M000019116

Chemotherapy in Psychiatry

Ross J. Baldessarini

Chemotherapy in Psychiatry

Pharmacologic Basis of Treatments for Major Mental Illness

Third Edition

 Springer

Ross J. Baldessarini
Harvard Medical School
McLean Hospital
Psychopharmacology Program
Belmont, MA, USA

ISBN 978-1-4614-3709-3   ISBN 978-1-4614-3710-9 (eBook)
DOI 10.1007/978-1-4614-3710-9
Springer New York Heidelberg Dordrecht London

Library of Congress Control Number: 2012939220

© Springer Science+Business Media New York 2013
This work is subject to copyright. All rights are reserved by the Publisher, whether the whole or part of the material is concerned, specifically the rights of translation, reprinting, reuse of illustrations, recitation, broadcasting, reproduction on microfilms or in any other physical way, and transmission or information storage and retrieval, electronic adaptation, computer software, or by similar or dissimilar methodology now known or hereafter developed. Exempted from this legal reservation are brief excerpts in connection with reviews or scholarly analysis or material supplied specifically for the purpose of being entered and executed on a computer system, for exclusive use by the purchaser of the work. Duplication of this publication or parts thereof is permitted only under the provisions of the Copyright Law of the Publisher's location, in its current version, and permission for use must always be obtained from Springer. Permissions for use may be obtained through RightsLink at the Copyright Clearance Center. Violations are liable to prosecution under the respective Copyright Law.
The use of general descriptive names, registered names, trademarks, service marks, etc. in this publication does not imply, even in the absence of a specific statement, that such names are exempt from the relevant protective laws and regulations and therefore free for general use.
While the advice and information in this book are believed to be true and accurate at the date of publication, neither the authors nor the editors nor the publisher can accept any legal responsibility for any errors or omissions that may be made. The publisher makes no warranty, express or implied, with respect to the material contained herein.

Printed on acid-free paper

Springer is part of Springer Science+Business Media (www.springer.com)

*For Frances, who thrice
encouraged and supported the
writing of this book*

Preface

This third edition of *Chemotherapy in Psychiatry* (*CTP-III*) has been extensively updated since the second edition of 1985, and is being produced by a new, international publisher. As in the first editions, the monograph addresses the basic aspects of modern psychopharmacology and the clinical applications of drugs that are useful for the treatment of many major psychiatric disorders, including psychoses, major mood disorders, and many anxiety disorders. The presentation covers descriptions of the main classes of psychotropic drugs, selected information concerning their known action mechanisms and metabolic disposition, and their clinical applications for acute illnesses and for prevention of recurrences or long-term morbidity, which are all too common among psychiatric disorders. In addition, limitations and adverse effects of each type of agent are covered.

The title of the book, by analogy to treatments used for cancer, was selected to emphasize the point that all psychotropic agents have adverse effects, ranging from annoying to potentially lethal. The art of clinical application of psychotropic medicines centers on the need to balance benefits and risks at the level of individual persons. Moreover, it is increasingly appreciated that use of psychotropic drugs has come to dominate clinical practice in psychiatry worldwide—perhaps owing largely to perceived simplicity, ease of use, and apparent efficiency, as well as apparent cost-effectiveness of such treatments. Nevertheless, it is important to emphasize that medicinal treatments for almost all psychiatric patients are but one component of comprehensive clinical care of complex human problems. Effective application of psychiatric chemotherapy requires more than rapid recognition of a syndrome and selection of an appropriate medicine and dose aimed at suppressing selected target symptoms. The now nearly universal domination of clinical psychiatry by a relatively narrow medicinal approach underscores the thesis that premature devaluing or abandonment of traditional psychological and social aspects of clinical psychiatry would be a grave error.

In addition to its profound impact on clinical practice since the 1950s, modern psychopharmacology has had a fundamental effect on contemporary academic and theoretical psychiatry. Despite limited advances in fundamental understanding of basic neurobiological and behavioral processes that may underlie major psychiatric disorders, the availability of effective medicinal treatments and partial understanding of their neuropharmacology have strongly stimulated a generation of research strategies that have moved modern psychiatry closer to the mainstream of general medicine. In part, as a contribution to more precise applications of specific treatments to particular disorders and individual patients, there is also a renewed interest in psychiatric nosology and in psychopathology after decades of neglect, particularly by American academic psychiatry.

This monograph developed initially from several years of seminars on psychopharmacology for psychiatric residents, medical students, and postgraduate training programs, and many additional years of consultative clinical practice and case conferences for inpatients and outpatients based at Massachusetts General and McLean Hospitals in Boston. It has also been informed by more than four decades of leading a laboratory program in preclinical neuropharmacology at the Mailman Research Center of McLean Hospital, aimed at better understanding of known psychotropic drugs and in developing improved agents. Additional insights have come from many years of industrial consulting to the pharmaceutical and biotechnology industries, as well as from authoring the chapters on psychopharmacology for a leading textbook of general pharmacology (*Goodman and Gilman's The Pharmacological Basis of Therapeutics;* New York: McGraw-Hill) from 1980 to 2010. Particularly valuable information has arisen from interactions with patients, trainees, and especially from experiences shared in a series of continuing medical education seminars for a broad range of mental health professionals in Boston, Cape Cod, Santa Fe and elsewhere, and from ongoing international research collaborations with academic and clinical colleagues throughout the world.

Since the first edition of the book appeared in 1977, at a time when textbooks in this field were uncommon, many specialized reviews and excellent comprehensive textbooks on various aspects of psychopharmacology have appeared. In the face of the growing complexity and specialization of the field, this monograph has attempted to provide a relatively brief, consistent, and hopefully readable text aimed at communicating basic principles as well as facts, and at contributing to safe and effective clinical application of chemotherapy in psychiatry. It may be limited in representing largely American, and possibly somewhat regional, perspectives and practices, but attempts to balance the risk of parochialism with the principles and major research findings that underlie modern theories and practices. Now that the field has been in existence for more than six decades, it also seemed timely to consider the current state of psychopharmacology, its evolution and promise, but also its theoretical and practical limitations, as the rate of innovation and of fundamentally new drug-products is declining, and its negative as well as many positive effects on clinical care of the mentally ill have become more apparent.

Many colleagues in psychiatry, pharmacology, and industry have shared expertise, scientific insights, and clinical experiences that have contributed importantly to this book. Special thanks is also given to Mrs. Rita Burke, for her tireless assistance in gathering research literature on which the book is based.

Belmont, MA, USA   Ross J. Baldessarini

Macbeth: *How does your patient, doctor?*

Doctor: *Not so sick, my lord, as she is troubled with thick-coming fantasies, that keep her from her rest.*

Macbeth: *Cure her of that: Canst thou not minister to a mind diseas'd: pluck from the memory a rooted sorrow; raze out the written troubles of the brain; and with some sweet oblivious antidote cleanse the stuff'd boson of that perilous stuff which weighs up on the heart?*

Doctor: *Therein the patient must minister to himself.*

—William Shakespeare, *Macbeth (67)*

Contents

Chapter 1
Modern Psychopharmacology and Psychiatric Treatment

In the fields of observation chance favors only the prepared mind

Louis Pasteur

The need for effective treatments of psychiatric disorders is indicated by the high prevalence of many of these disorders [47], particularly substance-use, anxiety, and mood disorders, as well as their high burden of direct and indirect costs to society (Tables 1.1 and 1.2). Psychiatric, substance-abuse, and primary brain disorders account for approximately 13% of the global disease burden (years of healthy life lost due to early death or disability). Depression alone, even excluding bipolar disorders, is the third leading contributor to the worldwide disability. Associated suicides number 1.5 million/year, and attempts are estimated at over 20 million/year [24]. Throughout the recorded history of medicine, efforts have been made to utilize chemical or medicinal means to modify abnormal behavior and emotional pain. Alcohol and opiates have been used for centuries not only by physicians and healers but also spontaneously for their soothing or mind-altering effects. Stimulant and hallucinogenic plant products also have been a part of folk practices for centuries. More recently, man has applied modern technology, first to "rediscovering" and purifying many natural products, later to synthesizing and manufacturing their active principles or structural variants with desired properties. Throughout the discussion that follows, the classes of chemicals used for their *psychotropic* effects (altering feelings, thinking, or behavior) are referred to by the somewhat arbitrary terms *antipsychotic*, *antimanic* or *mood-stabilizing*, *antidepressant*, and *antianxiety* agents. This system of terminology grows out of the *allopathic* tradition of modern scientific medicine, which treats with drugs producing effects opposite, or antagonistic to the signs and symptoms of a given illness.

The modern era of psychopharmacology can be dated from 1949, when the antimanic effects of the lithium ion were discovered, or 1952, when the psychotropic and antiadrenergic effects of reserpine were investigated, and when the special properties of chlorpromazine began to be recognized. The antidepressant monoamineoxidase (MAO) inhibitor iproniazid also was introduced in the early 1950s, and in the

R.J. Baldessarini, *Chemotherapy in Psychiatry: Pharmacologic Basis of Treatments for Major Mental Illness*, DOI 10.1007/978-1-4614-3710-9_1,
© Springer Science+Business Media New York 2013

Table 1.1 Needs and markets for psychotropic drugs

Prevalence of psychiatric illnesses (United States)
Substance abuse (alcohol and drug) (27%)
Anxiety disorders (25%)
Major depression and dysthymia (24%)
Bipolar disorders (5%)
Antisocial personality (3.5%)
Schizophrenia and related psychoses (1%)
Two or more disorders (27%)
Professionally treated (42%)

Illness costs
Mental illness + dementias (30%)
Cardiovascular (18%)
Brain + spinal injuries (13%)
Cancers (12%)
Mood disorders (11%)
Acquired immunodeficiency (HIV/AIDS) (8%)
Arthritis (5%)
Schizophrenia (4%)

Ranking by sales income
1. Psychotropics
2. Cancer chemotherapies
3. Anticholesterol agents
4. Diabetes drugs
5. Gastric proton-pump inhibitors
6. Antihypertensives
7. Analgesics

late 1950s, the "tricyclic" antidepressant imipramine was introduced. Use of the anxiolytic-sedative meprobamate began in 1954, and the first benzodiazepine, chlordiazepoxide, was being developed before 1960 as an antianxiety agent. By the end of the 1950s, general medicine and psychiatry had therapeutic agents available for the psychotic and major mood disorders—including schizophrenia, mania, and severe depression—and the anxiety disorders (formerly "neuroses"). Remarkably few fundamentally new kinds of psychiatrically therapeutic agents have been developed since that time. Instead, the past half-century has been marked by an accumulation of structural analogues of earlier agents or chemically dissimilar drugs with similar actions and clinical effects—all with adverse effects, some similar and others new. Nevertheless, there have been important advances in understanding the biological and clinical actions of the available psychotropic drugs and their appropriate clinical use. Currently available psychotropic drugs are the leading pharmaceutical products of all kinds, based on their annual sales (Table 1.1).

Psychotropic drugs that are currently available are summarized by generic and corresponding original or prominent brand names in Table 1.3. A recent market analysis of leading psychotropic drugs based on annual sales in the United States [30] indicated the following ranking: (1) alprazolam (*Xanax®*), (2) zolpidem (*Ambien®*), (3) S-citalopram (*Lexapro®*), (4) lorazepam (*Ativan®*), (5) gabapentin

Table 1.2 Prevalence of psychiatric and neurological disorders, sex risk, and disease burden statistics (Europe, 2010 [87])

Disorder	Prevalence (%) (median (IQR))	Female/male risk ratio	Disease burden (DALY/10,000)
Anxiety disorders	14.6 (6.80–20.4)	2.5	23.0
Sleep disorders	10.8 (8.10–16.8)	1.2	9.4
Juvenile behavioral disorders	9.60 (1.70–27.0)	1/2.4	—
Major depression	6.30 (3.54–8.18)	2.3	104
Somatoform disorders	5.60 (1.87–6.93)	2.1	—
Dementias	5.40 (0.30–1.00)	1.6	53.7
Substance abuse (alcohol + drug)	4.38 (0.90–6.05)	1/3.3	67.1
Personality disorders	1.30 (1.30–0.14)	1/2.9	—
Psychotic disorders	1.10 (0.22–1.21)	1/1.2	15.3
Mental retardation	1.00 (0.40–1.40)	1/1.2	0.10
Bipolar disorder	0.80 (0.80–1.03)	1.2	17.5
Eating disorders	0.62 (0.01–0.63)	6.2	—
Neurological disorders	—	1.2	71.5

DALY: disease-associated life-years lost
Reciprocal sex risk ratios are greater in males

(*Neurontin®*), (6) clonazepam (*Klonopin®*), (7) sertraline (*Zoloft®*), (8) duloxetine (*Cymbalta®*), (9) amphetamines (*Adderall®*), (10) venlafaxine (*Effexor®*), (11) quetiapine (*Seroquel®*), (12) trazodone (*Desyrel®*), (13) diazepam (*Valium®*), (14) R,S-citalopram (*Celexa®*), and (15) fluoxetine (*Prozac®*).

The impact of modern psychopharmaceuticals on the practice of psychiatry since the 1950s has been compared to the impact of the antibiotics on general medicine since the 1940s. Quantitatively, the utilization of chlorpromazine compares well with that of penicillin: in the first decade of its availability this antipsychotic drug was given to approximately 50 million patients throughout the world, and some 10,000 scientific papers were written about it [72]. At the present time, psychotropic drugs not only are among the leading pharmaceuticals of all types, but several command markets of several billion US dollars/year. These facts underscore the revolutionary impact of these drugs on modern psychiatry.

Prior to the 1950s, most severely disturbed psychiatric patients were managed in relatively secluded public or private institutions, usually with locked doors, barred windows, and other physical restraints. The few medical means of managing their symptoms included use of barbiturates, bromides, opioids, and anticholinergic drugs such as scopolamine for sedation. Other treatments included soothing baths and wet-packs, as well as "shock" treatments with insulin, atropine, or convulsant drugs, and later electrically induced convulsions, along with neurosurgical techniques including prefrontal leucotomy. Since then, most of these forms of treatment, except for electroconvulsive treatment (ECT) have virtually disappeared. Many locked doors have opened, except for severely disturbed, aggressive or suicidal patients, and both patients and psychiatric facilities have been returned to "the community," to general hospitals, open-door day-treatment centers, and to hospital-based or free-standing outpatient clinics and private offices. However, to conclude that modern psychotropic drugs have been solely responsible for these revolutionary

Table 1.3 Common psychotropic drugs: generic–*brand* names

Antidepressants (23)	Antipsychotics (19)	Mood-stabilizers (5)
Atomoxetine (*Strattera*)	Aripiprazole–*Abilify*	Carbamazepine–*Tegretol*
Amitriptyline–*Elavil*	Asenapine–*Saphris*	Divalproex–*Depakote*
Bupropion–*Wellbutrin*	Chlorpromazine–*Thorazine*	Oxcarbamazepine–*Trileptal*
Citalopram–*Celexa*	Clozapine–*Clozaril*	Lamotrigine–*Lamictal*
S-Citalopram–*Lexapro*	Fluphenazine–*Prolixin*	Lithium carbonate–*Lithobid*
Clomipramine–*Anafranil*	Haloperidol–*Haldol*	
Desipramine–*Norpramin*	Iloperidone–*Fanapt*	
Desvenlafaxine–*Pristiq*	Loxapine–*Loxitane*	
Doxepin–*Sinequan*	Lurasidone–*Latuda*	
Duloxetine–*Cymbalta*	Mesoridazine–*Serentil*	
Fluoxetine–*Prozac*	Olanzapine–*Zyprexa*	
Fluvoxamine–*Luvox*	Paliperidone–*Invega*	
Imipramine–*Tofranil*	Perphenazine–*Trilafon*	
Mirtazapine–*Remeron*	Quetiapine–*Seroquel*	
Nortriptyline–*Pamelor*	Risperidone–*Risperidal*	
Paroxetine–*Paxil*	Thiothixene–*Navane*	
Phenelzine–*Nardil*	Thioridazine–*Mellaril*	
Selegiline–*Emsam*	Trifluperazine–*Stelazine*	
Sertraline–*Zoloft*	Ziprasidone–*Geodon*	
Tranylcypromine–*Parnate*		
Trazodone–*Desyrel*		
Venlafaxine–*Effexor*		
Vilazodone–*Viibryd*		

Agents commonly employed in the United States. Atomoxetine is approved for attention disorders; oxcarbamazepine is used off-label for bipolar disorder

changes would be a gross exaggeration. In the same period, partly independent changes in the clinical management of psychiatric patients also were beginning. These included use of group and milieu techniques to complement individual psychotherapy, greater appreciation of the untoward regressive effects of institutions on behavior, and a strongly increased social consciousness throughout medicine and particularly in community-based psychiatry. A fair conclusion would be that social and administrative changes and the new drugs had mutually facilitating and enabling interactions, which resulted in a melioristic trend toward progress and change.

Observations that underscore the important impact on hospital practice associated with the new antipsychotic, antimanic, antidepressant, and antianxiety drugs introduced in and following the 1950s include the observation that in the United States the number of patients hospitalized in public mental institutions reached a peak of approximately 500,000 in the 1950s, with an initially rapid and now slower downward trend to less than 30,000 currently, despite a steady increase in the general population. This change has resulted not only from the beneficial effects of modern psychotropic drugs but also from policy decisions to alter the pattern of mental health-care delivery, notably including decisions to reduce the number of available beds in most public psychiatric institutions. Ironically, rates of new admissions and of readmissions have increased over time, particularly among the very young and very old, despite a misleading decline in the prevalence of psychiatric hospitalization,

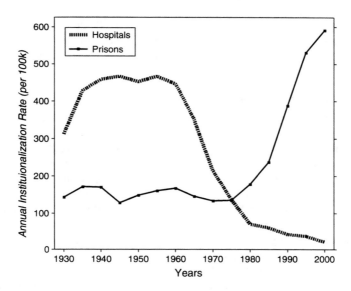

Fig. 1.1 Institutional trends in the United States, 1930–2000. Hospitalization in public mental hospitals peaked in the 1940s and 1950s and declined thereafter as numbers of persons in prisons increased, many of whom were severely mentally ill. Adapted from Harcourt [38]

as length-of-stay has declined markedly, largely through administrative demands driven by hopes of cost-containment [46]. In addition, there has been a major shift in the proportion of hospitalized seriously mentally ill persons in hospitals vs. in jails and prisons, whose numbers currently exceed those of the hospitalized mentally ill in the 1950s ([38], Fig. 1.1). A large proportion of patients formerly held in hospitals for many months now attain outpatient status within weeks or even days, owing to current philosophies and systems of care combined with the effects of modern chemotherapy. However, their aftercare, quality-of-life, and functional levels vary markedly among locales, diagnoses, and individuals [31].

In general, a number of serious problems remain despite striking improvements in the clinical treatment of patients with severe psychiatric disorders. Whereas many acute episodes of such disorders can be interrupted and shortened with modern treatments, and highly disturbed behavior is relatively infrequent, even in public mental institutions, the available chemotherapies have severe shortcomings, including general limitations of efficacy, particularly incomplete evidence of long-term preventive or prophylactic effectiveness, and incomplete or intermittent adherence to prescribed regimens, as well as sometimes medically significant adverse or toxic effects. Many patients with severe psychotic, mood, or anxiety disorders respond only marginally to available treatments, despite a tendency to continue their use or even to add increasingly complex combinations of drugs of largely untested benefit and safety.

An additional complication for the medical treatment of psychiatric disorders is the uncertainty or lack of specificity of diagnosis and classification that is characteristic of psychiatric illnesses. Diagnosis has become more important than ever, especially in order to optimize matching of clinical conditions, choice of treatment, and chances of a beneficial response with tolerable side effects. The advent of modern psychopharmacologic treatments, perhaps more than any other single factor, contributed to a vigorous renewal of interest in nosology and its ally, psychiatric epidemiology. There is also renewed interest in classical, descriptive psychopathology, which American psychiatry has tended to ignore more than other cultures. Nevertheless, etiology, whether biological or psychological, remains unknown for most psychiatric disorders, and classification or diagnosis rests on descriptions of clinical features of syndromes, as well as on their course or natural history and outcomes, familial associations, and to some extent, responses to treatment. Despite the fundamentally unsatisfactory basis of psychiatric nosology and the inescapable contributions of individuality even to the most classic syndromes, diagnosis continues to gain objectivity, coherence, and reliability. Associations between specific clinical syndromes and predictable responses to psychotropic agents continue to support efforts to improve psychiatric classifications. Indeed, examples of clinical syndromes that became much more widely recognized with the development of effective treatments for them include major depression, bipolar disorder, panic-agoraphobia, obsessive-compulsive disorder, and others [3, 36, 69].

The current clinical and academic standard for psychiatric diagnosis in the United States and many other countries is the *Diagnostic and Statistical Manual* (DSM) of the American Psychiatric Association, now entering its fifth revision. The World Health Organization sponsors a psychiatric component of its *International Classification of Diseases* (ICD); now in its tenth revision, it is widely used internationally as well as by many government agencies and insurance companies in the United States. Additional systems of diagnosis and gathering of clinical information have been developed for specialized research purposes, and clinical testing of novel psychotropic drugs depends heavily on rating scales that aim to capture the severity of characteristic clinical symptoms of common psychiatric disorders. However, even the best available systems of diagnostic classification and symptom rating for clinical or research applications require gathering and interpretation of data by and from individual persons, so that subjective and idiosyncratic, as well as culture-bound, elements of their application can scarcely be avoided.

Development of New Psychotropic Drugs

Drug Discovery

Most available psychopharmaceuticals have been discovered and introduced in one of three basic ways: (a) rediscovering and exploiting folk usage of natural products, usually with isolation of active principles and synthesis of similar molecules with comparable effects (e.g., reserpine, opioids, and centrally active sympathomimetics); (b) serendipitous observation that an agent developed for another purpose has a desirable but unexpected clinical effect (examples include chlorpromazine, haloperidol, iproniazid, imipramine, meprobamate, and trazodone); or (c) synthesis and functional

Table 1.4 Characteristics of psychotropic drug development

Use and later purification and synthesis of *natural products* (e.g., alcohol, amphetamines, cocaine, opiates, reserpine)

Serendipity combined with research uncovers leads (e.g., buspirone, chlordiazepoxide, chlorpromazine, clozapine, haloperidol, improniazid, imipramine)

Synthesis and screening of *chemical analogs* (e.g., benzodiazepines, most tricyclic antidepressants and antipsychotics)

Partial knowledge of pharmacodynamics supports *functional screening* (e.g., haloperidol, serotonin-reuptake inhibitors)

Cloning of genes for novel proteins and *rational, computer-assisted, design of small molecules* (functions not well defined; limited success for psychotropics)

Overall progress severely limited by *lack of pathophysiology* or etiology

screening of structural analogues of known drugs or of novel compounds in search of behavioral or molecular effects similar to those of known agents (examples include the piperidine phenothiazines, thioxanthenes, butyrophenones and diphenylbutylpiperidines, modern antipsychotic drugs modeled after the structure or activities of clozapine, and the serotonin-reuptake inhibitors and other modern antidepressants) [13]. These methods are summarized in Table 1.4, and examples of serendipity in the discovery of novel psychotropic drugs are provided in Table 1.5 [57].

An important reality that underlies the process of drug development in the psychopharmaceutical industry is the profit motive. Psychotropic drugs have gradually risen to the top of all classes of drugs in annual prescription counts and income from sales. As noted, individual psychotropic compounds represent tens of millions of prescriptions and several have generated several billions of dollars in sales annually. In a recent international drug market survey, drugs acting on the central nervous system (CNS) were the leading category of all types (Table 1.1), with annual international sales of nearly $120 billion, or more than 15% of the total world drug market, and four of the ten best-selling drugs recently were psychotropics (antipsychotics and antidepressants), all returning 4–5 billion dollars/year [54]. However, there is an emerging tendency for expected growth annual sales of psychotropic drugs to level off, with concern about a growing disparity between massively rising costs of research and development vs. a decline in the rate of new products marketed among all pharmaceuticals (Fig. 1.2). These trends are indications of a highly successful, but maturing, market and they reflect the enormous difficulties in developing new, and especially, innovative or superior drugs, and especially psychotropics.

The process of new-drug development in psychopharmacology is a fundamentally conservative and empirical process that appears to overvalue principles of drug action established or proposed for known agents. This process results in searches for more drugs with similar effects and limitations. For example, it remains hard to imagine investing tens or hundreds of millions of dollars in developing a potential antipsychotic agent that has no antagonistic action on central dopamine or serotonin receptors, or an antidepressant that does not limit the inactivation of serotonin or norepinephrine. Following such conservative models derived from the pharmacology of older, successful, agents may be an effective

Table 1.5 Examples of serendipity in psychopharmacological drug discovery

Agent	Background
Lithium carbonate	Not so good for *gout*
Amphetamines	Derived from ephedrine in Chinese herbal *ma huang*
Reserpine	Derived from Vedic herbal (*Rauwolfia*) for snakebite and madness
Chlorpromazine	Antihistaminic *preoperative sedative*
Imipramine	Putative *antipsychotic* ("analogue" of chlorpromazine)
Iproniazid	*Antituberculous* but also MAO inhibiting and mood elevating
Chlordiazepoxide	Possibly sedative *muscle relaxant*
Haloperidol	Nonanalgesic *meperidine analog*
Clozapine	Surprising *imipramine analog*
Carbamazepine	Not just another *anticonvulsant*
Valproic acid	A noninert *solvent*
Fluoxetine	Antidepressant and *anxiolytic*
Buspirone	An early *atypical antipsychotic*

business model, but is hardly likely to provide highly innovative or truly unique means of achieving desired clinical ends. More fundamentally, the process of psychopharmaceutical drug development over the past half-century reflects the severely limiting effect of a lack of knowledge of etiology of psychiatric disorders, and only fragmentary and unconvincing notions about their possible pathophysiology. Even the pathophysiological hypotheses that have been proposed are logically circular and based largely on known actions of available treatments. In short, drug development for psychiatry has been empirically effective, if basically repetitious, for several decades but true innovation remains extraordinarily elusive.

Current procedures for developing, testing, and seeking regulatory approval of new drugs in the United States have arisen by tradition as well as by the regulatory requirements of the US Food and Drug Administration (FDA), with similar procedures followed elsewhere by the activities of increasingly merged, international pharmaceutical corporations. In this process, once the potential clinical usefulness of a new molecule is suspected, initial animal experimentation is conducted to establish its apparent spectrum of pharmacological activities, to evaluate its metabolism and disposition, as well as its potential toxicity, and to estimate likely clinical doses and the ratio of its median toxic or lethal doses to its median effective doses (*therapeutic index*) or margin of safety. The typical course of discovery and development of new psychotropic drugs and the standard phases of drug development and typical times involved are summarized in Tables 1.6 and 1.7.

Phases of Drug Development

The early, preclinical steps in drug development and their refinement have become an increasingly sophisticated subspecialty within the field of psychopharmacology. They involve such technologies as molecular targeting of suspected primary sites of

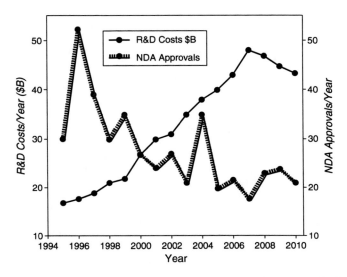

Fig. 1.2 Pharmaceutical research and development costs vs. new-drug approvals. Research and development (R&D) costs are in billions of US dollars/year and new-drug application (NDA) approvals by FDA are for all types of drugs (1994–2010). Adapted from Harris [39]

action (such as hormone or neurotransmitter receptor proteins, or cell membrane transporter proteins) in addition to more traditional modeling based on effects on the behavior of laboratory animals (discussed in more detail in the second edition of this book; [4]). It is now commonplace to design selective small molecules with high affinity for defined macromolecular target sites [25]. Moreover, many steps in the chemical synthesis of variants of a desired compound (to seek more ideal candidate molecules or potential follow-on agents for later development), as well as molecular screening procedures that seem to be highly efficient, are often automated and robotically controlled. These modern procedures involve high-capacity or combinatorial chemistry and high-throughput screening methods that are technically impressive but expensive and liable to produce large numbers of uninteresting candidate molecules [26, 27, 48, 49, 84]. In addition, pharmacokinetic modeling can identify oral bioavailability, elimination half-life and clearance rates as well as likely enzymatic routes of metabolic conversion, based usually on animal models [59]. Advanced techniques include the introduction and expression of human genes of interest into test animals so as to provide modeling that is more likely to represent human subjects [21]. Another increasingly employed technique is the early application in test animals and in human subjects of in vivo labeling of brain target sites of interest with radiolabeled tracer molecules (such as for positron-emission tomography [PET]) as an approach to estimating potency, dosing requirements, and pharmacokinetic measures [66, 83].

Following initial identification of a promising candidate molecule, the process of drug development usually splits into further pursuit of basic mechanisms of action and

Table 1.6 Historical evolution of modern psychopharmacology

Before 1950s	Long use of sedatives (antihistamines, barbiturates and newer nonbarbiturates, bromides, cannabis, chloral hydrate, ethanol, opiates, reserpine) as well as stimulants (amphetamines, camphor), antiparesis (cerebral syphilis) agents (mercurials, penicillin), and physical methods (baths, restraints, insulin, chemical and electroconvulsive treatments (ECT), leucotomy)
1950s	Serendipitous discoveries of lithium, phenothiazines and haloperidol, antidepressants, benzodiazepines
1960s	Gradual and contentious acceptance vs. psychotherapy; commercial potential emerges; regulatory requirements standardized; randomized-controlled trials become standard; rating-scales and simple statistics develop; initial efforts in pathophysiology inspired by early pharmacodynamics
1970s	Molecular and receptor-based preclinical screening compete with animal behavioral modeling; intensive pharmacocentric research aimed at pathophysiology; drugs and syndromes encourage each other as new diagnoses proliferate with new treatments (e.g., attention, bipolar, depressive, obsessive-compulsive, panic, phobic disorders)
1980s	High throughput, automatic synthesis and drug screening established; nosology expands to over 300 diagnoses, driven partly by drugs and insurability; few novel and not-better agents emerge; improved methods for analysis of therapeutic trials
1990s	"Decade of the brain"; genetics and brain imaging emerge; new molecular gene-product targets identified (often without clear functions); research and development costs rise; billion-dollar/year drugs; interest in "effectiveness" (practical utility) vs. efficacy emerges; markets expand to general medicine; adverse drug effects: different but unavoidable; effect-size declines as placebo-rates rise and multisite, off-shore trials become popular
2000s	Innovation limited despite continued advances in basic and clinical neuroscience; ratio of research and development costs-to-novel-products highly unfavorable; many patents run out and insurers encourage use of generics; markets begin to saturate; clinical practice dominated by limited contact and medicating; growing concern about adverse metabolic and behavioral effects; growth of postmarketing research; efforts at evidence-based therapeutics emerge

preclinical pharmacological characterization vs. initiation of human and clinical studies, as are summarized in Table 1.7. Basic pharmacology occurs in both academic and industrial laboratories, but clinical testing for safety and efficacy, as required by regulatory agencies, is largely the province of the pharmaceutical industry, owing mainly to the extraordinary costs involved, and its timing is driven by the limited patent-life of new drugs (typically 20 years in the United States since 1995). Developmental costs for a single new drug that acts on the CNS can run to hundreds of millions of dollars, although the actual costs vary with definitions of what is included in research and development costs as opposed to marketing activities [1, 2]. In turn, such costs drive efforts to develop "blockbuster" drugs with sufficiently large markets as to guarantee recovery of investment costs as well as substantial profit.

The first phase of human experimentation (Phase I) involves toxicological and pharmacokinetic studies in healthy human volunteers. Such subjects are increasingly difficult to access due to ethical constraints against enrolling persons whose

Table 1.7 Typical phases of drug development in the United States

Phase	Tasks	Years
Preclinical research	Identify leads, preliminary pharmacology; patents (delayed as long as possible)	2–5
Initial FDA review	IND, enter clinical testing	ca. 5
Clinical testing		
Phase I	Human tolerability, pharmacokinetics	4–6
Phase II (A and B)	Pilot and later rigorous small trials, dose-finding	
Phase III	Large controlled, pivotal trials	
FDA review	Review of all data, licensing (NDA)	2–3
Postmarketing ("Phase IV")	Refine uses, doses, adverse effects	Open-ended
End of patents	Generics appear, prices fall	20

voluntary status is questionable (such as prisoners, other institutionalized or cognitively compromised persons, or the poor anticipating financial compensation). Initial Phase I studies typically are carried out with small numbers of closely monitored subjects under clinical-laboratory conditions. If this phase of drug development is encouraging, preliminary clinical trials are undertaken with regulatory review and approval (typically, with an investigational new drug (IND) permit from the FDA).

These Phase II trials are typically relatively small and involve carefully diagnosed and closely clinically evaluated subjects under rigorous investigative conditions. It is also at this point that the drug development sponsor will have decided on a plausible target indication or diagnostic group of particular interest. Such planning is not always sustained, since candidate drugs sometimes have unexpected clinical effects that have not been anticipated, and may lead to alternative developmental paths: examples include the dopamine partial agonist pramipexole, which proved to be successful as an antiparkinsonism agent rather than an atypical antipsychotic drug, and the serendipitous discovery of the erectile-dysfunction activity of sildenafil [17, 68]. Phase II trials may or may not include placebo controls or comparisons with established treatments, and aim primarily to develop evidence of efficacy under highly controlled conditions ("proof of concept"). If the Phase II trials are encouraging, the next step are larger and broader Phase III trials, typically in more clinically representative samples and often involving multiple sites.

Such Phase III trials are sometimes considered early (IIIA) or late (IIIB, typically following submission of preliminary findings to a regulatory agency). Formerly, they were often conducted within academic medical centers, but have increasingly been shifting toward management by contract research organizations (CROs) working with individual clinicians, as well as toward large, international, multisite, collaborative trials, often in dissimilar cultures. In such settings, methods of diagnosis, clinical assessment, and symptom ratings may not be well standardized and validated. Placebo-controlled, randomized trials have long been considered optimal for testing new treatments. However, ethical and clinical controversy about their use in some clinical circumstances is now coupled with growing reluctance of potential patient-subjects to accept enrollment in trials involving a placebo condition when known treatments are available. Nevertheless, FDA approval typically requires at least two "pivotal" trials of substantial size, involving randomized, blinded, comparisons of the test drug against a placebo, and showing statistical superiority of the active agent. A major scientific reason to continue to require some placebo-controlled trials is that comparisons of new vs. established drugs are likely to yield little or no apparent difference, whereas clear superiority of a new agent over an established drug is rare in psychiatry. This circumstance can lead to the logically highly risky conclusion that a finding of "not different from" is equivalent to "about as good as." Such potentially false conclusions are especially likely if a trial is poorly designed or conducted, with high levels of random variation, or if the standard comparator does not show expected efficacy, as happens more often than one might expect of an established treatment.

If all three phases of clinical investigation are successful, the drug can be a candidate for licensing by a regulatory agency, such as by an FDA new-drug application (NDA). Licensing requirements and procedures in other countries differ in some details, including requirements for placebo-controlled trials. Drug monitoring and regulatory processes exist in many individual countries, including Health Canada, the British Medicines and Healthcare products Regulatory Agency (MHRA), and the scientifically-oriented British National Institute for Health and Clinical Excellents (NICE), and the European Medicines Agency (EMA). Approval processes also vary widely in their linking of the licensing of new drugs for clinical use with their pricing. Price regulation is not done in the United States, where brand-name drugs typically command high prices during their patent-life, and subsequent generic products are much less costly.

An important additional phase of drug development is "after-marketing" ("Phase IV"), in which the optimal clinical use and dosing of newer drugs is clarified, and perhaps extended to new indications with additional controlled trials. Of particular importance, this phase of drug development can provide much more information about potential adverse effects than are detected or quantified during Phases II or III, especially involving events of low prevalence, whose detection and quantification may require large numbers of exposed persons. Nevertheless, in general, current means of monitoring FDA-required safety of new drugs remain less well developed than tests of clinical efficacy. They continue to rely heavily on passive and incidental observation or reporting of adverse effects by patients to investigators or clinicians, with inconsistent reporting to regulatory authorities after licensing. There are growing efforts to incorporate explicit assessments of suspected areas of risk into clinical trials as well as in after-marketing monitoring programs, with the aim of improving timely detection and quantification of adverse effects and increasing the safety of marketed drugs. Another component of after-marketing investigations involves practical or clinical effectiveness trials that compare the therapeutic value, relative tolerability, and acceptability of licensed treatments in large samples of patients aiming to depict drug performance in broad clinical practice. Sometimes, such effectiveness trials incorporate the methods of randomization and blinding that are more typical of Phase II and III trials.

An important movement in recent decades involves efforts to place therapeutics on a more sound, scientific basis by pooling and comparing findings from available clinical trials, in what is referred to as *evidence-based therapeutics* [64, 70]. The aim is to rank specific treatments by evidence of their relative efficacy, safety, and tolerability. Such efforts have obvious potential clinical value, but are also encouraged by administrators, policy makers, and insurers, in order to maximize not only the clinical value of treatment choices, but also to limit costs. For most classes of psychotropic drugs, this effort has yielded very limited success, at least in determining compelling and consistent rankings of specific drugs within a class by efficacy and safety [50, 51, 77, 82].

Nature of Clinical Trials

It is important to have some appreciation of the nature of clinical therapeutic trials, their design, limitations, and interpretation, in order to evaluate new findings critically. This need is particularly important since treatment trials in general, and particularly for psychiatric disorders, vary markedly in quality. Basic requirements for a credible therapeutic trial include clearly defined and clinically reasonably similar, but hopefully representative, patient-subjects in substantial numbers (*N*), and with substantial morbidity, evaluated by clinically relevant, sensitive, and reliable measures of clinical change.

Effects of Trial Size

Sometimes, a large *N* can work against a successful outcome (support for the hypothesis that a drug is effective), particularly in trials involving not only large numbers of subjects, but also multiple collaborating sites, each of which may contribute only a few patients. In large, complex trials, it becomes very challenging to establish and maintain reliable and consistent diagnoses, clinical assessments, and quantitative ratings on standard symptom scales. Complexity surely increases when multiple regional, national, and cultural differences may affect the meaning of diagnostic terms and criteria and of items on standard symptom-rating scales, even after translation into local languages. In turn, compromises in quality control and inconsistent interpretations can increase heterogeneity and may result in high inter-site variability, which is rarely reported. These circumstances also can make pooling of data across sites risky, and commercial application of pooled data to all sites (such as for licensing across countries) questionable. Trial complexity and heterogeneity appear to have greater impact on responses during placebo treatments than with active test drugs, possibly as a reflection of the phenomenon of regression to the mean or chance outcomes, which are more likely with placebo [77, 82, 83]. It is also likely that heterogeneity and compromised control of the conduct of trials have contributed to a noteworthy trend in recent years toward falling drug–placebo contrasts in trials of various types of psychotropic drugs [77, 83]. In turn, it is tempting to combat this trend by use of ever-larger and more complex multisite trials in order to increase statistical power (N) as effect-size (drug–placebo contrast) diminishes, in a basically circular process [77, 78].

Subject Recruitment

As clinical psychopharmacology has evolved over recent decades, it has become increasingly difficult to enroll patients who are diagnostically and clinically typical, at least moderately symptomatic, and not already treated more or less optimally. This quest has become particularly difficult in academic medical centers of North America and Europe, where teaching and research clinics tend to accumulate ever-more difficult and treatment-unresponsive, but already more-or-less optimally treated, patients who may be not only unrepresentative of broader clinical samples, but also unlikely to improve sufficiently for plausible inclusion in clinical trials of new treatments, though often willing to participate in hopes of encountering a better treatment. Of note, in well-established academic centers, treatments tested earlier in the life of a research clinic may have shown superior outcomes compared to those tested later, as more and more difficult, less treatment-responsive, patients tend to be retained over time. Since minor clinical changes are now routinely encountered, outcome measures other than indices of clinical improvement are sometimes employed. These may, for example, include time to a decision to change treatment, usually assumed to reflect poor response or limited tolerability, but typically complex and incompletely defined [53]. In addition, the efficacy of many psychotropic drug treatments is modest, and drug–placebo differences sometimes quite small [6, 77]. In part for this reason, and because major clinical improvements may require several months, it is a common practice to define outcomes in terms of proportions of patient-subjects reaching a defined level of clinical change (commonly ≥50% improvement in a rating-scale measure).

Difficulties in recruiting appropriate trial subjects contribute to pressure to rely increasingly on CRO-organized trials based in private clinics and offices, or to seek far less extensively treated or studied patients in less developed countries—again with an uncertain impact on control of diagnosis and clinical assessments and on the reliability and generalizability of the results obtained. The trend toward larger, multinational trials is also motivated by the quest for less costly opportunities for carrying out work in underdeveloped countries with lower labor costs. However, risks involved, in addition to matters of heterogeneity and quality control already mentioned, include variable levels of required infrastructure and training of research clinicians in many sites, as well as risk of inadequate supervision and temptations to distort data collection and reporting to earn money. Countering such potential problems can become very expensive, and tend to limit anticipated cost-savings. Another emerging problem with the international quest for low-cost alternatives to developing drugs in developed countries is that the quality of drug products manufactured in underdeveloped countries can be highly variable, unreliable, and sometimes unsafe, perhaps particularly with generic drugs developed after original patents have expired [39, 73]. Again, interventions required to assure product quality can rapidly erode expected savings in drug production.

Clinical Heterogeneity, Randomization, and Blinding

A major source of potentially misleading outcomes in modern therapeutic trials is virtually irreducible clinical heterogeneity. Even with rigorous application of modern diagnostic criteria and formal systems of clinical assessment involving standardized and structured interviewing techniques, most clinical syndromes in psychiatry involve large inter-individual differences and change over time. Variability arises from individual differences in the quality and severity of illnesses, as well as in the timing of interventions in the course of acute, recurring, or chronic illnesses. Moreover, some disorders (e.g., acute psychoses and apparently unipolar major depression) may meet even rigorous diagnostic criteria at one time, but not later [65]. To a large extent, such variability can be managed more or less adequately by inclusion of substantial numbers of patients and effective randomization. Nevertheless, even with randomization, heterogeneity within and among trials yields average outcomes that may not differ markedly between treatments within individual trials, or after pooling of findings across trials. Currently, it is fashionable to pool data across trials by the methods of meta-analysis [34]. Meta-analytic summaries are descriptive clinical exercises, in which randomization among trials does not obtain, and statistical power is limited by the number of trials pooled, not by total numbers of subjects. Pooling data from trials involving psychiatric patients often yields average outcomes that fail clearly to distinguish one treatment from another, and so frustrate the aims of evidence-based therapeutics [28, 77, 82].

"Blinding" is an important method of controlling bias arising from observers or patients ("single-blind") or both ("double-blind") to keep those involved unaware of the identify of a presumably randomly assigned treatment option. Success of the technique varies, and can be compromised when characteristic or obvious drug side effects make it clear which treatment is involved. Some studies include a debriefing phase at the end of the trial, but before the blind is broken, in which both patients and investigators are asked to guess at the treatment given to individual patients. However, standards by which to use such information to assess the resulting primary outcome findings are rarely defined or applied.

Even with randomization and blinding, therapeutic trials are at risk of additional artifacts that are not always readily controlled. A common problem is subject-retention. The longer a trial continues, and the more prevalent are adverse effects of a test agent that limit its tolerability, or the greater the lack of obvious benefit (as with a placebo or an ineffective drug), the less likely a patient-subject will continue to the designed end of a trial. Premature "dropouts" in such nonrandom circumstances can introduce major sources of bias and limit statistical power through attrition, as well as losing potentially valuable information concerning reasons for early dropping out. Moreover, if an ineffective treatment or placebo condition leads to earlier discontinuation than with a test drug, with higher levels of symptoms, it is likely that drug–placebo differences will be exaggerated. Conversely, if dropouts occur earlier with an active drug than a placebo, the opposite conclusion may be drawn, and the experimental treatment be considered ineffective, particularly if some degree of spontaneous improvement occurs over time to favor placebo treatment.

Analysis of Trial Outcomes

Various analytical methods have been developed to limit such risks [41]. They include statistical modeling techniques that make use of all data available at all time points, such as mixed-effects modeling, generalized estimating equation (GEE) regression modeling, and survival analysis. Such methods are far more powerful than older, simpler, but potentially-biased methods of comparing outcomes between subjects who complete a trial ("completer" analyses) or by making contrasts of initial and final ratings obtained at any time (as in "intent-to-treat" analyses), as well as based on considering a last observation as an end-point (last-observation-carried forward method). Such methods risk introducing biases that can have unpredictable effects on conclusions drawn [88], but even modern statistical modeling techniques involve assumptions and simplifications. A summary of methodological concepts associated with the design and analysis of clinical trials is provided in Table 1.8.

Survival analysis has become particularly popular for use in evaluating outcomes in clinical trials, by considering the latency to a defined outcome, such as the time from the start of randomized treatment to a particular level of clinical improvement or to a first relapse, although it was originally introduced to compare times to death under different clinical conditions or treatments [23]. The technique assumes that latency to a defined outcome is a suitable proxy measure for a general clinical outcome. For example, in the long-term treatment of a recurrent major mood disorder, a longer delay of a next recurrence, following initial recovery, is often considered evidence of a superior prophylactic treatment. A tacit assumption involved is that delay of an initial recurrence is a fair surrogate for long-term wellness, but this assumption has rarely been tested empirically [10, 11].

The use of randomization to an active test drug vs. an inactive but similar-appearing dummy pill (placebo) continues to be considered about the best research design for experimental therapeutic trials. This conclusion is probably valid scientifically, but use of placebos can lead to complicated sociological and even ethical dilemmas. Moreover, different cultures vary in their insistence on data from placebo-controlled trials for licensing of new drugs. Use of placebo controls continues to be a policy of the U.S. FDA in its requirement for a minimum number (usually two) of so-called "pivotal" trials with placebo controls. In large part, the controversy involved reflects

Table 1.8 Methodological concepts for clinical psychopharmacology

Term	Definition
RCT	Randomized, controlled trial vs. placebo, standard agent, or other controls; usually under double-blind conditions
Efficacy	Typically, superiority to placebo at $p < 0.05$, but effect-size can be large or small (commonly based on "response" as <50% reduction of severity ratings)
Blinding	*Single*: patient unaware of treatment provided; *double*: neither patient nor clinician or rater is aware of treatment; double-blinding is standard but not always guaranteed
Enrichment	Selecting subjects for a second-phase of a trial based on responsiveness in a preliminary phase (typically in acute illness); popular for efficiency, but can bias and be misleading
Effectiveness	Beneficial and tolerable in clinical settings, typically long-term
Probability (p)	Likelihood of statistical superiority vs. control (typically null or no-difference) is <5% likely (smaller p-value is generally better)
Continuation	Treatment continued after initial response in acute illness, ideally (but not always) to full clinical recovery
Prophylaxis or maintenance	Reduction of risk of future morbidity or recurrences of acute episodes with long-term treatment; proof of such benefits remain inadequate in much of psychopharmacology
Discontinuation	A trial design that removes an initially effective treatment; sometimes proposed to indicate prophylaxis but at great risk of discontinuation-associated stress, especially when discontinuation is rapid and patients incompletely recovered
Power	Ability to identify difference between treatments; greater with larger effect-size, greater precision, and larger sample (N) to yield lower p-value
Effect-size (d)	Response-difference ([treatment–control]/[measurement variance])
RR and RD	Relative response rate (RR), typically proportion attaining a criterion outcome (typically 50% reduction in symptom-rating scores) with drug/control, or the difference (RD) in responses or improvements between drug and placebo
CI	Confidence interval (typically reported 95% CI includes 95% of measurement variance; nonoverlap of 95% CIs indicates $p < 0.05$, by inspection)
Meta-analysis	Statistical method for pooling results across trials (N=trials, not cases); results can be expressed as RR (response rate-ratio), odds ratio (OR), or RD (response difference), usually weighted by trial size and variance measurement
NNT or NNH	Number of cases needed to treat or harm, for superior benefit or risk of an adverse effect vs. a control (ideally NNT<NNH); computed as reciprocals of RD; NNT >10 indicates limited benefit
Noninferiority	Similar outcome with two treatments compared head-to-head with sufficient statistical power to rule out defined levels of potential difference
EBM or EBT	Evidence-based medicine or therapeutics: research-supported treatment, commonly arising from meta-analyses + clinical experience + commonsense (risks over-interpretation); difficult to achieve in psychopharmacology due to typically similar efficacy assessments within drug-classes

the tension between scientific aims and clinical considerations. Now that effective and tolerated medicinal agents are available for most serious mental illnesses, some observers and many patients and their families argue that depriving a suffering or potentially suicidal patient of an established treatment is clinically very questionable and perhaps unethical, and would prefer to rely on comparisons of older vs. newer drugs. However, as noted above, such comparisons may mislead by failing to distinguish between active treatments and encouraging the risky conclusion that they are similar in efficacy. Moreover, testing aimed at supporting equal efficacy is challenging and typically requires large numbers of subjects.

Ideally, it would be good to seek innovative treatments that are demonstrably superior in efficacy or safety to older, standard, treatments. However, at least in clinical psychopharmacology, such aims are usually excessively idealistic, as clear outcomes favoring one treatment over another are uncommon or rare, as already noted. Most often, outcome involves similar final clinical status or levels of improvement with any two active treatments of similar kind. However, even well-established standard treatments sometimes fail to perform in any given trial for a variety of reasons, and leave the logical conundrum of deciding whether "similar to" implies "about as good as, or not." Without an inactive treatment condition (placebo), it is hard to know if the new drug is indeed about as good as an older standard agent, or if the trial simply failed. Some trial designers attempt to limit the clinical and possible ethical liability involved by arranging for smaller numbers of subjects to be randomized to placebo than to an active treatment. If "quantitative ethics" is a legitimate concept, it can also be argued that trials not involving a placebo typically require much larger numbers of participants since differences between active drugs of similar type are typically small (i.e., effect-size is small), so that larger N is needed to demonstrate a statistical difference. Conversely, drug–placebo contrasts are usually easier to detect with smaller numbers of subjects, so that overall exposure to potential risks is more limited.

A further dilemma in interpreting trial outcomes arises from the interplay among effect-size (difference between treatments assigned randomly), N, and variance (variation that reflects the precision of measurements) is whether to make more of effect-size or the probability of difference from a null condition (p-value indicating that a null [typically, no-difference] can be rejected). Many clinicians and some investigators are highly impressed by tiny p-values and pay less attention to effect-size. Indeed, some regulatory agencies consider a novel drug to be "effective" if it outperforms a comparator or placebo at the 5% probability level ($p < 0.05$). However, it is not uncommon in clinical psychopharmacology to encounter levels of effect-size that strain credulity as to clinical value. For example, differences between antipsychotic drugs and placebo in schizophrenia patients or between antidepressants and placebo in depressed patients may be on the order of 10% or less, even with probabilities below 5%. Moreover, although it can be expensive to do so, it is usually possible to "force" a low p-value by increasing N, even with very modest effect-sizes. That is, probability values (p) are smaller with (as desired, or inversely proportional to): larger effect-size (difference between treatments), larger N, and smaller measurement variance. Some regulatory agencies also set standards of efficacy or safety of new drugs based on differences in observed outcomes that meet the statistical criterion of

$p < 0.05$, despite sometimes clinically trivial effect-size. Exaggerating the significance of probabilities can also lead to pharmacologically illogical, and clinically potentially dangerous conclusions, including concepts such as "safe vs. risky" doses, or "effective vs. ineffective" doses. Notably, if an adverse effect occurs at $p > 0.05$ at a particular dosing level, it is not reasonable to conclude that the observed adverse effect is clinically unimportant. Instead, adverse effect risks, as well as benefits relate to drug dose in a continuous, logarithmic, statistical manner, such that beneficial and adverse effects are almost universally nearly linearly proportional to the logarithm of dose. Moreover, risks or benefits at particular doses and tissue concentrations of a drug can vary markedly among individuals or groups (such as juvenile vs. geriatric patients).

Drug Dosing

Scientifically sound information on relationships of doses and beneficial and adverse effects of most psychotropic drugs remains notoriously meager. Very few prospective, randomized trials seeking to assess dose-effects are reported, and estimates of clinically appropriate or approximately equivalent doses of specific drugs remain largely clinical, informal, and impressionistic [32]. Doses that have regulatory approval and appear in product information bulletins appear to balance the need to be large enough to assure substantial efficacy in typical patients, and yet to limit risks of unpleasant or toxic adverse effects, both of which are usually more or less linearly related to the logarithm of daily dose of most drugs. Given such limited information, dosing is largely left to clinical, empirical trial-and-error assessments with individual patients, starting at the low end of the recommended dosing range, and gradually increasing to gain benefits and avoid intolerable side effects. Exceeding recommended dosing limits, clinical indications or diagnoses in individual patients, as well as use of drug combinations that are not formally recognized by regulatory agencies are common practices. They are most safely carried out with the understanding that they represent informal $N = 1$ trials for individual patients for whom more standard treatments have proved to be inadequate or poorly tolerated, ideally with some indication to the patient of the "off-label" status of a recommendation, and with extra documentation in the clinical record to support such decisions. Assessments of relationships among dose, circulating serum concentrations, and clinical effects of drugs are best carried out with due consideration of typical pharmacokinetic characteristics, such that stable or steady-state tissue levels are reached in multiples of five or six times elimination half-life.

Carryover and Withdrawal Artifacts

A final consideration about the design of experimental therapeutic trials is the importance of time, both regarding duration of treatment, the timing of outcome assessments with respect to previous treatments, and the impact of treatment discontinuation. Duration of trials is typically decided in efforts to balance chances of finding robust clinical improvement while limiting early dropouts and the cost of the trial. Some conditions show beneficial effects surprisingly quickly, for example within days in cases of juvenile attention-deficit/hyperactivity disorder, and within a few weeks in acute mania, whereas major depression tends to show more improvement and better separation of drug and placebo with trials that last more than a month [74].

A particularly important aspect of the timing of therapeutic trials is the relationship of what occurs during the trial to what has preceded it. That is, many trials are affected by carry-over effects due to gradual elimination of previous treatments, especially when these are long-acting, such as with highly lipophilic agents (whose elimination half-life may be as long as weeks rather than days) or injected pro-drugs, which are hydrolyzed in the body to liberate the active parent compound and can require months to be cleared from tissue pools [18, 86]. Another commonly encountered situation is rapid discontinuation of ongoing treatment to enter an experimental trial or to shift from a short-term to long-term phase of an extended trial (treatment discontinuation trial). Both situations can lead to markedly increased symptomatic expression of the illness being treated or major relapses that may be delayed for some weeks, with significant adverse clinical effects and ethical implications as well as profoundly interfering with sound interpretation of findings from trials. Such treatment discontinuation-associated clinical worsening has been reported with mood-stabilizing [29], antipsychotic [80], and antidepressant drugs [10], and is to be distinguished from physiological withdrawal reactions that usually occur within days of discontinuing some psychotropic drugs, including benzodiazepines [62] and short-acting serotonin-reuptake inhibitors [43]. All of these reactions can be markedly limited by gradual discontinuation of psychotropic drugs, usually over several weeks, especially after long-term use of short-acting agents in relatively high doses.

Long-Term or Prophylactic Trials

Much more research is available to support relatively short-term treatment of acute phases of illness than for long-term treatment, even though most major psychiatric disorders are recurrent-episodic, or chronic with variable levels of symptomatic intensity. In part, this disparity reflects the major complexities encountered in attempting to prove long-term, preventive, or prophylactic effectiveness. These include high costs, recruiting and retaining research subjects, especially when a placebo condition is involved and as dropout rates rise with longer trial duration, maintaining blinding to treatment conditions, inclusion or control of multiple treatments or not, long-term burdens of adverse effects, and many logistical difficulties. More fundamentally, there is a lack of agreement on standards by which such trials should be designed, conducted, and interpreted [7, 35, 61, 71]. An older tradition was to enter patients in any phase of complex illnesses and to follow them with various treatment options, often including placebos, and for very long periods that usually exceeded a year. Outcome measures were often quite simple: counting recurrences or time-ill between treatments. More recently, long-term trials have relied on the methods of survival analysis, largely as a means of limiting trial duration by measuring time to a first recurrence of illness with vs. without active treatment. As noted above, the assumption that such latencies are robust indications of long-term morbidity or wellness remains poorly tested [11].

It has become common to require passing through a short-term trial for treatment of acute illness (often with a placebo control) and then to continue observations for at least several months, often after discontinuing active treatment at some point. Many acute psychiatric illnesses or exacerbations or recurrences of chronic or episodic disorders require many weeks or even months for full clinical recovery, even though major symptomatic improvement can often be documented within several weeks [75]. However, there is striking lack of agreement about when it is clinically and ethically appropriate to discontinue treatment for experimental purposes, or how to provide adequate information to participants as to qualify them for informed consent. Drug-discontinuation-associated recurrences are especially likely during incomplete clinical recovery from an acute illness, but may occur after discontinuing treatment even many months or years after full remission and sustained clinical stability [9, 81]. It is also not clear to what extent even very slow and gradual removal of a drug can limit such risks [12].

The preceding considerations indicate that incomplete clinical recovery and even late risk of reaction to relatively rapid treatment discontinuation severely complicate the design of long-term trials or extensions of short-term trials. Moreover, extension trials are often biased by requiring short-term responsiveness to a particular treatment (usually one produced by the study-sponsor's product, so-called "enrichment" designs) before being continued or re-randomized into a continuation phase that involves treatment discontinuation. Discontinuation trials of this kind may add support for short-term efficacy of a particular treatment by documenting the impact of its removal, but not provide evidence of true, long-term prophylaxis. In short, scientifically optimal and clinically appropriate, ethical, and feasible design and analysis of trials aiming to test for long-term preventive effects remain major challenges for experimental psychiatric therapeutics.

Reporting and Pooling of Trials Data

In recent years, corporate-sponsored therapeutic trials of psychotropic agents have come to dominate the research literature. Even reports of trials nominally arising from academic centers often have corporate sponsorship. In such circumstances, it would not be surprising if purely objective and scientific purposes of testing efficacy and safety of new agents were not entirely isolated from commercial aims, including licensing and product placement in markets [2, 40]. Given the close monitoring by regulatory agencies, it is unlikely that overtly misleading information is reported in corporate-sponsored findings. It is, however, likely that the design, analysis, and reporting of trials by manufacturers who know it best can be crafted so as to favor a particular product. In recent years, there have been vigorous efforts to create systematic and publicly accessible registries of treatment trials that are planned, in progress, or recently completed (e.g., [22, 89]) as well as standards accepted by journal editors for the publication of reports of trials (e.g., [45]). These procedures are designed to improve access to information by healthcare professionals and, to some extent, the general public and to limit selective and incomplete reporting [44]. Nevertheless, there is growing evidence of selective reporting of some trial findings in the peer-reviewed research literature, often in directions that tend to favor pharmaceutical manufacturers' products, even in head-to-head comparisons, which are potentially valuable but uncommon and put particular products at risk of inferior performance [42, 58, 63, 76, 79].

Biological Hypotheses in Psychiatry

The introduction of relatively effective and somewhat selective medicinal agents for the treatment of psychotic, manic-depressive, depressed, and anxious patients in the 1950s and 1960s, as well as other agents that could mimic or worsen symptoms of some illnesses encouraged formulation of biological hypotheses about the possible pathophysiology of various major mental illnesses. Such biomedical hypothesizing was further encouraged as knowledge of drug actions increased in subsequent years. For example, an association of depressed mood in some vulnerable persons given reserpine or other antiadrenergic agents to treat hypertension and in others exposed to diets deficient in aromatic amino acids that are precursors of monoamine neurotransmitters [60], as well as beneficial effects of exposure to antidepressants or stimulants, encouraged speculation that depressive illness may involve a deficiency of neurotransmission in the brain mediated by monoamines including norepinephrine and serotonin (5-hydroxytryptamine). In addition elevation of mood or worsening of mania or psychotic illness with dopaminergic stimulants, and the opposite effects of antipsychotic–antimanic drugs with potent central antidopaminergic actions encouraged a dopamine-excess hypothesis concerning mania and schizophrenia. Additional alternatives, also strongly encouraged by psychopharmacological findings, included searches for endogenous "psychotoxins" or hallucinogens in psychotic disorder patients, encouraged by the striking psychotropic actions of hallucinogens found in nature or synthesized in laboratories, including LSD (d-lysergic acid diethylamide) and alkylated tryptamines [8].

Such theorizing seemed plausible and rational, and it strongly encouraged a generation of intensive investigations of what might be considered basically "pharmacocentric" theories in a modern phase of biological psychiatry [5]. Years of efforts were made to test the implications of such hypotheses, especially by seeking evidence of corresponding biological abnormalities in psychiatric patients. Nevertheless, to date, the results have not, on balance, led to a compelling and coherent body of evidence for specific metabolic, pathophysiological changes in patients with a range of major psychiatric disorders. Many changes that have been identified may represent state-dependent conditions secondary to physiological changes associated with acute or chronic psychiatric illnesses—that is, effects, not causes. Alternative strategies as well as failures to find strong support have largely led to the abandonment of the pharmacocentric approach. Nevertheless, applications of modern metabolic, genetic, and brain imaging strategies also have led to few clear conclusions, despite growing evidence of major genetic contributions to the risk of several psychiatric disorders [14–16, 20, 33, 56].

There are several difficulties in efforts to support biological hypotheses of the pathophysiology of mental illnesses. One is that actions of psychotropic drugs which are both necessary and sufficient to account for their clinical benefits remain elusive. In addition, immediate pharmacodynamic actions of such drugs are only a small part of increasingly complex, and later-emerging actions that involve deep alterations in cellular functioning, even at the level of gene expression and regulation [19, 37, 52, 55]. There is also a risk that treatments themselves induce metabolic and physiological changes in the nervous system that can lead to artifactual research findings in patients who have been treated, especially for prolonged periods with multiple psychotropic drugs.

Moreover, many drugs in general medicine are effective but act at sites far removed and only indirectly associated with the pathophysiology of diseases. Examples include the effects of antipyretics in infections, or of diuretics in congestive heart failure. It is also an uncomfortable fact that most psychotropic drugs are not highly disorder-specific, but instead tend to be broadly useful palliative agents that produce beneficial effects in a variety of syndromes. Examples include both antipsychotic and antimanic and sometimes antidepressant effects of antipsychotic drugs, as well as antidepressant, anxiolytic, and even analgesic effects of many antidepressants. Such nonspecificity is not ideal for spinning biological hypotheses based on limited knowledge of drug actions. Moreover, decades of essentially repetitious development of increasingly chemically diverse drugs in a particular class (such as antidepressants or antipsychotics), based on conservative and limited understanding of their actions, had tended to set up and reinforce a logically circular conceptual system. This conservative cycle (Fig. 1.3) can lead not only to misleading biomedical speculation, but also impede progress in developing innovative drugs based on novel principles.

Despite the lack of compelling evidence for pathophysiological factors in major psychiatric disorders derived from speculations concerning clinical and pharmacological effects of drugs that either worsen or ameliorate symptoms of the disorders, major advances have been made in methods and technology, and in advancing basic understanding of the brain, as well as encouraging much deeper interest in brain and behavioral neurosciences, and the training of a growing cadre of scientifically competent and imaginative investigators with an interest in psychiatric disorders. Additional secondary benefits from the explosive upsurge in interest in biomedical aspects of psychiatry, strongly encouraged by the beneficial effects of psychotropic drugs, is renewed interest in descriptive psychiatry and nosology,

Pharmacocentric Cycle of Biological Psychiatry

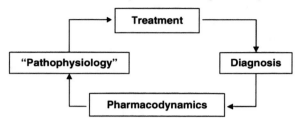

Examples

Treatment	Diagnosis	Pharmacodynamics	"Pathophysiology"
Haloperidol	Schizophrenia	Antidopaminergic	Dopamine excess
Lithium	Mania	Anticatecholaminergic	Catecholamine excess
Imipramine	Depression	Noradrenergic	Norepinephrine deficiency
Alprazolam	Panic	GABAergic	GABA deficiency
Fluoxetine	OCD	Serotonergic	Serotonin deficiency

OCD: obsessive-compulsive disorder.

Fig. 1.3 Pharmacocentric cycle of biological psychiatry. Interrelationships among drug treatments, diagnosis, partial understanding of drug actions (pharmacodynamics), and speculations about putative pathophysiology

in efforts to address the "phenotype problem" of needing more secure and credible diagnostic groups with whom to search for anatomical, metabolic or genetic anomalies. This fundamental problem is far from resolved by currently available diagnostic systems, and requires much closer collaboration among biological scientists and clinical investigators. As a final and repeated comment, it is likely that the lack of solid clinical findings to support pharmacocentric biological theorizing arises not from technical shortcomings of modern medical biology, but from oversimplifications and false expectations arising from pharmacology, particularly as applied to disorders of essentially unknown etiology.

Conclusions

The development of modern psychopharmacology has had a striking impact on psychiatry in the past half-century. Effective and relatively safe medicinal treatments are available for most of the major psychiatric disorders. These treatments have had beneficial interactions with changes in the philosophy and administration of healthcare delivery programs, and have contributed to markedly decreased importance of prolonged psychiatric hospitalization. In addition, the development of psychopharmacology has contributed to a heightened awareness of the medical and scientific traditions of psychiatry and encouraged greater reliance on clinical studies of diagnosis, psychopathology, and epidemiology as well as of pathophysiology and treatment, especially for the psychotic and major mood disorders. The scientific design and conduct of objective, controlled clinical trials of new drugs in psychiatry have become increasingly sophisticated and serve as a model for other medical specialties. An important limitation to the discovery of new psychiatric medicines is that many of the preclinical methods for screening compounds with potentially useful effects have led to more compounds with previously known effects as well as adverse actions. Psychotropic drug development remains highly empirical and remarkably conservative conceptually. Progress remains hampered by a lack of knowledge of biological causes or of precise pathophysiological bases of the major mental disorders, most of which remain idiopathic. Indeed, it remains an act of faith to believe that such cause-related biological mechanisms will be discovered. The partial success of older treatments, paradoxically, contributes further to the problem of testing new agents, in part because patients available for study are increasingly treatment-resistant, complex, and otherwise atypical, as more of the broader spectrum of patients is successfully managed with available treatments.

References

1. Adams C, Brantner V. Estimating the cost of new drug development: is it really 802 million dollars? Health Aff. 2006;25:420–8.
2. Angell M. The Truth About the Drug Companies. New York: Random House; 2004.
3. Baldessarini RJ. Frequency of Diagnoses of Schizophrenia vs. Affective Psychoses from 1944-1968. Am J Psychiatry. 1970;127:759–63.
4. Baldessarini RJ. Chemotherapy in Psychiatry. 2nd ed. Cambridge, MA: Harvard University Press; 1985.
5. Baldessarini RJ. American Biological Psychiatry and Psychopharmacology 1944–1994 (Chap. 16). In: Menninger RW, Nemiah JC, editors. American Psychiatry After World War II (1944–1994), a Volume Celebrating the 150th Anniversary of the Founding of the American Psychiatric Association. Washington, DC: APA; 2000. p. 371–412.
6. Baldessarini RJ. Drug Therapy of Depression and Anxiety Disorders (Chap. 17). In: Brunton LL, Lazo JS, Parker KL, editors. Goodman and Gilman's the Pharmacological Basis of Therapeutics. 11th ed. New York: McGraw-Hill; 2005. p. 429–59.
7. Baldessarini RJ. Long-term drug efficacy in major psychiatric disorders: problems and opportunities. In: Proceedings of an International Society for Clinical Trials and Methodology (ISCTM) Annual Meeting, Chicago, IL. 2006. http://www.isctm.org/public_access/r_baldessarini-jul06-session1present.pdf.
8. Baldessarini RJ, Stramentinoli G, Lipinski JF. Methylation hypothesis. Arch Gen Psychiatry. 1979;36:303–7.
9. Baldessarini RJ, Suppes T, Tondo L. Lithium withdrawal in bipolar disorder: Implications for clinical practice and experimental therapeutics research. Am J Ther. 1996;3:492–6.
10. Baldessarini RJ, Tondo L, Ghiani C, Lepri B. Discontinuation rate vs. recurrence risk following long-term antidepressant treatment in major depressive disorder patients. Am J Psychiatry. 2010;167:934–41.
11. Baldessarini RJ, Tondo L, Lepri B. Empirical test of a basic assumption underlying survival analysis as applied to experimental therapeutics in psychopharmacology. J Clin Psychopharmacol. 2010;30:72–5.
12. Baldessarini RJ, Tondo L, Viguera AC. Discontinuing psychotropic agents. J Psychopharmacol. 1999;13:292–3.
13. Ban TA. The role of serendipity in drug discovery. Dialogues Clin Neurosci. 2006;8:335–44.

14. Bondy B. Genetics in psychiatry: are the promises met? World J Biol Psychiatry. 2011;12:81–8.
15. Brown GG, Thompson WK. Functional brain imaging in schizophrenia: selected results and methods. Curr Topics Behav Neurosci. 2010;4:181–214.
16. Burmeister M, McInnis MG, Zöllner S. Psychiatric genetics: progress amid controversy. Nat Rev Genet. 2008;9:527–40.
17. Campbell SF. Science, art and drug discovery: a personal perspective. Clin Sci (Lond). 2000;99:255–60.
18. Campbell A, Baldessarini RJ, Teicher MH, Kula NS. Prolonged antidopaminergic actions of single doses of butyrophenones in the rat. Psychopharmacology (Berl). 1985;87:161–6.
19. Chen G, Huang LD, Zeng WZ, Manji HK. Mood stabilizers regulate cytoprotective and mRNA-binding proteins in the brain: long-term effects on cell survival and transcript stability. Int J Neuropsychopharmacol. 2001;4:47–64.
20. Chen CH, Suckling J, Lennox BR, Ooi C, Bullmore ET. Quantitative meta-analysis of fMRI studies in bipolar disorder. Bipolar Disord. 2011;13:1–15.
21. Chen AA, Thomas DK, Ong LL, Schwartz RE, Golub TR, Bhatia SN. Humanized mice with ectopic artificial liver tissues. Proc Natl Acad Sci USA. 2011;108:11842–7.
22. Clinicaltrials.gov. Clinical trials registration. 2011. http://www.clinicaltrials.gov/ct2/info.
23. Collett D. Modeling Survival Data in Medical Research. 2nd ed. Boca Raton, FL: Chapman & Hall/CRC; 2003.
24. Collins PY, Patel V, Joestl SS, March D, Insel TR, Daar AS. Grand challenges in global mental health. Nature. 2011;475:27–30.
25. Dahl SG, Sylte I. Molecular modeling of drug targets: the past, the present and the future. Basic Clin Pharmacol Toxicol. 2005;96:151–5.
26. Dedeurwaerdere S, Wintmolders C, Straetemans R, Pemberton D, Langlois X. Memantine-induced brain activation as a model for the rapid screening of potential novel antipsychotic compounds: exemplified by activity of an mGlu2/3 receptor agonist. Psychopharmacology (Berl). 2011;214:505–14.
27. Dolle RE, Bourdonnec BL, Worm K, Morales GA, Thomas CJ, Zhang W. Comprehensive survey of chemical libraries for drug discovery and chemical biology: 2009. J Comb Chem. 2010;12:765–806.
28. Eddy DM. Evidence-based medicine: a unified approach. Health Aff. 2005;24:9–17.
29. Faedda GL, Tondo L, Baldessarini RJ, Suppes T, Tohen M. Outcome after rapid vs. gradual discontinuation of lithium treatment in bipolar mood disorders. Arch Gen Psychiatry. 1993;50:448–55.
30. Forbes.com. America's most popular psychiatric drugs. 2010. http://www.forbes.com/2010/09/16/prozac-xanax-valium-business-healthcare-psychiatric-drugs.html.
31. Fransen V, editor. Mental Health Services in the United States and England: Struggling for Change. Princeton, NJ: Robert Wood Johnson Foundation; 1991.
32. Gardner DM, Murphy AL, O'Donnell H, Centorrino F, Baldessarini RJ. International consensus study of antipsychotic dosing. Am J Psychiatry. 2010;167:686–93.
33. Gershon ES, Alliey-Rodriguez N, Liu C. After GWAS: searching for genetic risk for schizophrenia and bipolar disorder. Am J Psychiatry. 2011;168:253–6.
34. Glass GV. Integrating findings: the meta-analysis of research. Rev Res Educ. 1977;5:351–79.
35. Glick ID. Understanding the results of CATIE in the context of the field. CNS Spectr. 2006;11 Suppl 7:40–7.
36. Greenberg PE, Stiglin LE, Finkelstein SN, Berndt ER. Depression: a neglected major illness. J Clin Psychiatry. 1993;54:419–24.
37. Haenisch B, Bönisch H. Depression and antidepressants: insights from knockout of dopamine, serotonin or noradrenaline re-uptake transporters. Pharmacol Ther. 2011;129:352–68.
38. Harcourt BE. The mentally ill, behind bars. New York Times, 15 Jan 2007, p. A19.
39. Harris G. Deal in place for inspecting foreign drugs. New York, NY: New York Times; 13 Aug 2011. http://www.nytimes.com/2011/08/13/science/13drug.html?scp=2&sq=Gardiner%20Harris&st=cse.

40. Healy D. The Creation of Psychopharmacology. Cambridge, MA: Harvard University Press; 2002.
41. Hennen J. Statistical methods for longitudinal research on bipolar disorders. Bipolar Disord. 2003;5:156–68.
42. Heres S, Davis JM, Maino K, Jetzinger E, Kissling W, Leucht S. Why olanzapine beats risperidone, risperidone beats quetiapine, and quetiapine beats olanzapine: exploratory analysis of head-to-head comparison studies of second-generation antipsychotics. Am J Psychiatry. 2006;163:184–94.
43. Himei A, Okamura T. Discontinuation syndrome associated with paroxetine in depressed patients: a retrospective analysis of factors involved in the occurrence of the syndrome. CNS Drugs. 2006;20:665–72.
44. Hirsch L. Trial registration and results disclosure: impact of US legislation on sponsors, investigators, and medical journal editors. Curr Med Res Opin. 2008;24:1683–9.
45. International Committee of Medical Journal Editors (ICJME). Uniform requirements for manuscripts. 2011. http://www.icmje.org/urm–main.html.
46. Isaac RJ, Armat VC. Madness in the Streets. New York: Simon & Schuster; 1990.
47. Kessler RC, McGonagle KA, Zhao S, Nelson CB, Hughes M, Eshleman S, et al. Lifetime and 12-month prevalence of DSM-III-R psychiatric disorders in the United States. Arch Gen Psychiatry. 1994;51:8–19.
48. Kiselyuk A, Farber-Katz S, Cohen T, Lee SH, Geron I, Azimi B, et al. Phenothiazine neuroleptics signal to the human insulin promoter as revealed by a novel high-throughput screen. J Biomol Screen. 2010;15:663–70.
49. Lazo JS, Wipf P. Combinatorial chemistry and contemporary pharmacology. J Pharmacol Exp Ther. 2000;293:705–9.
50. Leucht S, Corves C, Arbter D, Engel RR, Li C, Davis JM. Second-generation versus first-generation antipsychotic drugs for schizophrenia: meta-analysis. Lancet. 2009;373:31–41.
51. Leucht S, Heres S, Kissling W, Davis JM. Evidence-based pharmacotherapy of schizophrenia. Int J Neuropsychopharmacol. 2011;14:269–84.
52. Lesch KP. Serotonergic gene expression and depression: implications for developing novel antidepressants. J Affect Disord. 2001;62:57–76.
53. Lieberman JA, Stroup TS, McEvoy JP, Swartz MS, Rosenheck RA, Perkins DO, et al. Clinical Antipsychotic Trials of Intervention Effectiveness (CATIE) Investigators. Effectiveness of antipsychotic drugs in patients with chronic schizophrenia. N Engl J Med. 2005;53:1209–23.
54. Maggon K. Pharmaceutical market intelligence monograph. 2010. http://knol.google.com/k/global-pharmaceutical-market-intelligence-monograph.
55. Manji HK, Moore GJ, Chen G. Lithium up-regulates the cytoprotective protein Bcl-2 in the CNS in vivo: a role for neurotrophic and neuroprotective effects in manic depressive illness. J Clin Psychiatry. 2000;61 Suppl 9:82–96.
56. Mantripragada KK, Carroll LS, Williams NM. Experimental approaches for identifying schizophrenia risk genes. Curr Topics Behav Neurosci. 2010;4:587–610.
57. Meyers M. Happy Accidents: Serendipity in Modern Medical Breakthroughs. New York: Arcade; 2007.
58. Moreno SG, Sutton AJ, Turner H, Abrams KR, Cooper NJ, Palmer TM, et al. Novel methods to deal with publication biases: secondary analysis of FDA trial registry database and related journal publications. BMJ. 2009;339:2981–6.
59. Pang KS, Durk MR. Physiologically-based pharmacokinetic modeling for absorption, transport, metabolism and excretion. J Pharmacokinet Pharmacodyn. 2010;37:591–615.
60. Parker G, Brotchie H. Mood effects of the amino acids tryptophan and tyrosine. Acta Psychiatr Scand. 2011;124:417–26.
61. Pecenak J. Relapse prevention in schizophrenia: evidence from long-term, randomized, double-blind clinical trials. NeuroEndocrinol Lett. 2007;28 Suppl 1:49–70.
62. Podhorna J. Experimental pharmacotherapy of benzodiazepine withdrawal. Curr Pharm Des. 2002;8:23–43.

63. Rief W, Nestoriuc Y, von Lilienfeld-Toal A, Dogan I, Schreiber F, Hofmann SG, et al. Differences in adverse effect reporting in placebo groups in SSRI and tricylcic antidepressant trials: systematic review and meta-analysis. Drug Saf. 2009;32:1041–56.
64. Sackett DL, Rosenberg WM. The need for evidence-based medicine. J R Soc Med. 1995;88:620–4.
65. Salvatore P, Baldessarini RJ, Tohen M, Khalsa HMK, Perez Sanchez-Toledo J, Zarate Jr CA, et al. McLean-Harvard International First-Episode Project: Two-year stability of DSM-IV diagnoses in 500 first-episode psychotic disorder patients. J Clin Psychiatry. 2009; 70:458–66.
66. Schou M, Sóvágó J, Pike VW, Gulyás B, Bøgesø KP, Farde L, et al. Synthesis and positron emission tomography evaluation of three norepinephrine transporter radioligands: [¹¹C]desipramine, [¹¹C]talopram and [¹¹C]talsupram. Mol Imaging Biol. 2006;8:1–8.
67. Shakespeare W. Macbeth (published ca. 1603–1611). In: Boyce (editor). Encyclopedia of Shakespeare. New York: Roundtable Press.
68. Sit SY. Dopamine agonists in the treatment of Parkinson's disease past, present and future. Curr Pharmaceut Design. 2000;6:1211–48.
69. Stoll AL, Tohen M, Baldessarini RJ. Increasing diagnostic frequency of obsessive-compulsive disorder. Am J Psychiatry. 1992;149:38–640.
70. Straus SE, Richardson WS, Glasziou P, Haynes RB. Evidence-Based Medicine. 3rd ed. Edinburgh: Churchill-Livingstone; 2005.
71. Swann AC. Long-term treatment in bipolar disorder. J Clin Psychiatry. 2005;66(Suppl):7–12.
72. Swazey J. Chlorpromazine in Psychiatry: A study in Therapeutic Innovation. Cambridge, MA: MIT; 1974.
73. TalentMash. Generic drugs: yes or no? 2011. http://talentmash.com/generic-drugs-yes-or-no.
74. Tedeschini E, Fava M, Papakostas GI. Placebo-controlled, antidepressant clinical trials cannot be shortened to less than 4 weeks' duration: a pooled analysis of randomized clinical trials employing a diagnostic odds ratio-based approach. J Clin Psychiatry. 2011;72:98–118.
75. Tohen M, Zarate Jr CA, Hennen J, Kaur Khalsa HM, Strakowski SM, Gebre-Medhin P, et al. The McLean-Harvard First-Episode Mania Study: prediction of recovery and first recurrence. Am J Psychiatry. 2003;160:2099–107.
76. Turner EH, Matthews AM, Linardatos E, Tell RA, Rosenthal R. Selective publication of antidepressant trials and its influence on apparent efficacy. N Eng J Med. 2008;358:252–90.
77. Undurraga J, Baldessarini RJ. A 30-year meta-analytic review of antidepressant efficacy in acute major depression. Neuropsychopharmacology. 2012;37:851–64.
78. Vázquez G, Baldessarini RJ, Yildiz A, Tamayo J, Tondo L, Salvatore P. Multi-site international collaborative clinical trials in mania: commentary. Int J Neuropsychopharmacol. 2011;14:1013–6.
79. Vedula SS, Bero L, Scherer RW, Dickersin K. Outcome reporting in industry-sponsored trials of gabapentin. N Engl J Med. 2009;361:1963–71.
80. Viguera AC, Baldessarini RJ, Hegarty JD, van Kammen DP, Tohen M. Clinical risk following abrupt and gradual withdrawal of maintenance neuroleptic treatment. Arch Gen Psychiatry. 1997;54:49–55.
81. Viguera AC, Baldessarini RJ, Friedberg J. Risks of interrupting continuation or maintenance treatment with antidepressants in major depressive disorders. Harv Rev Psychiatry. 1998;5:293–306.
82. Yildiz A, Vieta E, Leucht S, Baldessarini RJ. Efficacy of antimanic treatments: meta-analysis of randomized, controlled trials. Neuropsychopharmacology. 2011;36:375–89.
83. Yildiz A, Vieta E, Tohen M, Baldessarini RJ. Factors modifying drug and placebo responses in randomized trials for bipolar mania. Int J Neuropsychopharmacol. 2011;14:863–75.
84. Wang J, Maurer L. Positron emission tomography: applications in drug discovery and drug development. Curr Topics Med Chem. 2005;5:1053–75.
85. Weaver DF, Weaver CA. Exploring neurotherapeutic space: how many neurological drugs exist (or could exist)? J Pharm Pharmacol. 2011;63:136–9.

86. Wistedt B, Wiles D, Kolakowska T. Slow decline of plasma drug and prolactin levels after discontinuation of chronic treatment with depot neuroleptics. Lancet. 1981;1(8230):1163–4.

87. Wittchen HU, Jacobi F, Rehm J, Gustavsson A, Svensson M, Jönsson B, Olesen J. Size and burden of mental disorders and other disorders of the brain in Europe in 2010. Eur Neuropsychopharmacol. 2011;21:655–79.

88. Woolley SB, Cardoni AA, Goethe JW. Last-observation-carried-forward imputation method in clinical efficacy trials: review of 352 antidepressant studies. Pharmacotherapy. 2009;29: 1408–16.

89. World Health Organization (WHO). International clinical trials registry platform (ICTRP). 2011. http://www.who.int/ictrp/en/.

Chapter 2
Antipsychotic Agents

There is no insanity so devastating in man's life as utter sanity.

William Allen White

Introduction

The start of the era of modern psychopharmacology can be dated from 1949, with the introduction of lithium carbonate in Melbourne, Australia, as a selective and effective antimanic agent. However, the earliest antipsychotic drugs appeared soon after that, and exerted immediate and truly revolutionary changes in the care of the severely mentally ill. The first modern antipsychotics were the phenothiazines, starting with the discovery of antimanic and antipsychotic properties of chlorpromazine in Paris in 1952. In the following year, reserpine was isolated and chemically identified [180] as a major alkaloid of *Rauwolfia serpentina* (snakeroot), a crude plant material used for centuries in traditional Vedic folk medicine in India for psychosis and agitation and known to the Western medicine since the 1930s [184]. *Reserpine* (contained in the alkaloid mixture, Serpasil®; Fig. 2.1) and several other *Rauwolfia* alkaloids were found to interfere with the intraneuronal storage of monoamine neurotransmitters, and induce their rapid depletion from tissue stores. These drugs maintained a role in the treatment of hypertension for some time, but were abandoned for use in psychiatry since the relatively high doses required to obtain apparent antipsychotic effects were poorly tolerated due to the profound sedating, antiadrenergic, and gastrointestinal effects of these agents, and their much less favorable clinical effects than the emerging phenothiazines [25, 113]. A surviving monoamine-depleting agent is *tetrabenazapine* (Nitoman®; Fig. 2.1), a much simpler molecule than reserpine, that was considered briefly as a potential antipsychotic drug and has been used as an antidyskinetic agent to treat tardive dyskinesia induced by neuroleptic agents, as well as Huntington's chorea [133, 136].

R.J. Baldessarini, *Chemotherapy in Psychiatry: Pharmacologic Basis of Treatments for Major Mental Illness*, DOI 10.1007/978-1-4614-3710-9_2,
© Springer Science+Business Media New York 2013

Fig. 2.1 Structures of
reserpine and tetrabenazine

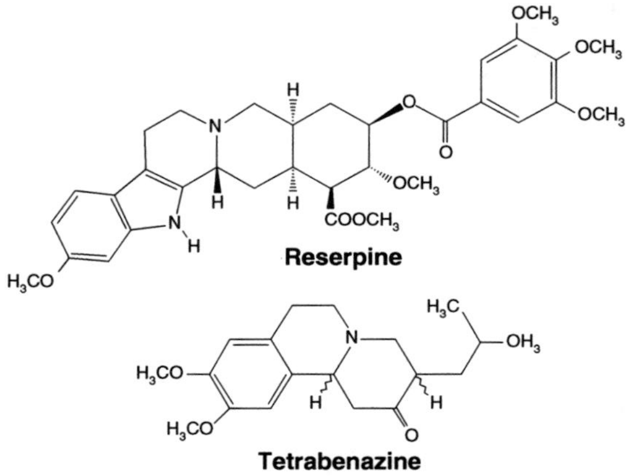

Reserpine

Tetrabenazine

The first clinically successful antipsychotic agent was the phenothiazine, *chlorpromazine* (Thorazine®). It was developed by the Rhône-Poulenc Laboratories in France. The phenothiazine nucleus of the molecule had been synthesized in the late nineteenth century with the development of dyes derived from aniline (aminobenzene), such as the thionine or phenothiazine compound methylene blue, synthesized in 1876 [65]. The history of the study of these dye-stuffs is intimately related to the early development by Paul Ehrlich and others of the theory of specific drug–tissue interactions, a cornerstone of general pharmacology [18]. Like several of the thionine dyes, the molecule phenothiazine itself was for a time used clinically as an antimicrobial agent.

In the late 1930s, a phenothiazine derivative, *promethazine* (Phenergan®), was noted to have antihistaminic and sedative properties by Paul Charpentier at Rhône-Poulenc. Further chemical modifications of promethazine were made in hopes of developing additional unique molecules with potentially clinically useful effects on the nervous system. The eventual stunning success of chlorpromazine in this developmental process was unexpected since several centrally-active antihistamine drugs had been tried in psychiatric patients in the 1940s, with little evidence of particularly useful beneficial effects compared with other sedatives. Chlorpromazine was first tried clinically in 1951 as a preanesthetic sedative by the Parisian military surgeon Henri Laborit, who described some of its peculiar effects on behavior ("artificial hibernation") that included retention of consciousness associated with striking indifference to the surroundings. In addition to possible value as a preanesthetic sedative, it was hoped to limit physiological responses to surgical shock ("vegetative stabilization"), and it disrupted physiological regulation of body-temperature (poikilothermic effect) that facilitated induction of hypothermia in general medicine.

In 1951–1952 in Paris, several psychiatrists noted the ability of the new phenothiazine to increase the efficacy with which barbiturates sedated manic and agitated psychotic patients. In 1952–1953 at Ste. Anne's Hospital in Paris, psychiatrists Jean Delay and Pierre Deniker reported further experience with the new drug, and used it alone, initially for acute mania, and gradually for schizophrenia and other psychotic disorders to control agitation. Early clinical experience with chlorpromazine established its regular association with useful clinical effects and with abnormalities of posture and movement (including dystonias, restlessness, bradykinesia, and dyskinesias), that supported the appellation "neuroleptic" (seizing the nervous system) for it and its later congeners with similar broad effects on the central nervous system (CNS). Gradually, it also became clear that chlorpromazine was more than a potent sedative, and that it exerted effects on emotional responses associated with delusions and hallucinations to decrease their intensity, as well as reducing general excitement and agitation, so as to suggest particular potential utility in the treatment of psychotic disorders typified by schizophrenia [194, 204]. Chlorpromazine was given as early as 1954 in Canada and the United States [123, 229]. Based on the initial experience with chlorpromazine, many pharmaceutical companies recognized an important innovative discovery and rapidly developed a series of chemically analogous compounds in the late 1950s, several of which became established antipsychotic drugs by the 1960s [21, 88, 113].

Since the discovery of chlorpromazine in the early 1950s, a large and growing number of compounds have been proved to be effective in the clinical management of a broad range of syndromes marked by agitation and psychosis, and found to be particularly useful in the treatment of schizophrenia and mania [1, 18, 117]. There is a current tendency to separate available antipsychotic drugs, somewhat arbitrarily, into those introduced before and since the 1980s, although the chemistry, pharmacology, and clinical effects of older and newer drugs vary greatly. Since most of the older agents, in addition to their antipsychotic actions, produce abnormalities of movement and posture (so-called "extrapyramidal syndromes" [EPS] despite limited knowledge of their pathophysiology), they have been termed "neuroleptic." They are also known as "typical" or "first-generation agents." Most such neuroleptic antipsychotic agents are chemically classified as *phenothiazines* (e.g., chlorpromazine, fluphenazine, perphenazine, thioridazine, trifluperazine, and others), thioxanthenes (such as thiothixene), *phenylpiperidines* (including the butyrophenone haloperidol and closely related diphenylbutylpiperidines such as pimozide), or *dibenzapines* (e.g., clozapine, loxapine and analogous asenapine). More recently introduced agents include a variety of heterocyclic compounds (e.g., aripiprazole, iloperidone, lurasidone, olanzapine, paliperidone, quetiapine, risperidone, ziprasidone), most of which were developed by mimicking the chemical structure or some of the complex pharmacodynamic actions of clozapine.

Clozapine is a tricyclic dibenzodiazepine compound with a seven-member central ring. It is an unusually effective, but potentially toxic agent. Though initially patented as early as 1960, it is usually included among the modern or so-called "atypical" or "second-generation" antipsychotics for two main reasons. First, it has far lower risks of inducing some forms of adverse neurological effects, notably dystonia, bradykinesia, and dyskinesias, than other older neuroleptic-type antipsychotics. In addition, it fell into clinical disfavor during the 1970s owing to a high risk of potentially lethal bone-marrow dysfunction leading to agranulocytosis, but was reintroduced with FDA-approval in 1989 with specific restrictions. These included use for chronically psychotic patients diagnosed with schizophrenia, who failed at least two vigorous trials of alternative antipsychotic drugs, and a system of required, regular monitoring of white blood cell counts (WBC) for safety [15].

Pharmacology

Types of Antipsychotic Drugs

Currently clinically used, effective, antipsychotic agents include several chemical classes. The first discovered were the *phenothiazines*—tricyclic compounds containing two benzene rings ("pheno"), as well as sulfur ("thio"), and nitrogen ("azo") atoms in a six-atom central ring. Antipsychotic phenothiazines include *chlorpromazine* (Thorazine®), *fluphenazine* (Prolixin®), *mesoridazine* (Serentil®), *perphenazine* (Trilafon®), *thioridazine* (Mellaril®), and others. Most of these compounds differ only in the chemical structure of their side-chains attached at the central ring (position 10) or in electronegative ring-substituents (position 2). The side-chains can be aliphatic or aminoaklyl moieties with straight carbon-chains, or can include nitrogen-containing cyclic components (piperidine or piperazine rings). They consistently follow the structure-activity rule of: central ring-nitrogen, three carbons, and an evidently functionally critical, doubly substituted or secondary-amino nitrogen atom (Fig. 2.2). The piperazine derivatives are usually of higher potency, but not greater clinical efficacy, and have relatively lesser sedative and hypotensive effects (Table 2.1).

Later, more varied structures were considered, including compounds that substituted the sulfur or nitrogen atom of the central ring of phenothiazines with a carbon atom to produce additional tricyclic analogues with similar neuropharmacological and behavioral properties. These have included a series of experimental *acridanes* (retain nitrogen), and *thioxanthenes* (retain sulfur but not nitrogen in the central ring; not to be confused with the xanthine alkaloids including caffeine, theobromine, and theophylline). Several thioxanthenes reached clinical application (including chlorprothixene with an aliphatic side-chain like that of chlorpromazine, and the piperazine thiothixene, as well as clopenthixol and its *cis*-isomer zuclopenthixol, and flupenthixol). These compounds have side chains attached to a carbon atom in the central ring by a double-bond, which provides geometric isomers, only one of which (*cis*) is active as a neuroleptic [195]. This molecular arrangement affords a close conformational relationship of an electronegative ring-substituent (at position 2: a chlorine atom in chlorprothixene and clopenthixol, CF_3 in flupenthixol, $SO_2N[CH_3]_2$ in thiothixene) to a side-chain secondary amino or piperazine nitrogen atom (Fig. 2.3). Of the thioxanthenes, only *thiothixene* (Navane®) continues in clinical use as an antipsychotic drug in the United States.

Other tricyclic compounds include the *dibenzepines* and related agents, initially represented by *clozapine* (Clozaril®) and *loxapine* (Loxitane®), and later to include *asenapine* (Saphris®), *olanzapine* (Zyprexa®), and *quetiapine* (Seroquel®) as well

Tricyclic Neuroleptic-Antipsychotic Drugs

Phenothiazines

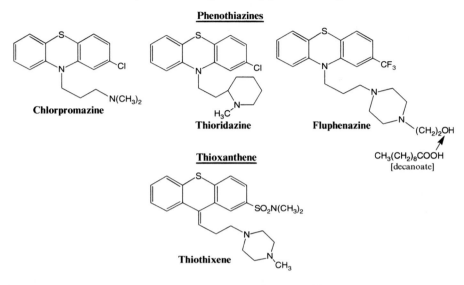

Fig. 2.2 Structures of tricyclic neuroleptic-antipsychotic agents. Of the phenothiazines, chlorpromazine has an aliphatic side-chain, thioridazine a piperidine-containing side chain, and fluphenazine, and thiothixene (a thioxanthene) have piperazine-containing side-chains. Fluphenazine can be esterified at the free OH group, to form the long-acting injectable decanoate

(Fig. 2.3). All of these agents are used as antipsychotic drugs. All of these antipsychotic tricyclic molecules have structural conformations that allow them to superimpose on the preferred conformation of the neurotransmitter dopamine, suggesting a common theme of an active center ("pharmacophore") with structural analogy to dopamine [18, 66].

Additional antipsychotic molecules of dissimilar chemical structure include a series of *butyrophenones* (haloperidol, droperidol) and *other heterocyclic* compounds (e.g., amisulpride, aripiprazole, molindone, iloperidone, lurasidone, paliperidone, pimozide, risperidone, and ziprasidone; Fig. 2.3). Haloperidol is important historically as it represented a break from the line of development of tricyclic antipsychotic molecules modeled after chlorpromazine. This butyrophenone (or phenylbutylpiperidine) is a derivative of the potent phenylpiperidine, narcotic analgesic agent *meperidine* (or pethidine; Demerol®), prepared initially in a search for novel analgesics by Paul Janssen in Belgium in 1959 [95]. Haloperidol arose from pursuing activities in intermediate compounds that included analgesic effects as well as unexpected effects on animal behavior that resembled the actions of phenothiazines and thioxanthenes. Haloperidol turned out to be a highly potent and effective antipsychotic agent without analgesic activity, but with neurological effects on posture and movement typical of those of the older tricyclic neuroleptics. It became one of the most commonly employed antipsychotic agents, worldwide, until the 1990s.

Table 2.1 Characteristics of antipsychotic agents

| Drug | Brand | Half-life (h) | Doses (mg/day) | | Injectable | Preferred CYP | Adverse effect risks | | |
			Typical	Extreme			EPS	Sedation	Hypotension
Phenothiazines									
Chlorpromazine	Thorazine	24	200–600	30–1,500	Yes	1A2, 2D6, 3A4	++	+++	++/+++
Fluphenazine	Prolixin	24	2–20	0.5–30	Yes+Depot	2D6	++++	+	+
Mesoridazine	Serentil	30	75–300	30–400	Yes	1A2, 2D6, 3A4	+	+++	++
Perphenazine	Trilafon	12	8–32	4–64	Yes	2D6	++	++	+
Thioridazine	Mellaril	22	150–600	20–750	No	1A2, 2D6	+	+++	+++
Trifluoperazine	Stelazine	20	5–20	2–30	Yes	1A2, 2D6	+++	+	+
Triflupromazine	Vesprin	–	50–100	10–150	Yes	–	++	+++	++/+++
Thioxanthene									
Thiothixene	Navane	18	5–30	2–30	Yes	1A2	+++	+/++	++
Phenylpiperidines									
Droperidol	Inapsine	2	2.5–5.0	1.25–10	Yes	–	++++	++++	+++
Haloperidol	Haldol	20	4–20	1–50	Yes+Depot	3A4	++++	+	+
Pimozide	Orap	55	2–6	1–10	No	3A4	++++	+	+
Benzepines									
Clozapine	Clozaril	12	150–450	12.5–900	No	1A2, 2D6, 3A4	+/–	+++	+++
Loxapine	Loxitane	16	60–100	20–250	Yes	–	++/+++	+	+
Olanzapine	Zyprexa	30	5–20	2.5–30	Yes+Depot	1A2	+	++	++
Quetiapine	Seroquel	6.5	300–600	50–800	No	3A4	+/–	+++	++
Other heterocyclics									
Aripiprazole	Abilify	75	10–20	5–30	Yes	3A4, 2D6	+/–	+/–	+/–
Asenapine	Saphris	24	10–20	5–20	No	1A2, 2D6	+/–	++	++
Iloperidone	Fanapt	26	12–24	2–32	No	2D6, 3A4	+/–	+/–	++
Lurasidone	Latuda	18	40–80	40–80	No	3A4	+	++	+
Molindone	Moban	2	50–200	15–225	No	2D6	+/++	++	+
Paliperidone	Invega	22	6–12	3–15	No	–	++	++	++
Risperidone	Risperidal	3+22	6–8	0.25–16	Yes+Depot	2D6, 3A4	++	++	++
Ziprasidone	Geodon	7.5	80–160	20–180	Yes	3A4	+/–	–	–

Limited to drugs used clinically in the United States. Common brand-names are provided. Half-life is nominal plasma elimination half-life (second values are for major active metabolites). CYP are cytochrome-P450 hepatic microsomal metabolizing enzymes. Adverse effect risks (from + to ++++) are for extrapyramidal neurological (EPS), sedative, and hypotensive effects. *Depot*: long-acting, injectable.

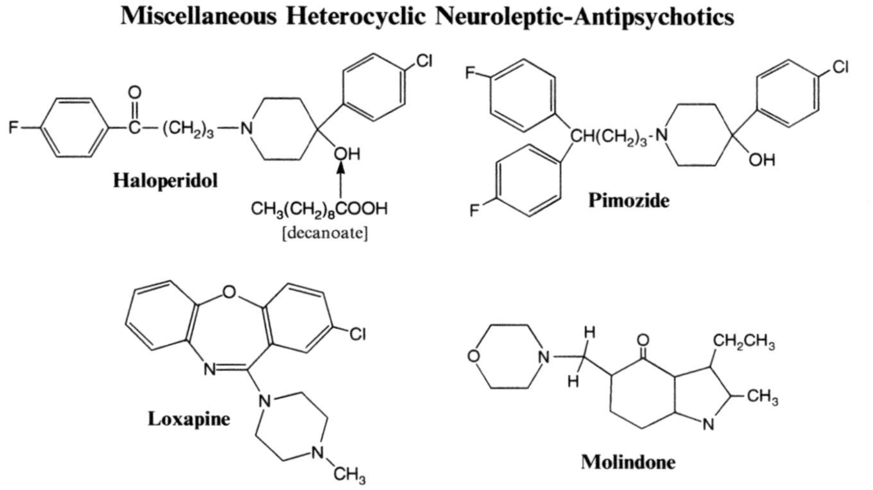

Miscellaneous Heterocyclic Neuroleptic-Antipsychotics

Fig. 2.3 Structures of miscellaneous neuroleptic-antipsychotics. Haloperidol (a butyrophenone) and pimozide (a diphenylbutylpiperadine) are phenylbutylpiperadines derived originally from the analgesic meperidine. The free OH group of haloperidol can be esterified to produce the long-acting injectable decanoate. Loxapine is structurally similar to clozapine except that its side-chain piperidine nitrogen atoms are closer to the ring-chlorine, preserving homology to dopamine and making it a typical neuroleptic. Molindone is one of the few indole antipsychotics

Other phenylbutylpiperidine compounds chemically related to haloperidol include the butyrophenone *droperidol* (Inapsine®), and several dephenylpiperidines, including *pimozide* (Orap®), *penfluridol* (Semap®), and *fluspiriline* (Imap®). All of these agents are potent neuroleptic-type compounds with antipsychotic and neurological effects. Droperidol is used mainly as a co-anesthetic agent and in combination with the potent narcotic analgesic *fentanyl* (Duragesic®), another substituted phenylpiperidine. Droperidol has sometimes been used as a potent sedative-neuroleptic for psychiatric emergencies, or as an antiemetic drug [48]. Pimozide is a potent, long-acting, and effective antipsychotic agent that is licensed in the United States only for the treatment of Gilles de la Tourette's disorder (Tourette syndrome) of tics and coprolalia [187].

The older dibenzazepine clozapine as well as newer ("second-generation") antipsychotics modeled after the chemical structure of clozapine or designed to mimic some of its pharmacological characteristics, are often considered *"atypical"* for having relatively low risk of most extrapyramidal side effects (EPS), particularly dystonia and bradykinesia (Fig. 2.4). Following many years without notable innovation in antipsychotic drug development, the following "second-generation" drugs were licensed for clinical use in the United States since *clozapine* was FDA-approved in 1989 with restrictions: *risperidone* (1993), *olanzapine* (1997), *quetiapine* (1998), *ziprasidone* (2001), *aripiprazole* (2003), *paliperidone* (2006), *asenapine* (2009), *iloperidone* (2009), and *lurasidone* (2010). The approximate clinical potencies of antipsychotic drugs relative to chlorpromazine range from 60% less potent (quetiapine) to 120 times more potent (droperidol) (Table 2.2).

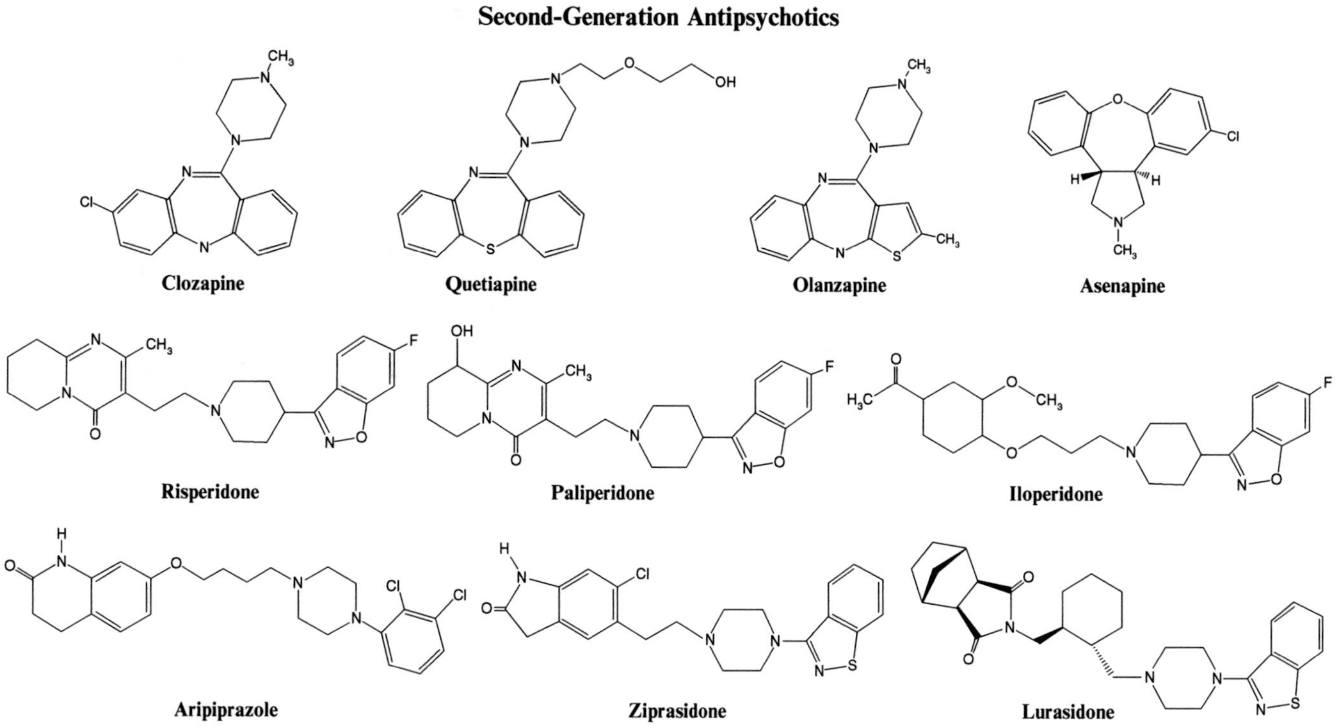

Second-Generation Antipsychotics

Fig. 2.4 Structures of second-generation antipsychotic agents. Note that the configuration of clozapine has the piperidine nitrogen atoms in the side chain maximally separated from the ring-chlorine, in contrast to loxapine, and limiting its homology to dopamine

Table 2.2 Equivalent doses of antipsychotics

Drug	Median dose (mg/day)	Relative potency	CPZ-eq factor
Phenothiazines			
Chlorpromazine	300	Low	1.0
Fluphenazine	7.5	High	40
Mesoridazine	250	Low	1.2
Perphenazine	24	Intermediate	12.5
Thioridazine	300	Low	1.0
Trifluoperazine	15	High	20
Triflupromazine	80	Intermediate	3.8
Thioxanthene			
Thiothixene	15	High	20
Phenylpiperidines			
Droperidol	2.5	High	120
Haloperidol	7.5	High	40
Pimozide	3.0	High	100
Benzamides			
Clozapine	200	Low	1.5
Loxapine	30	Intermediate	10
Olanzapine	15	High	20
Quetiapine	500	Low	0.6
Other heterocyclics			
Aripiprazole	20	High	15
Asenapine	15	High	20
Iloperidone	18	High	16.7
Lurasidone	60	Intermediate	5.0
Molindone	30	Intermediate	10
Paliperidone	6.0	High	50
Risperidone	4.0	High	75
Ziprasidone	120	Intermediate	2.5

CPZ-eq factor converts dose to the approximate daily equivalent of 300 mg of chlorpromazine as a standard [71]; e.g., 10 mg/day of fluphenazine is similar in clinical effect to 400 mg of chlorpromazine

In addition to a growing variety of chemical structures with antipsychotic activity, several agents are available as preparations given by injection for long-acting effects. Most are esters of highly lipophilic, long-chain fatty acids (most are ten-carbon decanoates or seven-carbon enanthates), covalently attached at a free hydroxyl group of the parent molecule. These so-called "depot" preparations (injected intramuscularly in an oily vehicle for slow absorption) are also "pro-drugs" in that they are hydrolyzed in the body by nonspecific esterase enzymes to liberate an active parent molecule (such as fluphenazine, haloperidol or paliperidone). Other long-acting preparations include physical-chemical complexes of the unmodified parent drug molecule, such as in carbohydrate microspheres for risperidone, or with the lipophilic, polycyclic, naphtholic acid derivative pamoate for olanzapine (Fig. 2.5).

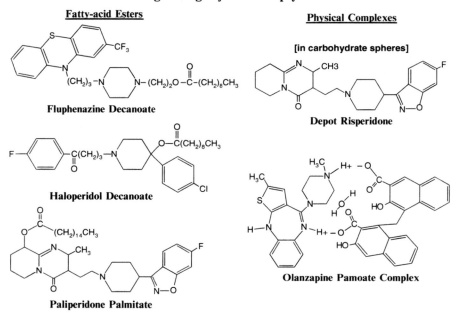

Long-Acting Injected Antipsychotics

<u>Fatty-acid Esters</u>

Fluphenazine Decanoate

Haloperidol Decanoate

Paliperidone Palmitate

<u>Physical Complexes</u>

[in carbohydrate spheres]

Depot Risperidone

Olanzapine Pamoate Complex

Fig. 2.5 Long-acting injected antipsychotic agents. Preparations include esters of fatty-acids such as decanoate (for fluphenazine, haloperidol, and paliperidone), or physico-chemical complexes that include carbohydrate microspheres for risperidone and pamoate for olanzapine

Basic Pharmacology of Antipsychotics

Early neurochemical studies of the effects of various the first psychotropic agents on the metabolism of the catecholamines (dopamine and norepinephrine) and serotonin (5-hydroxy-tryptamine), notably by Arvid Carlsson in Sweden, drew attention to the consistent ability of antipsychotic drugs to increase metabolic turnover of dopamine in rodent brain tissue relatively selectively, whereas stimulant agents, which can induce or worsen psychosis, had the opposite effect [32]. Such seminal observations led to the hypothesis that antipsychotic agents may block synaptic neurotransmission mediated by dopamine in the forebrain, probably by blocking the then-unidentified dopamine receptors, and that increased production and release of dopamine represented compensatory responses to its neuroleptic-induced decreased functioning. In the 1970s, an association was made by Paul Greengard and his colleagues to stimulation of adenylyl cyclase by dopamine, and its blockade by some (phenothiazines and thioxanthenes) but not all (butyrophenones) neuroleptics. These drug interactions supported the concept of a first dopamine receptor ("D_1") function to be identified [43]. A second receptor type ("D_2") that was affected by haloperidol and chemically similar agents and tonically suppressed output of prolactin from the anterior pituitary was strongly suspected by John Kebabian and later identified as the binding site of radioactively-labeled haloperidol and other analogous compounds [107]. Subsequently, other selective small-molecule radioligands were developed to probe D_1 or D_2 receptor proteins in brain tissue and to label them by autoradiographic localization and laboratory quantification, and eventually in clinical brain-imaging. Receptor affinity assays based on competition with radioactive labeling agents showed that all neuroleptic drugs bind to striatal D_2 receptors with affinities closely correlated with their antipsychotic doses and clinical plasma concentrations (clinical potency). This relationship became a compelling, if logically somewhat circular, element in the dopamine theory of antipsychotic action of neuroleptics and, more speculatively, consideration of dopamine as a potential contributor to the pathophysiology of idiopathic psychotic or manic illness (Table 2.3).

From these beginnings, there has been remarkable progress in understanding the molecular biology and pharmacology of dopamine receptors in mammalian brain, in turn, stimulating clinical studies of the neurogenetics, neuropathophysiology, and neuroradiology of dopamine receptors. Knowledge of receptor and effector components of brain dopaminergic neurotransmission systems has yielded revolutionary insights into the complexity of molecular neurotransmission [12, 17]. The most prevalent central dopaminergic receptors, types D_1 and D_2, have been cloned, and their gene and peptide sequences and structures proposed. Genes for all major dopamine receptors (types 1–5) have been localized on human chromosomes. Less prevalent or more anatomically restricted, novel gene products which resemble either D_1 (D_5) or D_2 (D_{2S} [a short variant of D_2], D_3, D_4) receptors also have been identified, their tissue distributions described, and their pharmacology characterized, mainly in gene-transfected cultured cells, and selective small-molecule ligands for them have been developed. The dopamine receptors belong to a super-family of membrane proteins with seven relatively hydrophobic membrane-spanning regions and an intracytoplasmic segment which interacts with guanosine nucleotide-binding proteins (G-proteins) and other molecular elements of neurotransmission effector systems. The super-family includes receptors for the catecholamines and serotonin, as well as the light-sensitive visual pigment, retinal rhodopsin. The original categorization of D_1 and D_2 receptor types has been sustained in that all known dopamine receptors are molecular variants of these two major types. Close molecular similarities of D_1 and D_2 receptors and their subtypes make it difficult to design highly selective drug molecules to target them; notably, many antipsychotic drugs have effects at both D_2 and D_3 receptors. Despite the formidable obstacles presented by the molecular similarities of receptors within the D_1 and D_2 subfamilies and the low natural abundance of D_{2S}, D_3, D_4, and D_5 receptors, additional agonist, antagonist, and receptor-labeling agents with even greater selectivity for specific dopamine receptors continue to emerge [22]. Clinical development of such compounds as diagnostic neuroradiopharmaceuticals that produce signals suitable for clinical brain imaging, or as neuropsychiatric drugs remains empirical and unpredictable. There have been some disappointing preliminary clinical trials of D_1/D_5 antagonists, as well as D_4 antagonists as potential atypical antipsychotics modeled after the partially D_4-selective drug clozapine, and

Table 2.3 Potencies (K_i, nM) of antipsychotic agents at central neurotransmitter receptors

Drugs	DA receptors			$5HT_{2A}$	$5HT_{2A}/D_2$ ratio	ACh_m	NE receptors		H_1
	D_1	D_2	D_4				α_1	α_2	
cis-Thiothixene	340	0.45	77.0	130	289	2500	11.0	200	6.00
Fluphenazine	15.0	0.80	9.30	19.0	23.8	2000	9.00	1600	20.8
Perphenazine	–	1.40	–	5.60	4.00	1500	10.0	510	–
Asenapine	2.30	1.40	1.30	1.40	0.06	21.7	1.20	1.20	1.00
Thioridazine	22.0	2.30	12.0	41.0	17.8	10.0	1.10	–	–
Pimozide	–	2.50	30.0	13.0	5.20	–	–	–	–
Risperidone	750	3.30	16.5	0.15	0.05	>10,000	2.00	55.6	58.8
Haloperidol	45.0	4.00	10.0	36.0	9.00	>20,000	6.20	3800	1890
Ziprasidone	339	4.79	39.0	0.40	0.08	≥10,000	10.5	–	46.8
Mesoridazine	–	5.00	13.5	6.30	1.26	–	–	–	–
Sulpiride	≥1000	7.40	52.0	≥1000	135	≥1000	≥1000	–	–
Olanzapine	31.0	11.0	9.60	4.00	0.36	1.90	19.0	230	7.15
Chlorpromazine	56.0	19.0	12.5	1.40	0.07	60.0	0.60	750	9.10
Loxapine	–	71.4	12.0	1.69	0.02	62.5	28.0	2400	5.00
Molindone	–	125	–	5000	40.0	–	2500	625	>10,000
Quetiapine	455	160	1164	294	1.84	120	62.5	2500	11.0
Clozapine	38.0	180	9.60	1.60	0.01	7.50	9.00	160	2.75

Affinity constants (K_i) are inversely proportional to potency, as determined by radioreceptor assays with radioactively labeled index agents, selective for each receptor type in mammalian brain tissue or in genetically transfected cells expressing human receptor genes [18]. Drugs are ranked in descending order of affinity to dopamine D_2 receptors, a presumably critical characteristic. Note that thiothixene has the highest D_2 affinity, and clozapine the lowest, whereas asenapine and risperidone have the highest $5HT_{2A}$, and the indole, molindone, and sulpiride the lowest. Sulpiride (Dogmatil®) is a racemic, substituted benzamide (not available in the US), of a chemical type that includes remoxipride (Roxiam®, withdrawn for toxic aplastic anemia)

ACh_m muscarinic acetylcholine; *DA* dopamine (D_1, D_2, or D_4 receptors); *5HT* serotonin receptors; $5HT_{2A}/D_2$ ratio of potency at dopamine D_2 vs. $5HT_{2A}$ receptors; H_1 histamine type-1 receptors highly selective D_3 receptor agonists and antagonists require further development and testing [106, 118, 190].

The molecular neurobiology of dopamine receptors and their effectors stimulated renewed interest in many traditional problems for the physiology and pharmacology of catecholamine neurotransmission, including molecular control of the synthesis, storage, release, actions, and inactivation of dopamine, as well as potentially rational development of novel CNS drugs. New insights from molecular and cell biology have revealed many details of processes that modulate and control production and functioning of dopamine and other neurotransmitters, including adaptive changes in receptor sensitivity and plasticity during normal development and in response to drug-treatments. Some of these processes may or may not prove to be relevant to the pathophysiology of major idiopathic neuropsychiatric disorders but are likely to contribute to understanding the benefits, limitations, and adverse effects of many psychotropic drugs. However, attempts to find simple relationships between alterations in the physiology or genetics of dopamine receptor subypes, and clinical disorders or drug responses have not been especially fruitful. In particular, predictions that dopamine receptor abundance in postmortem brain tissue would be increased in psychotic disorders remain inconclusive, with suggestions of diagnostic nonspecificity as well as artifacts associated with antemortem drug treatment. Visualization of dopamine receptors in life by positron-emission tomographic (PET) or single-photon emission computed tomographic (SPECT) brain scanning has also yielded inconsistent results, not fully explained by the pharmacology of the radioligands employed or artifacts arising from previous antipsychotic treatment [233]. More substantial and practical neuroradiological findings include quantitative estimates of D_2, $5HT_{2A}$, or other receptor occupation by specific doses or circulating concentrations of antipsychotic and other psychotropic drugs, as well as indications of their elimination half-lives from presumably critical cerebral sites of pharmacodynamic action [228]. In general, since identification of new genes and putative gene products moves much faster than understanding of their physiological roles, parallel studies of the physiology and pharmacology of cells, nervous systems, and behaving organisms are essential for clarifying the functional and possible clinical significance of the many newly emerging gene products of the brain.

Current understanding of the actions of known antipsychotic drugs is highly concentrated on knowledge of the neurobiology of dopaminergic neurotransmission, although whether alternative mechanisms may yield clinically effective antipsychotic treatments remains unknown. Indeed, it remains virtually inconceivable to pursue development of potential antipsychotic agents that have no interactions with cerebral dopaminergic systems [182]. Concentrating on dopamine has been highly successful as an empirical strategy for drug-development, arising largely from mimicking the structures and then the pharmacodynamic actions of drugs such as chlorpromazine, discovered by serendipity rather than biological theory. The antidopamine approach remains the most developed theory of antipsychotic drug action, even as modified in recent years. Current strategies include seeking to balance relatively potent effects on serotonin-mediated neurotransmission with only moderate D_2 antagonism, as is characteristic of clozapine, as a means of identifying potential "atypical" antipsychotics with low risk of adverse extrapyramidal neurological effects (EPS [146]). Affinities (K_i values in nanomolar [nM] concentrations) of various antipsychotic agents at dopamine, serotonin, and other cerebral neurotransmitter receptors are summarized in Table 2.3. Of the agents tested, cis-thiothixene had the highest D_2 affinity, and clozapine the lowest, whereas clozapine and loxapine had the highest ratio of $5HT_{2A}$ affinity relative to D_2, and cis-thiothixene the lowest. However, loxapine is chemically and clinically more like a typical neuroleptic and not as effective or free of risk of adverse extrapyramidal effects as clozapine. Interactions with norepinephrine (especially α-adrenergic) receptors may contribute to antipsychotic effects of some antipsychotic drugs, and antimuscarinic potency probably contributes to some adverse cognitive and autonomic effects as well as limiting risk of some forms of EPS (notably, bradykinesia), whereas potent central antihistaminic action probably contributes to sedative actions [14, 18].

Theories concerning the pharmacological actions of antipsychotic agents have concentrated on four major cerebral systems that involve dopaminergic neurotransmission or neuroendocrine effects. A scheme of their organization is provided in Fig. 2.6. They include: (a) a projection from the midbrain substantia nigra to the basal ganglia (caudate-putamen) that degenerates for unknown reasons in Parkinson's disease to impair posture and initiation of movement; (b, c) projections from the midbrain tegmentum to targets in the limbic forebrain that include the amygdala as well as to cerebral cortex, particularly frontal and temporal regions; and (d) the tuberoinfundibular system of secretory neurons arising largely from the arcuate nucleus of the hypothalamus to deliver dopamine to the hypophyseoportal microvascular system at the median eminence to reach the anterior pituitary, where dopamine acts, notably, as a prolactin inhibitory factor by stimulating sensitive D_2-like dopamine receptors on mammotrophic cells. This anatomy has provided fertile ground for theorizing about various actions of antipsychotic drugs as dopamine receptor (especially D_2) antagonists, including an association between bradykinetic (Parkinson's disease-like) effects mediated by dysfunction of the nigrostriatal system; possible antipsychotic effects mediated by altered function of limbic and cortical dopamine projections; and probable induction of at least moderate hyperprolactinemia by many antipsychotic agents (especially potent neuroleptics. risperidone, and paliperidone) by interfering with the prolactin-inhibitory effects of dopamine acting as a neurohormone in the anterior pituitary gland.

An accounting of known or proposed actions of antipsychotic drugs at dopaminergic nerve terminals can be guided by the schematic summary provided in Fig. 2.7. This scheme depicts a dopamine nerve terminal above and a receptive neuron or other cell below. Actual dopamine neurons are designed to deliver their neurotransmitter in a less spatially specific manner than is depicted in the schematic figure, since their terminal axons divide into fine terminal arborizations with swellings along their near-terminal portions; these swellings (varicosities) represent the "terminals" depicted in Fig. 2.7. Similar arrangements occur in the central noradrenergic and serotonergic neuronal systems. All of these monoaminergic systems resemble the anatomy of the peripheral sympathetic autonomic nervous system,

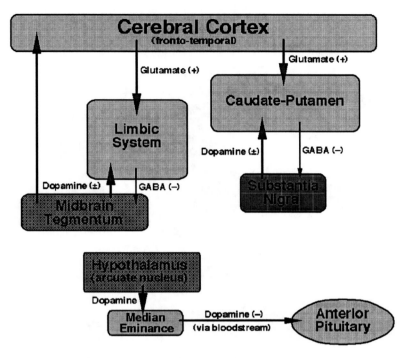

Fig. 2.6 Major dopaminergic systems in mammalian brain. The four major subsystems involving dopaminergic neurons (with mixed inhibitory and some excitatory actions) are regulated by prominent feedback systems mediated by glutamic acid (excitatory) and γ-aminobutyric acid (GABA; inhibitory). A major system projecting from the midbrain substantia nigra to the basal ganglia regulates posture and initiation of movement (degenerates in Parkinson disease). Another arises from the midbrain tegmentum to project to elements of limbic forebrain and portions of the frontal and temporal cerebral cortex. Dopamine produced largely in the arcuate nucleus of the hypothalamus is secreted at the median eminence into the tuberoinfundibular microvascular system to reach the anterior pituitary as an inhibitory neurohormone acting on sensitive dopamine D_2-type receptors, notably to limit release of prolactin from mammotrophic cells

with branching terminal axons which make synapses *en passant* as they course through smooth-muscle, glandular, or other targets of autonomic nervous system control. Such systems are effective in regulating the background tonality of neuronal function rather than carrying precise information from point-to-point. The presynaptic apparatus in dopaminergic neurons includes cytoplasmic enzymatic mechanisms for synthesis of dopamine from the essential aromatic amino acid L-tyrosine, obtained from food and provided by the bloodstream. Tyrosine is first 3-oxidized to form L-dihydroxyphenylalanine (L-DOPA) mediated by the enzyme tyrosine hydroxylase, This rate-limiting ("bottleneck") enzymatic step is highly regulated, in part by activation by phosphorylation mediated by the action of cyclic-adenosine-monophosphate (cyclic-AMP). Production of presynaptic cyclic-AMP, in turn, is regulated by D_2 autoreceptors, which decrease production of cyclic-AMP

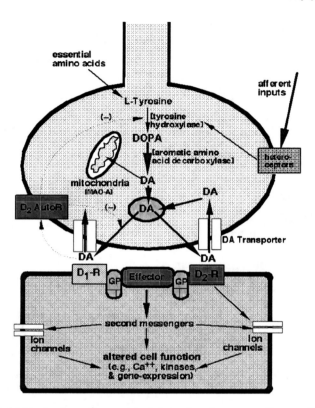

Fig. 2.7 Organizational scheme for a cerebral dopaminergic synapse. Presynaptic components (above) include enzymatic steps required for the synthesis of dopamine (DA, by tyrosine-hyroxylase, which is rate-limiting, and aromatic amino acid decarboxylase), its storage in presynaptic, membrane-enclosed vesicles (in a complex with adenosine-triphosphate (ATP), chromogranin proteins, and cations), which fuse with the cell membrane during presynaptic neuronal depolarization and neurotransmission (exocytosis) in the presence of calcium ion, to release dopamine into the synaptic cleft. The synthesis and release processes are regulated by dopaminergic D_2-like autoreceptors, as well as by inputs from other types of neurotransmitters and their nondopaminergic heteroreceptors (including adenosine A_2 receptors, which increase production of cyclic-adenosylmonophosphate (cyclic-AMP) to activate tyrosine hydroxylase, whereas D_2 autoreceptors have the opposite effects). Regulation of DA release probably involves effects on calcium ion (Ca^{++}) or potassium (K^+) currents. Residual dopamine in the presynaptic cytoplasm, not re-stored in presynaptic vesicles, is removed by oxidative deamination by monoamine oxidase-A (MAO-A, characteristic of nerve endings, as well as MAO-B in nearby glial cells). Released dopamine is inactivated mainly by active reuptake in exchange for sodium ions, by dopamine transporter proteins (DAT) of the presynaptic neuronal cell membrane (which are characteristic of dopamine neurons and are inhibited by stimulant drugs). Neurotransmitter effects of dopamine are mediated by at least five subtypes of dopamine receptors (D-R) that belong to D_1 (D_1 and D_5) and D_2 (D_2, D_3, D_4) families, of which D_1 and D_2 are most abundant, and exert opposite effects to either stimulate (D_1) or decrease activation (D_2) on the complex of adenylyl cyclase or other effector proteins including PLC, both regulated by facilitatory (G_s) or inhibitory (G_i) G-proteins (GP, guanine nucleotide-binding proteins with three (α, β, γ) subunits). "Second-messengers" involved include cyclic-AMP produced from ATP, as regulated positively by D_1 receptors acting with G_s proteins, and negatively

and phosphorylation-activation of tyrosine hydroxylase and reduce production of dopamine, as well as by adenosine A_2 heteroceptors that stimulate production of cyclic-AMP. DOPA is then rapidly deaminated by a common aromatic amino acid decarboxylase to yield dopamine. This monoamine neurotransmitter product is rapidly incorporated into intracellular, membrane-enclosed storage vesicles in a complex with adenosine-triphosphate (ATP), chromogranin proteins, and magnesium ion (Mg^{++}), where it is protected until required for neurotransmission.

Release of dopamine is effected by events initiated by the arrival of nerve impulses at dopamine nerve terminals that include rapid fusing of the vesicle membranes with the neuronal cell-membrane in a process that requires free calcium ions (Ca^{++}) and may be sensitive to potassium ions (K^+), known as *exocytosis*. This process is a ubiquitous method for gaining the release of various neurotransmitters or hormones from intracellular storage sites. Exocytosis is also regulated by the negative feedback influence of D_2-type autoreceptors to limit the rate of release of neurotransmitter, probably in part by limiting availability of Ca^{++}. In turn, both forms of negative-feedback control on dopamine synthesis and its release respond to the levels of free dopamine released into the synaptic cleft, so as to establish a degree of homeostatic control over the levels of dopamine provided for neurotransmission. The major mechanism for inactivating released dopamine is by active re-uptake in exchange for intracellular sodium ions (Na^+), mediated by dopamine-transporter proteins (DAT). These cell-membrane proteins are highly characteristic and uniquely expressed on dopamine terminals, and can be used as a target of radiopharmaceuticals aimed at labeling dopamine neurons. Such radiolabeling in brain-imaging can support the diagnosis of Parkinson's disease by major losses of dopamine neurons, and is used in studies of psychiatric patients [33, 142]. The dopamine transport mechanism and components of intracellular storage of dopamine are interfered with by typical stimulant drugs including amphetamine, methamphetamine, and methylphenidate, which increase the biological activity of dopamine (and other neurotransmitter amines) and can induce or worsen psychosis [230].

Once released, dopamine acts through D_1-type and D_2-type postsynaptic, membrane receptor proteins (Fig. 2.7). They operate in concert with particular types of G-proteins, notably, to alter activity of adenylyl cyclase as a major effector molecule: increasing activity by D_1 receptors operating with stimulatory G_s proteins and decreasing it with D_2 receptors and inhibitory G_i proteins. D_2 receptors have other actions that include inhibition of phospholipase-C (PLC), another major effector mechanism that converts phosphatidyl inositol-bisphosphate (PIP_2) to inositol trisphosphate (IP_3).

Fig. 2.7 (continued) by D_2 receptors acting with G_i proteins, and inositol trisphosphate (IP_3) and DAG produced by PLC from PIP_2, with secondary effects on protein kinases and distribution of Ca^{++} in receptive cells. The dopamine receptor complex also regulates the function of ion channels in the receptive cells. All of these components, as well as additional regulatory processes in receptive cells, are affected by antipsychotic drugs, and the functional organization of other monoamine neurotransmission systems mediated by norepinephrine (NE) or serotonin (5-hydroxytryptamine, 5-HT), which are affected by other psychotropic drugs, is very similar [17, 18]

The second-messenger IP_3 liberates Ca^{++} and diacylglycerol (DAG), which activates protein kinases to alter the functional states of other proteins through phosphorylation. The receptor-effector products cyclic-AMP and IP_3 are considered "second messengers" in that they carry the initial effect of a neurotransmitter at its cell-membrane receptor to alter cellular physiology of receptive cells. D_1 and D_2 receptors probably also exert opposite effects on the control of intracellular Ca^{++} availability, and may also influence the functioning of ion-transporter proteins in the cell membranes of receptive neurons [150]. All of these metabolic influences on the functioning of receptive cells may include long-term adaptations of cellular mechanisms that probably include alterations in gene expression [139, 183]. Such adaptations imply long-lasting physiological changes after long-term exposure to antipsychotic treatment.

The actions of antipsychotic drugs are believed to include immediate, direct, and sustained blockade of D_2 (and other) receptors, although the mechanisms involved in subsequent actions are best defined for the older neuroleptics. Responses to the blockade of postsynaptic D_2 receptors include effects mediated by the D_2-like autoreceptors on the presynaptic terminals of dopamine neurons. When these autoreceptors are blocked, their negative-feedback effects on dopamine synthesis and release are disinhibited, and more dopamine is made available, as if in an attempt to restore homeostasis. This early signal of increased dopamine production and release is readily detected in animal models. As the process of presynaptic disinhibition continues, over some days (or soon after large injected doses of potent antipsychotics), the presynaptic neurons become depolarized and inactive. As production and release of the neurotransmitter falls, the post-synaptic blockade of D_2 receptors continues, to yield highly inhibited dopaminergic neurotransmission [151]. The timing of this process strikingly parallels the emergence of Parkinsonian bradykinesia, which does not occur immediately, but evolves gradually over at least several days, as the risk of acute, transient dystonias declines, suggesting that the dystonias may represent a transient excess of dopaminergic function. An additional consequence of the sustained dysfunction of dopaminergic neurotransmission are changes in the postsynaptic dopamine receptors to become more abundant ("upregulate") and more responsive ("supersensitive") to the neurotransmitter or its agonists such as $R(-)$-apomorphine. Such adaptations may contribute to the pathophysiology of later-appearing (tardive) dyskinesia (TD [199]).

Clinical Use of Antipsychotics

All of the newer and older antipsychotic drugs offer relatively effective treatment of psychotic disorders or mania with acceptable safety and practicality, with generally similar efficacy but dissimilar risks of particular adverse effects. This class of drugs is not specific for treating schizophrenia and probably is more efficacious in other, especially acute or affective disorders with psychotic features, and in acute mania or mixed-manic-depressive states. Antipsychotic drugs also may have useful effects in organic mental syndromes marked by agitation and psychotic features, and some second-generation antipsychotics may have useful adjunctive, apparent, antidepressant effects in severe, major depression. Of overall usage of antipsychotic drugs, over 60% is currently "off-label" for FDA-unapproved conditions other than schizophrenia or major affective disorders with psychotic features [3]. This striking pattern can be viewed either as underscoring the diagnostic nonspecificity of this class of psychotropic agents or as indicating their broad range of utility. A few agents have performed particularly poorly in controlled trials, and most of these usually are not employed clinically as antipsychotics; they include prochlorperazine (Compazine®), promazine (Sparine®), mepazine (Pacatal®), molindone (Moban®), and pimozide (Orap®), as well as reserpine and barbiturates. Together, these agents have outperformed placebo in only about half of their controlled trials, in contrast to rates of statistical superiority to a placebo in more than 90% of commonly clinically employed antipsychotic drugs [102, 104].

Formulations of Antipsychotics

Dosing with antipsychotics in short-term treatment of acute phases of psychotic illnesses typically involves daily doses up to the equivalent of 15–25 mg of haloperidol or fluphenazine, 300–500 of chlorpromazine, 2–6 mg of risperidone or paliperidone, or 10–25 mg of olanzapine. Dosing is typically given in divided daily distribution to allow time for nonspecific adaptation to occur, but moderate doses can usually be given once-daily after the first few days. Most newer antipsychotics are limited in dosing range and dosage forms. Rapidly orally dissolving preparations are available for some agents, notably aripiprazole, clozapine, olanzapine, and risperidone; such preparations usually are given to facilitate rapid responses or for patients who are too confused to cooperate reliably with swallowed tablets or capsules. Several antipsychotics are also available as short-acting, intramuscularly injectable preparations, notably aripiprazole, olanzapine, risperidone, and ziprasidone; these preparations as well as fast-dissolving oral preparations can be used to gain rapid clinical effects. Long-acting preparations for intramuscular injection, typically in intervals of 2–4 weeks, include older depot preparations of slowly hydrolyzed long-chain fatty-acid esters of several neuroleptic agents with a free hydroxyl group, notably fluphenazine and haloperidol (*decanoates*). Long-acting, injectable preparations of modern antipsychotics include risperidone incorporated into biodegradable *carbohydrate microspheres* (Risperdal Consta®), as well as the ester paliperidone palmitate (Invega Sustenna®), and olanzapine in a *pamoate complex* (Zyprexa Relprevv®) (Fig. 2.5).

Drug Monitoring

Assays of plasma or serum concentrations of most antipsychotics have not proved useful to guide routine clinical practice, although there is some convergence of clinical dose-response and plasma drug concentration data and results of PET brain scanning, such that plasma concentrations of haloperidol found effective in a majority of patients (5–20 ng/mL) also occupy D_2 dopamine receptors by 75–85%, as do other neuroleptics at daily doses equivalent to 200–400 mg of chlorpromazine ("CPZ-eq" doses), with lesser occupation by agents with lower affinity for dopamine receptors, including clozapine, even at clinically effective doses with serum concentrations of approximately 200–300 ng/mL [55, 63, 160, 212]. Modern pharmacokinetic studies with assays of high sensitivity, as well as application of PET brain scanning indicate that some neuroleptics, especially butyrophenones, have slower, multiphasic, tissue elimination kinetics than are indicated by typical, published, nominal plasma elimination half-life values of about 24 h. Such values for clearance from plasma probably apply validly to phenothiazines, even though many neuroleptics tend to accumulate in brain and other tissues [44]. Recent efforts to identify clinically predictive, individual genetic differences related to hepatic and other enzymes that metabolize antipsychotic and other psychotropic drugs have not yet proved useful [152, 220]. The coherence of neuroradiological findings concerning dopamine receptor-occupancy and the known antidopaminergic pharmacology of antipsychotic drugs is impressive, but logically circular, and does not prove that antidopaminergic actions are necessary for antipsychotic benefits.

Expected Effects and Selection of Antipsychotic Drugs

Typically, antipsychotic drugs palliate acute psychotic symptoms and they are widely considered to limit risk of future acute exacerbations when used over months or years in maintenance treatment for chronic or frequently recurring psychosis. However, evidence that they markedly improve the overall prognosis and long-term functional outcomes in schizophrenia is limited and inconsistent. Placebo response rates in psychotic illnesses treated for at least several weeks commonly average 10–20% if mild remaining symptoms are the outcome considered, and even less if virtual remission is required, compared to 40–50% of schizophrenia patients showing major, clinically important, improvement after being treated with most antipsychotic agents. Although over 90% of controlled trials involving commonly clinically employed antipsychotic drugs have shown their statistical superiority over placebo, since many of the controlled trials involved in such high success rates were aimed at supporting regulatory approval and licensing, this outcome is hardly surprising. However, the degree of improvement has been highly variable. Drug–placebo contrasts with antipsychotic treatments vary greatly with the target outcomes considered. The contrast is very high (perhaps 10–20-fold) with respect to virtual clinical remission, but the proportion of patients achieving such outcomes with active drugs is low (≤20%) and the proportion with placebo is far lower (1–2% [70, 102, 104]). Rates of major clinical improvement with antipsychotic drugs vs. placebo are typically 40–50% vs. 20–25%, for an approximately twofold difference [102, 104]. More than three-quarters of patients with schizophrenia show some measurable symptomatic improvement, although nearly one-third, though improved, remain substantially ill and 5–10% show little or no improvement. These estimated rates of clinically meaningful improvement vary markedly among disorders other than schizophrenia, typically being higher in acute psychotic or affective disorders with major psychotic features, and even lower in dementia and many other neuropsychiatric disorders.

Based on findings from both short- and long-term, controlled therapeutic trials, when data are pooled by meta-analytic methods, the 95% statistical confidence-intervals among antipsychotics (and most other psychotropic drugs) generally overlap, making it impossible to rank specific agents securely by efficacy. With the exception of the relatively toxic agent clozapine [15, 143, 222], and suggestive findings for olanzapine and risperidone [47, 132, 192], there is little evidence that any commonly used antipsychotic agent is clearly and consistently more effective, on average, than others for particular illnesses. The same impression applies also to positive (agitation, delusions, hallucinations), negative (emotional withdrawal, autism, abulia), cognitive, or functional features of psychotic illnesses. Such findings underscore the impression that hopes for a rational or evidence-based differential therapeutics for psychotic disorders remains elusive.

There is also a lack of research evidence that more than one antipsychotic agent is better than one for either short- or long-term treatment, even though antipsychotic combinations ("polytherapy," preferred to the widely employed term "polypharmacy") are commonly used empirically, probably largely in response to typically limited benefits in chronic and disabling illnesses that are difficult to treat, and perhaps owing to clinical sensitivity to the obligation to limit risks commonly associated with unusually high doses of individual drugs [35, 37, 60, 67]. Polytherapy has rarely been tested for added benefits or safety, but can be expected to add to overall patient side-effect burden, complexity of regimens, risk of nonadherence, and costs, whether or not clinically significant drug-interactions or major adverse effects increase [37].

In general, differences in adverse-effect risks can be considered in selecting treatment regimens for individual patients. These include a relatively high risk of acute and later EPS with older neuroleptics, especially those of high potency, in contrast to relatively more sedative, hypotensive, and peripheral autonomic or metabolic adverse effects with many older and some newer agents, especially drugs of relatively low potency (Table 2.1).

Modern antipsychotic drugs are both widely used and relatively expensive before the end of their patent protection and appearance of generic versions. They recently have come to account for the largest proportion of sales of drugs of all kinds. Most of their usage is for psychotic and psychotic-affective disorders, which account for nearly 20% of prescriptions in an approximately 16-billion US dollar annual market [89, 175]. Leading agents currently rank quetiapine, risperidone, aripiprazole, and olanzapine (all with ≥10% of market-share). Between 1995 and 2011, the proportion of antipsychotic drug-treated patients given first-generation drugs declined precipitously from 80–90% to less than 10%, as modern, second-generation agents rose to account for more than 90% of current usage. Markedly dissimilar acquisition costs among specific antipsychotics can be considered in

Table 2.4 Short-term indications for antipsychotic drug treatment

Acute psychotic episodes of virtually any type

Acute exacerbations in (chronic) schizophrenia

Acute mania or mixed manic-depressive states

Adjunctive use in major depressive episodes, especially with psychotic features
     or prominent agitation

Severe agitation in delirium or dementia while seeking specific, remediable causes

Agitation and abnormal movements in Tourette's and Huntington's syndromes

Childhood psychotic disorders, often of uncertain type

Miscellaneous medical uses, including nausea and vomiting and intractable hiccups (singultus)

Most of these indications, other than for exacerbations of schizophrenia and acute mania and bipolar mixed-states, are largely empirical and lack extensive research support, but are both plausible and supported by decades of practical clinical application

designing treatment programs for individual patients. Costs can be 10–50-times more for recently marketed, brand-name products than for older generic agents or generic risperidone, and costs are falling as patent-life is ending for a growing number of antipsychotic and other psychotropic drugs and more generic products appear [175]. Many managed-care systems encourage selection of antipsychotic and other drugs based on costs, especially when differences in efficacy or safety between inexpensive and costly products are hard to demonstrate.

Short-Term Indications for Antipsychotic Drugs

Treatment with antipsychotic drugs can be divided for convenience into short-term applications for up to several months, and long-term or maintenance treatment for many months or indefinitely. Short-term uses are broad, and include applications in general medical and neuromedical conditions as well as in acute psychotic illnesses, mania, and exacerbations of chronic psychotic disorders, notably schizophrenia, as well as serving as adjunctive treatments in severe, psychotic, or bipolar depression (Table 2.4). However, the research support for many short-term applications is remarkably limited. Moreover, a surprising number of basic clinical questions remain about the nature of responses to antipsychotic drug treatments, even in schizophrenia—the paradigmatic indication, including the timing of clinical changes, individual factors that predict response or tolerability, effects of dosing, and expected responses and outcomes. In part, this lack of secure information arises from the complex variety of clinical conditions for which antipsychotic drugs have been used and their inter-individual heterogeneity, as well as the nature of most therapeutic trials. Most trials focus on relatively narrowly defined indications aimed at seeking regulatory approval and anticipation of future marketing efforts. That is, short-term treatment trials in schizophrenia or mania typically aim at demonstrating statistically significant contrasts between an active drug and a placebo-control condition, rather than determining the time-course and sequencing of steps toward attaining major clinical improvements or remission, or considering clinical characteristics of potential subgroups with high or low responsiveness or tolerability to particular treatments.

Time-Course of Short-Term Responses to Antipsychotic Treatment

Some insights have arisen from clinical observations and unpublished studies carried out at our center over the past several decades. It is our impression that rapid and probably largely nonspecific sedating or quieting effects occur early in the treatment of highly agitated, sleepless, anxious, and sometime aggressive psychotic or manic patients given antipsychotic agents. Pharmacological nonspecificity of such early quieting is strongly suggested by similar effects of both antipsychotic and sedative drugs including potent benzodiazepines in acute psychosis or mania [49, 125]. It is less clear how symptoms believed to be more fundamental to psychotic disorders, including irrational thinking, abnormal sensations, cognition, affective responsiveness, and functional status respond. Improvements in impaired cognition and in so-called "negative symptoms" of emotional and affective withdrawal typically are quite limited, even with modern treatments [59].

From the first descriptions of responses of psychotic patients to chlorpromazine, it appeared that an early effect was quieting of emotional reactions to delusions and hallucinations, with the gradual fading of such "positive symptoms" into the background and sometimes disappearing, or at least becoming less compelling and disruptive [194, 204]. Usually several weeks are required for major improvements in such phenomena as delusions and hallucinations, somewhat longer to regain rational thinking, attention, and concentration, and even longer times to achieve improvements in functional measures such as social relations, ability to care for one's self, or to regain former functional independence [181]. Even in acute mania or acute psychotic episodes, times to such important levels of clinical remission may require many weeks, and some psychotic or manic patients never do recover functionally even if major symptoms are substantially modified or reduced in intensity [102, 104, 208]. Response rates tend to parallel the acute vs. insidious onset of psychotic illnesses, but time-courses to particular outcomes in acute psychotic illnesses are not well-defined. Nevertheless, it has been standard clinical practice to continue antipsychotic treatment at tolerated doses for many weeks and often for several months before considering major, gradual dose-reductions or even cautious and tentative discontinuation of treatment, based on both current levels of recovery and on critical assessments of strengths and vulnerabilities. For many chronically psychotic patients, it may never be feasible to discontinue or to rely on intermittent treatment [68].

Short-Term Dosing of Antipsychotic Drugs

The question of effects of antipsychotic doses for use in acute episodes of psychotic illness has been considered in some experimental detail. An important source of information arose from studies in the 1970s and 1980s that compared the then-prevalent, very
high doses of antipsychotic agents with standard doses. The contrasts typically involved
agents of high-potency which could be given at daily doses equivalent to several grams
of chlorpromazine compared to more usual doses equivalent to several hundred mg/day.
Such studies indicated very little gain in efficacy by the use of median doses equivalent
to over 3,600 mg/day of chlorpromazine vs. only about 300 mg/day, but that risks of
adverse effects were increased. An exception was slightly greater responses with very
high doses within the first 24 h of treatment, which may well represent nonspecific central depressant effects and sedation. However, with both dosing ranges, the degree of
symptomatic improvement in these studies averaged 50–60% within 2 months, with
perhaps two-thirds of these gains realized within 1 month. Moreover, responses were
similar with high and low doses over times ranging from 2 days to 2 months, making it
unlikely that the rate (percentage change in symptom ratings per day) or latency for
reaching similar endpoints differed appreciably by drug dose [13].

A common strategy for limiting antipsychotic doses in acute psychosis or mania is to
add sedatives temporarily, typically a high-potency benzodiazepine such as lorazepam
or clonazepam, and later to remove them gradually. Many psychotic patients can be
managed safely with adequate medication plus skillful interpersonal and environmental
management, ideally in a well-staffed and highly controlled clinical setting such as an
inpatient or partial-hospital unit. A risk associated with the application of multiple central depressant drugs in such circumstances ("polytherapy") is that more conservative,
but adequate treatment, will be delayed and that total daily drug doses increase, along
with risks of adverse effects, drug-interactions, and costs [35, 37, 219].

Critical consideration of the plausible but unproved proposition that unusually
large doses of antipsychotics hasten clinical response is especially important in the
era of modern antipsychotics. These agents are generally better-tolerated than older
neuroleptics, and so encourage aggressive dosing. There is currently a great deal of
administrative, financial, and clinical pressure to seek more rapid improvement by
using greater doses of individual antipsychotic drugs, or by the use of combinations
of dissimilar agents, such as antipsychotics, sedatives, and anticonvulsants, with
overall increases in daily medication loads [37, 219]. Nevertheless, it is essential to
emphasize that evidence is lacking that such practices are predictably more rapidly
successful than conservative dosing with fewer drugs.

Impact of Shorter Hospitalization

A related aspect of responses to antipsychotic and antimanic treatments is the hope
that reducing hospital length-of-stay by administrative fiat, itself, would lead to
greater rates of improvement (percentage change per day), with similar endpoints

achieved more rapidly than with more open-ended hospitalizations. This hope can be considered an application of "Parkinson's law" of the inverse relationship of resources to work-efficiency [162]. We had an unusual opportunity to test this plausible expectation during an era of administratively required gradual shortening of time in hospital, coupled with repeated symptomatic ratings with the Brief Psychiatric Rating Scale (BPRS) required for a research study. In this experience, the hospital length-of-stay during the 1990s fell from an average of 6–8 weeks, to 1–2 weeks. During the same years average symptomatic improvement from admission to discharge fell from 40–50% to only 10–20%, indicating a lack of increase in the *rate* of change with forced shortening of hospitalization. That is, on average, patients were much more symptomatic when discharged after short hospitalizations. Fortunately, there was enough flexibility in the system as to avoid potentially dangerously premature discharges. Moreover, we also found that clinical outcomes at 6 months or later following discharge from the initial hospitalization were indistinguishable based on length of index hospitalization, suggesting that clinical improvement continued in the weeks or months following hospital discharge [208].

This experience did not address differences in outcomes at discharge or during follow-up between even earlier years when hospitalization sometimes persisted for months, but our impression was that mean reductions of lengths-of-stay from months to weeks had less impact than reductions from weeks to days. We also did not assess the clinical progress of recently discharged patients, or the impact of continued illness on their families and communities. One means of dealing with such effects of modern medical economics might be to rely on alternatives to full hospitalization, such as residential-care, day-treatment, or other partial-hospitalization, or the use of intensive outpatient care with frequent visits in specialized centers. Unfortunately, such plausible and effective alternatives, too, are not readily available to many patients, and they are typically viewed as added costs by insurers and program administrators.

An additional consequence of these increasingly constrained contemporary policies and practices regarding the treatment of acutely psychotic or manic patients, especially in hospitals, is that it is very difficult to justify days in hospital that do not include evidence of "active treatment." This concept, in turn, appears to encourage inflexibly continued or enhanced treatment regimens, with rising average drug-doses and increasingly complex drug-combinations in "polytherapy." It is rare currently to evaluate patients for possible benefits of decreased or changed medications, owing to the pressure of time and requirement to demonstrate "active treatment," which often leads to "more treatment." Sometimes, such practices are rationalized by hopes that outpatient clinicians will sort out treatment requirements during aftercare. However, the pressures in outpatient systems of care also are rising to seek to achieve more with less time, staffing, or financial support. Although such efforts to improve efficiency of clinical care are understandable, they arise largely from assumptions that are not supported by adequate empirical evidence. That is, hypothetical benefits of brief and aggressive treatment remain not only unproved but poorly tested, whereas the risks of side-effect burdens and potentially medically serious additive or interactive drug reactions probably are increasing [35, 37].

Table 2.5 Long-term indications for antipsychotic treatments

Primary indications

Schizophrenia (and broadly defined chronic psychotic disorders)

Schizoaffective disorders (poorly defined and little-studied)

Primary maintenance treatment in bipolar-I (with mania) disorder (olanzapine, possibly
    aripiprazole) or as a co-treatment with other mood-stabilizing agents (mainly quetiapine with
    lithium or valproate (used long-term off-label))

As an adjunctive option in unipolar, recurrent major depressive disorder especially with severe
    agitation or psychotic features (short-term mood-elevation with an antidepressant is claimed
    with aripiprazole, quetiapine, and olanzapine-plus-fluoxetine); sometimes second-generation
    antipsychotics may be effective in depression in low doses

Delusional disorder (paranoia; typically with limited efficacy)

Childhood psychoses (their categorization and therapeutics remain underdeveloped)

Tourette's and Huntington syndromes (for motor and psychiatric manifestations)

Secondary indications

Unstable bipolar disorder patients not adequately managed otherwise (including clozapine,
    off-label)

Otherwise unmanageable agitation in mentally retarded or pervasive developmental (autistic
    spectrum) disorder patients

Otherwise unmanageable agitation in dementia (with increased risk of cerebrovascular,
    infectious, and other causes of death with both older and modern antipsychotics)

Schizotypal personality disorder not adequately managed otherwise

Questionable or poorly evaluated indications

Most personality (Axis-II) disorders (including schizoid and borderline)

Adjunctively in obsessive-compulsive disorder and other anxiety disorders

The only FDA-approved long-term indications are for broadly defined schizophrenia and, for a few
agents, usually as an adjunct to lithium or divalproex, in type-I bipolar disorder (with mania),
including olanzapine. Most other indications are poorly studied

Long-Term Antipsychotic Treatment

A large research literature appears to support the impression that patients with
chronic or frequently relapsing psychotic disorders are less symptomatic and more
stable when antipsychotic medication is continued long-term, for more than several
months. A range of conditions have been considered appropriate indications for
long-term antipsychotic treatment (Table 2.5). However, except for disorders diagnosed within the clinically heterogeneous category of schizophrenia, research support for long-term effectiveness in the control of psychotic symptoms and
improvement of functioning is surprisingly limited. In an earlier review of nearly 30
randomized, controlled, long-term trials in patients diagnosed with schizophrenia
carried out over 6–18 months, we found mean rates of clinically important exacerbations ("relapses") of psychotic illness to differ by 3.5-fold (58.0% vs. 16.4%)
with vs. without continued antipsychotic medication [11]. It is probably important
that the range of outcomes in these early comparisons of antipsychotic drugs to
placebos showed striking differences between trials, such that relapse rates ranged
from 0 to 52% with antipsychotic drugs (mostly first-generation), and even more

strikingly, from 18 to 100% with placebo. Such variance is consistent with clinical heterogeneity among patients diagnosed with schizophrenia, particularly prior to wide acceptance of narrower diagnostic criteria since 1980, as represented by the DSM-IV of the American Psychiatric Association [6].

In contrast to studies of patients diagnosed with schizophrenia, most other long-term applications of antipsychotic drugs remain largely empirical and clinically based, without extensive research support, and usually "off-label," or not officially approved by the FDA or other regulatory agencies. In recent years, as summarized in Table 2.5, there have been efforts to extend the approved indications of antipsychotic drugs to include not only mania and its recurrences in bipolar disorders, but also to limit risks of the recurrences of depressive phases of bipolar-I disorder (with mania), and further, as adjunctive treatments for acute depression, especially major depression with prominent agitation or psychotic features, or cases not responding well to an antidepressant with or without psychotherapy [54, 115, 135, 178]. For bipolar-I disorder, long-term applications of antipsychotics are best-established for olanzapine, despite its often limiting, adverse weight-gain and metabolic effects, and also FDA-approved for aripiprazole [51, 144]. Antipsychotic drugs are also a mainstay in the treatment of patients diagnosed with schizoaffective disorders. The conceptualization and diagnostic criteria for this syndrome vary and its treatment has rarely been studied separate from trials aimed at patients diagnosed with schizophrenia [153].

Even in schizophrenia, many characteristic manifestations are questionably or little-benefited by long-term antipsychotic treatment, particularly the so-called "negative" symptoms such as abulia, passivity, and lack of affective responsiveness, as well as impaired cognition and logical incoherence [50, 79, 116, 148]. Such symptoms appear to be more central to the psychopathology and disability in schizophrenia than "positive" symptoms including agitation, delusions, and hallucinations [216]. In schizophrenia, modern long-term trials sometimes are barely able to demonstrate beneficial effects on symptom-ratings over those of a placebo, or between baseline assessments and follow-up evaluations months later [134]. In part such outcomes reflect extensive treatment already given to patients who become subjects in research trials in advanced, industrialized cultures, which tends to limit effects of experimental treatments.

Drug-Discontinuation Trials

Another source of support of long-term beneficial effects of sustained antipsychotic treatment in chronic psychotic illness is based on comparing the impact on morbidity of randomized continuation vs. discontinuation of treatment following initial improvement and sustained stabilization. Even with continuous antipsychotic treatment, rates of falling significantly ill again approach 50% of patients within 2 years, either as a manifestation of limitations in therapeutic effectiveness, or very likely, due to incomplete or inconsistent adherence to prescribed treatment

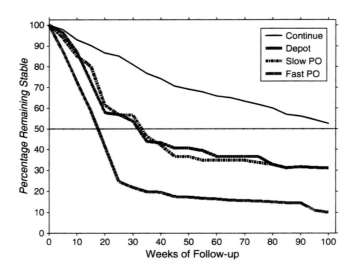

Fig. 2.8 Clinical stability with continued vs. discontinued antipsychotic treatment in patients diagnosed with schizophrenia. Based on pooling and survival analysis of data from long-term trials of continued vs. discontinued antipsychotic treatment [221]. The *horizontal dotted line* indicates 50% risk of becoming ill again (attained within 2 years even with medication continued). Similar numbers of patients (approximately 1,000) were followed for 2 years with continued treatment and relatively abruptly discontinued, orally (PO) administered drugs following initial treatment and weeks or months of stabilization, with marked differences (approximately 3.5-fold) in rates of clinical worsening/month within the initial 6 months. Note that the gradual dose-reduction to discontinuation over several weeks, or abrupt discontinuation of long-acting, injected, "depot" preparations (*n*=83) involving smaller numbers of patients, reduced the 6-month recurrence risk by 2.3-fold compared to rapid discontinuation of oral medication (*n*=106). Note also that the survival rates (relapse risk/month) are virtually parallel at times later than 6 months, suggesting that most of the impact of discontinuation of treatment occurred within the initial months. The survival function after rapid discontinuation of oral antipsychotics remained remarkably stable after 80–90% relapse had occurred within the initial 6 months, suggesting clinical heterogeneity

(Fig. 2.8). Comparisons of continued vs. discontinued antipsychotic treatment, consistently, indicate marked differences in the rate of recurrences of major clinical manifestations of psychotic illness over time, usually most clearly differentiated by differences in rates (percentage falling ill per month) within the initial 6 months after discontinuing treatment, with little difference thereafter ([221]; Fig. 2.8). Risks of illness at 6 months were approximately 13% with antipsychotic medicines continued, 62% after stopping long-acting preparations or after slow reduction of oral doses, and 75% after rapid discontinuation of oral drug [221]. The risk-by-time survival functions were virtually parallel (similar relapse rate/ month) after the first 6 months of continuing or discontinuing treatment, either rapidly or gradually (Fig. 2.8). This pattern again suggests substantial clinical heterogeneity among patients diagnosed with schizophrenia, at least in terms of vulnerability to relapse or responsiveness to antipsychotic treatment.

The observed differences in relapse risk with differences in rates of discontinuing antipsychotic medication (Fig. 2.8) raises the possibility that treatment-discontinuation, itself, may act as a stressor in the drug-adapted state, and contribute in a confounding way to a less favorable outcome. This hypothesis appears to be supported by still-limited evidence that rates of new illness over time are much less after stopping long-acting injected antipsychotic preparations, which can require several months to clear metabolically and physiologically [90, 186], or after slow, gradual reductions of oral doses (Fig. 2.8). The possibility that discontinuation of ongoing, long-term medication, especially abruptly or rapidly, can lead to clinical worsening or relapses within several weeks or months may be a common property of most psychotropic drugs [16, 20]. If it is correct that discontinuation adds to early risk of clinical worsening of chronic psychotic illness, then the difference between continued and slowly discontinued antipsychotic medication may be a closer representation of benefits to be ascribed fairly to long-term treatment than differences between continued and rapidly discontinued medication (Fig. 2.8).

Long-Term Antipsychotic Dosing

In long-term maintenance treatment, antipsychotic doses are best individualized, and can usefully be varied in response to changes in levels of symptoms or stress. Typically, daily doses in the range of 10–20 mg of aripiprazole, 5–15 mg of olanzapine, 300–600 mg of quetiapine, 2–4 mg of risperidone or paliperidone, 80–120 mg of ziprasidone, 2.5–10 mg of fluphenazine or haloperidol, or 100–300 mg of chlorpromazine, clozapine or their equivalent, are adequate and tolerated by most patients.

Usually it is clinically unrealistic, probably risky, and potentially expensive (due to increased utilization of emergency resources), to stop medication completely, and especially rapidly, in chronically psychotic patients. Decreases in maintenance doses are best and most practically made gradually and as slowly as circumstances allow. Rapid discontinuation of high doses of potent neuroleptics (such as fluphenazine and perhaps risperidone) can lead to temporary withdrawal-emergent dyskinesias, and rapid discontinuation of low-potency agents (such as chlorpromazine or thioridazine) can lead to nonspecific autonomic withdrawal symptoms and malaise. Stopping either type of agent rapidly or abruptly (as occurs frequently with pregnancy, often without adequate justification) is associated with excess early risk of relapse or exacerbation of psychosis, and this risk may be dose-dependent. Such risks are particularly striking after stopping clozapine abruptly, with poor cross-protection by gradual conversion to other antipsychotics.

Tolerance, or loss of efficacy, after several years of continuous antipsychotic drug treatment has been suspected in a small minority of chronically psychotic patients, but this is not a routinely encountered phenomenon. However, even with close clinical monitoring, it is common for chronically psychotic patients to take antipsychotics only intermittently, or to discontinue them altogether within a year, often without discussion with their families or physicians. Nonadherence to prescribed treatment is encouraged by clinically or subjectively unsatisfactory responses associated with fluctuation of symptom-severity of severe psychotic illness, as well as poor tolerability of antipsychotic treatment, and financial concerns. Treatment-nonadherence is often encouraged by the prolonged burden of side effects that may not be obvious clinically, but be troublesome to patients. Nonadherence is a major contributor to poor treatment response over time. As with short-term use of antipsychotic drugs, long-term treatment often is adequate and better-tolerated, with greater adherence, when doses are moderate.

In an earlier review of differential effects of unusually large and more ordinary doses of antipsychotic drugs for long-term use in nearly 30 trials, we found that daily doses averaging the daily equivalent of over 5,000 mg vs. 400 mg of oral chlorpromazine for an average of 6 months yielded similar beneficial effects in over 60% of chronically psychotic patients, and were more effective in only one-third, but less effective in approximately one-quarter. In notable contrast, approximately 95% of patients experienced more adverse effects with the higher doses. These adverse effects ranged from excessive sedation or mild extrapyramidal responses, including bradykinesia, to hypotension, falls and injuries, and epileptic seizures. These findings indicate that the long-term risk:benefit ratio of unusually high doses of antipsychotic agents is unfavorable [13]. If moderately higher-than-usual doses are explored empirically with individual patients who are not doing well with current, standard, doses, it is essential to evaluate critically and objectively the positive and negative effects encountered at higher doses, and to reduce them again and consider other changes in the treatment program if there is little evidence of further improvement within 1–3 months, even if higher doses seem to be well-tolerated. A related problem is that many effects of antipsychotic drugs, including such expected effects as EPS and decreased alertness and motility, can exert an adverse impact on mental status, cognition, and general functioning—again calling for critical, ongoing, assessments of risks as well as benefits of increased, sustained, or even reduced doses.

An important source of information concerning long-term dosing with antipsychotic agents was a long-term dose-ranging study led by Kane [101]. In this study, patients with modern (DSM-III) diagnoses of schizophrenia were treated with fluphenazine decanoate injections, given at usual average doses or doses reduced by five or tenfold. The results indicated strongly dose-dependent responses as the percentage of patients remaining stable for at least 12 months (Fig. 2.9). Remarkably, doses equivalent to perhaps only 30 mg/day of oral chlorpromazine [71] produced benefits that were 2.5-times greater than placebo-associated responses found earlier in the same clinic among similar patients. The apparent half-maximally effective dose (ED_{50}) of fluphenazine decanoate was only about one-quarter that of typically administered doses, indicating that many chronically psychotic patients can be managed successfully at remarkably conservative doses.

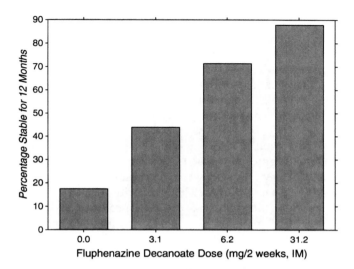

Fig. 2.9 Antipsychotic dose-response in schizophrenia: Depot fluphenazine decanoate. Rates of 1-year stability (percentage of DSM-III schizophrenia patients) vs. the mean dose of fluphenazine decanoate injected (intramuscularly [IM]) every 2 weeks. The highest dose (31.2 mg/2 weeks, IM) is similar to doses typically employed clinically, which are equivalent to approximately 500 mg/day of orally administered chlorpromazine. However, the findings indicate that half of the patients could be maintained adequately for a year at doses only about one-quarter of those typically given. The placebo data (0 mg/2 weeks) were obtained from earlier trials at the same center with similar patients, shown for comparison, and the data are adapted from Kane [101]. Note that doses only one-tenth those typically given were 2.5-times more effective than placebo

Long-Acting, Injected Antipsychotic Preparations

Discontinuous treatment with antipsychotic and other psychotropic drugs has much more limited benefits than sustained treatment [68]. It is a truism that most patients being maintained on long-term treatments vary in their adherence to prescribed medication. Repeated shifts in average daily doses ingested or in days on and off medication may well contribute to clinical instability. Clinical stability may be increased somewhat by use of long-acting, intramuscularly injectable antipsychotic agents (Fig. 2.5). These include the decanoate esters of fluphenazine (typically at doses of 12.5–25 mg) or haloperidol (at 25–50 mg) every 2–4 weeks, the palmitate ester of paliperidone (100–200 mg/month), long-acting carbohydrate microspheres containing risperidone (25–50 mg/2 weeks), or the pamoate complex of olanzapine (150–400 mg/2–4 weeks). Since it is nearly impossible to assess treatment-adherence reliably among outpatients, a major reason to consider long-acting, injected antipsychotic treatment is failure to respond well to presumably adequate doses of oral medication, particularly in sufficient nominal doses as to induce objective side effects.

Despite their theoretical advantages, use of long-acting, injected or depot preparations of antipsychotic drugs with relatively assured delivery, compared to orally administered preparations usually has shown surprisingly limited average gains in clinical stability [2, 114, 130]. In a recent meta-analysis of ten randomized comparisons of injected, long-acting to oral antipsychotic treatments given for an average of 64 ± 22 weeks, reported between 1979 and 2010, there was a pooled, overall significant reduction of relapse risk with depot agents by 30% (95% confidence interval [CI]: 13–43% less), a nonsignificant smaller effect on rehospitalization (22% reduction; CI: 43% less to 5% greater), and a nonsignificantly greater risk of stopping treatment due to intolerance of side effects (by 34%; CI: 30% less to 158% greater [130]). However, outcomes across individual trials varied markedly, from 50% superiority to 92% inferiority of injected medication with respect to relapse risks. Similarly, in an earlier, unpublished, review of 13 such trials, my colleagues and I found an overall average sparing of relapse with depot vs. oral antipsychotic treatment of 11%, ranging from 50% superiority to 4% inferiority, with marked differences across sites.

Such findings suggest that improving clinical outcomes with injected, long-acting antipsychotic agents may be limited in some clinical circumstances. Notably, treatment by well-trained clinicians and in specialized centers that are adequately staffed and funded, orally based treatments can be expected to have their best chance of helping, and added benefits of injected medication to be low. Future assessments of long-acting antipsychotic treatments might well be sited in more tenuous and vulnerable circumstances, such as emergency services or poorly supported clinics, and ideally continued for very long times (see Fig. 2.8, where relapse risk approached 50% within 2 years), so that the certainty of injected and sustained medication might have a greater impact. In general, selection of long-acting, injected antipsychotic drugs is highly patient-specific, may vary over time, and appears to depend greatly on the circumstances of the treatment program.

Comparative Efficacy Among Antipsychotic Drugs

As the variety of available antipsychotic agents continues to expand, an important topic is whether there is credible evidence that specific drugs differ substantially in efficacy for either short-term or long-term applications. This question falls within the currently influential approach of *evidence-based medicine* or evidence-based therapeutics [159, 177, 191]. Typically, this approach to developing rational therapeutics policies and practices relies heavily on combining findings across related therapeutic trials by the method of *meta-analysis* [45, 76, 77]. Ideally, meta-analysis relies primarily on results of well-designed, randomized, controlled therapeutic trials, and attempts to combine findings from studies that appear to be comparable. Typically, the results of individual trials are weighted, usually by the numbers of subjects and sometimes also for measurement variance, to produce an overall, pooled, estimate of effect size that can be a rate ratio (RR; such as response rates based on attaining a criterion [typically ≥50%] of improvement with drug vs. placebo) or an outcome difference, such as difference in response rate (RD) between a drug and placebo control (the reciprocal of which is the number-needed-to-treat [NNT] to produce an effect superior to that associated with placebo, or the number-needed-to-harm [NNH] in the case of an adverse effect as the outcome), or difference in percentage change on a standard symptom rating scale [110].

In practice, use of meta-analyses is potentially misleading in that such data-pooling exercises represent observational experiments (the number of observations is the count of trials pooled, not the numbers of subjects), and treatment-randomization does not apply to the meta-analysis itself. Moreover, some trials are not truly comparable, even if they appear to meet nominal selection and exclusion criteria, since potential confounding variables may not be considered or taken into account. Notably, there has been a striking trend toward falling drug–placebo differences in trials of many psychotropic drugs in recent decades. To compensate, some investigators seek ever-larger samples and pool data across multiple, often culturally and clinically heterogeneous, sites. This option can provide larger numbers of subjects seemingly more efficiently, but standards of diagnosis, clinical assessment, and staff-training may be compromised, and greater variance in outcomes introduced from variance in local, cultural differences in the interpretation of diagnostic and symptomatic concepts [215]. Such efforts to increase numbers may be able to force statistical significance from modest drug–placebo differences, but may well increase risk of misleading findings. There also appear to be important secular effects associated with the rising numbers of subjects and collaborating sites per trial of various psychotropic agents.

Such secular effects are particularly noticeable in rising responses associated with placebo-treatment as trial-size has increased over the years. In turn, increased placebo-responses may reflect compromises in quality-control as the numbers of subjects and heterogeneity of sites has increased, probably with more variable levels of expertise and experience in trial-staff providing diagnostic and symptomatic assessments [213, 215, 235, 236]. Alternative or additional factors may include nonspecific effects of trial-duration, which also has increased in recent decades [213]. Longer trials provide more, potentially helpful, clinical contact with patients and more time for spontaneous improvement. In many advanced countries and in academic clinical research centers, the mix of patient-types also has changed over the years, with more treatment-resistant, co-morbid, or otherwise clinically complex cases tending to encourage willingness to try new treatments but also limiting responses to them. Contrarily, there also appears to be a tendency for less severely ill, and often partially treated, patients to accept the risk of randomization to a placebo in experimental trials, with greater likeliness of their responding during placebo-treatment.

Additional limitations of pooling data from multiple trials include growing evidence of selective submission, peer-reviewer and editorial acceptance, and publication of "positive" findings with significant drug–placebo differences [91, 211]. Resolution of this problem is complicated by difficult access to unpublished findings, especially those from non-commercial sources, with great likelihood of incomplete sampling of trials outcomes. Although even unsuccessful pharmaceutically-sponsored trials, with nonsignificant differences in outcomes between drugs and placebos, are increasingly likely to be posted on-line, the information provided can be incomplete and not based on adequate peer-review.

With respect to the findings of reviews and meta-analyses of antipsychotic drug trials, most indicate only modest and inconsistent differences among specific agents, with the apparent exception of the relatively toxic drug clozapine, which is consistently superior in efficacy to most older neuroleptics, and more variable in comparisons to various modern antipsychotics [15, 75, 103, 126, 129, 131, 132, 143, 222]. Despite clozapine's generally superior antipsychotic efficacy, its risks of medically severe or even fatal adverse effects (leukopenia, seizures, ileus, and metabolic syndrome) have held it in reserve until other treatments have failed. Among most other antipsychotic drugs, old and new, overlapping 95% confidence intervals found in meta-analyses of their efficacy vs. placebo or vs. other antipsychotics indicate that most are similarly effective and that it is not possible to rank them with confidence by relative efficacy (Fig. 2.10). A recent meta-analysis of 23 trials of antipsychotic agents lasting an average of 15.4 months found that pooled results across all types of second-generation antipsychotic drugs yielded slightly superior effectiveness to older, first-generation neuroleptic drugs [112]. However, the superiority was minor (e.g., 8.5% regarding relapse-prevention) and the estimated NNT was large (17). Moreover, the appropriateness of pooling data from markedly pharmacologically dissimilar drugs into "classes" is questionable.

Additional efforts have been made in recent years to evaluate various antipsychotic drugs in large, randomized, practical, and often prolonged clinical trials, usually in patients diagnosed with schizophrenia. Notable examples include the US NIMH-sponsored CATIE project (*Clinical Antipsychotic Trials of Intervention Effectiveness* [105, 134, 143]) and the British trial CUtLASS (*Cost Utility of the Latest Antipsychotic Drugs in Schizophrenia Study* [52, 155]). These clinical effectiveness trials were not conducted or supported by the pharmaceutical industry. Both indicated few advantages (and much higher cost) of modern antipsychotic agents over first-generation neuroleptics, but also revealed quite modest changes in clinical symptoms with any treatment—at least among patients in developed countries who had already been extensively treated and so perhaps poorly representative of recently-ill and little-treated patients. In addition to limitations imposed by the types of patients available for such clinical trials, the marginal differences in efficacy among antipsychotic drugs, as well as the clinical heterogeneity of chronically psychotic patients diagnosed with schizophrenia even by contemporary research methods, are likely to limit the amount of clinical change that can be expected with any treatment.

Finally a recent analysis of a national Finnish cohort of nearly 2,600 psychotic-disorder patients also indicated only minor differences in long-term rehospitalization risks among treatments with orally administered, older or modern antipsychotic drugs. An exception, again, was clozapine, as well as substantial apparent superiority of treatment with long-acting injected agents ([205]; Table 2.6).

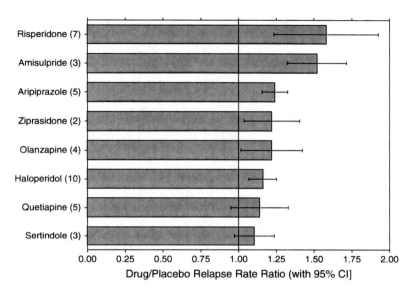

Fig. 2.10 Efficacy of representative antipsychotic drugs vs. placebo (relapse rate ratio with 95% confidence interval). Drugs are ranked in descending order of apparent efficacy in 1–12 months (with number of randomized, controlled trials reviewed). Note that all confidence intervals overlap, indicating lack of significant differences in drug efficacy (Adapted from Leucht et al. [128]). The vertical *dotted line* indicates no differences in relapse risk with drug vs. placebo

Long-Term Care of Psychotic-Disorder Patients

In the clinical management of chronically ill psychotic-disorder patients, it is essential to plan for the long-term, taking into account the patient's symptoms, course and treatment response, as well as his and the family's resources. The challenge is to support the family, help minimize discouragement and abandonment of the patient, encourage habilitation and a tolerable quality of life, avoid social isolation and depletion of the family's emotional and economic resources, and to counsel patients and families to avoid unwise or wasteful searches for unproved treatments [98].

Possible beneficial effects of alternatives to antipsychotic medication have been suggested for some anticonvulsants as well as adjunctive use of antidepressants, especially to combat negative symptoms and depressed mood in psychotic-disorder patients [42, 121, 170, 203, 206, 207, 219, 238]. Most of the experience underlying such empirical practices remains anecdotal and scientifically limited. Electroconvulsive treatment (ECT) is still tried occasionally when other medical treatments fail in chronic psychotic disorders, and it may be particularly useful with catatonia and perhaps in young patients early in the course of relatively acute psychotic illnesses, but evidence for clear or sustained benefits in chronic schizophrenia is meager [237]. However, there is more compelling research support for favorable interactions of commonsense, cost-effective psychosocial interventions

Table 2.6 Risk of rehospitalization in schizophrenia patients vs. type of antipsychotic treatment

Treatment	Hazard ratio (HR with 95% CI)
Depot haloperidol	0.21 (0.03–1.60)
Oral clozapine	0.48 (0.31–0.76)
Oral olanzapine	0.54 (0.40–0.73)
Depot risperidone	0.57 (0.30–1.08)
Depot perphenazine	0.59 (0.31–1.12)
Oral risperidone	1.00 (comparator)
Oral perphenazine	1.11 (0.57–2.19)
Oral quetiapine	1.11 (0.75–1.64)
Oral haloperidol	1.79 (0.63–5.09)
Any depot vs. oral	0.36 (0.17–0.75)
Continued ≥30 days (%)	45.7% (43.7–47.6%)
Deaths (with/without treatment)	0.38 (0.19–0.74)

Data are hazard ratios (relative risk; adjusted for sex, age, prior antipsychotic use, use of other psychotropic medicines, and duration of the first-lifetime hospitalization, in descending order of apparent efficacy) with their 95% confidence intervals for six orally administered, and three long-acting, injected preparations of first- and second-generation antipsychotic drugs, based on a Finnish national sample of 2,588 cases in the years 2000–2007. HR values are based on oral risperidone as the standard comparator; the null value is 1.0, lower values indicate superiority, and higher values, inferiority to the comparator (CI >1.0 indicates statistical nonsignificant difference from the comparator). Adapted from Tiihonen et al. [205].

with sound antipsychotic treatment of patients with chronic psychotic disorders. Specific forms of behavioral, educational, and family interventions with a supportive or rehabilitative orientation, in particular, can add substantially to the stability of antipsychotic-treated chronically psychotic patients, often reducing relapse rates by more than 50% compared to drug-alone for up to 2 years [109, 120, 165, 217, 234]. Overall, typical 1-year outcomes in patients diagnosed with schizophrenia average approximately 85% relapse risk without specific treatment, perhaps 5–10% less with a placebo, about 70% relapse risk with psychosocial treatment only, 25% with an antipsychotic drug alone, and perhaps 10% risk with both medicine and psychosocial treatment.

Prevalence of long-term custodial hospitalization has decreased greatly since the 1950s, but rehospitalization rates and incarcerations of the mentally ill in jails and prisons have increased several-fold, in parallel with diminishing hospital-access and premature discharges due to economic and administrative factors (Fig. 1.1). Even with optimal treatment, most gains among psychotic-disorder patients are found in reduction of morbidity in relatively acute phases of illness, whereas studies of long-term outcome in schizophrenia over the past century provide inconsistent evidence of a substantially improved average, long-term functional, and occupational prognosis in schizophrenia diagnosed by modern criteria (such as DSM-IV of the American Psychiatric Association or ICD-10 or the World Health Organization) since the introduction of modern antipsychotic treatments [86].

Adverse Effects

Older and modern antipsychotic drugs carry substantial risks of adverse effects that range from those that are uncomfortable and annoying to patients, to those with medically significant risks, and some that can be life-threatening. The term "adverse effects" is preferred to the euphemistic term "side effects" to draw attention to the medically significant as well as unpleasant nature of many of these, as well as to their strong association with non-adherence to recommended treatment [40, 78]. Common adverse effects include excessive sedation, fatigue, decreased alertness, decreased or slowed motility, and an overall burden of nonspecific effects of the "medicated state." It is also not uncommon for antipsychotic agents, especially those of relatively low potency, to induce postural hypotension or fainting, particularly soon after starting a new drug. Nausea, altered appetite (increased or decreased), and dry-mouth or blurred vision also are not uncommon. Iloperidone has been associated with febrile reactions of uncertain type or significance.

Weight-Gain and Metabolic Syndrome

Importantly, some first- and second-generation antipsychotic drugs are associated with weight-gain, with or without the metabolic syndrome of hypertension, hyperlipidemia, and type 2 diabetes mellitus [24, 61, 97, 161, 163]. Such reactions are most closely associated with older neuroleptics of low potency (such as chlorpromazine and thioridazine) and several modern agents (notably, clozapine, olanzapine, and quetiapine). Some psychotic-disorder patients may be particularly vulnerable to risk of weight-gain and adverse metabolic effects of antipsychotic treatment, but specific and clinically useful genetic markers for such risks remain elusive [29, 82, 127]. With respect to the pharmacology of the drugs themselves as a contributing factor in weight-gain and metabolic syndrome, inhibition of serotonin receptors, particularly type 2C by antipsychotic drugs may be involved, although additional effects on the functioning of insulin or on carbohydrate metabolism also are suspected [146].

Body-mass-index (BMI; weight/height ratio) values in the overweight range ($25–29$ kg/m^2) are commonly found among chronically psychotic and bipolar disorder patients, and clinical obesity, indicated by BMI ≥ 30 kg/m^2, is found increasingly often and at younger ages. BMI and waist-circumference are considered useful predictors or precursors of metabolic syndrome in antipsychotic-treated patients, although metabolic changes can occur without significant weight-gain, including in juveniles [58]. Treatment options include selection of lower but adequate doses of antipsychotic agents, particularly those less likely to increase weight, efforts at exercise and dietary control, and possibly early introduction of hypoglycemic agents such as metformin [38, 69, 167].

Adverse Neurological Effects

Weight-gain and adverse metabolic effects associated with some modern antipsychotics have to an important extent replaced the formerly very common neurological adverse effects of first-generation antipsychotic drugs, and they may be even more serious and potentially life-threatening. First-generation antipsychotic drugs were often termed "neuroleptic" ("seizing the nervous system") due to their regular association with typical abnormalities of posture and movement. Such reactions still occur with modern antipsychotic agents, though most are much less frequent [158, 166, 201, 231]. Characteristic adverse neurological effects, or so-called "extrapyramidal syndromes" of older neuroleptic agents include several discrete syndromes that vary in timing, severity, distress, knowledge of their mechanisms, and the effectiveness of their treatment (Table 2.7). They include: (a) akathisia or sustained restlessness, (b) acute dystonias, (c) gradually evolving and sustained parkinsonism marked by bradykinesia with variable tremor, and (d) late-emerging or tardive dyskinesias or dystonias (TD). In addition, a formerly not-uncommon syndrome of general cerebral intoxication was the neuroleptic malignant syndrome (NMS), whose former and current clinical presentations can differ markedly.

Akathisia ("inability to sit still") remains a common form of motor disturbance associated with antipsychotic drug treatment, including both older neuroleptics and most modern agents as well, particularly aripiprazole, ziprasidone, and even clozapine [102, 104, 119, 239]. Its clinical manifestations include strong subjective, anxiety-like, distress associated with inability to sit still and restless movements; movement or pacing may temporarily limit subjective distress to some extent. The distress involved is sometimes associated with increased aggressive or suicidal inclinations, and with worsening of mental status [57, 124]. The disorder arises almost immediately with the start of antipsychotic treatment, and usually continues indefinitely. It can be confused with increased primary agitation associated with most psychotic disorders. The pathophysiology of akathisia is poorly understood, but may involve central noradrenergic mechanisms since lipid-soluble β-adrenergic receptor antagonists (such as propranolol (Inderal®), nadolol (Corgard®), and atenolol (Tenormin®)) can be helpful clinically in this but not other neurological disturbances associated with antipsychotic treatment [137, 173]. Antiparkinsonism agents are usually ineffective, and anxiolytic benzodiazepines have limited benefits. Additional treatments that may have beneficial effects include the α_2-adrenergic agonist clonidine, the atypical antidepressant mirtazapine, and the tryptan serotonin agonists used to treat migraine [9, 92, 209]. Reducing the impact of akathisia by lowering the dose of the suspected offending agent, or changing to a dissimilar antipsychotic drug, often exerts noticeable clinical benefits on subjective discomfort and on mental status.

Dystonias are commonly acute, but can occur late in treatment (tardive dystonias), sometimes as a component of tardive dyskinesias. Acute dystonias involve sometimes painful and often frightening muscular contractions, commonly of the face, neck, and back. Occasionally, acute dystonias involving laryngeal or

Table 2.7 Neurological syndromes characteristic of antipsychotic drugs

Syndrome	Clinical features	Timing	Proposed mechanisms	Treatments
Acute dystonias	Spasms of muscles of tongue, face, neck, back; terrifying; rarely fatal from asphyxia	1–5 days or with each injection of depot agents; worse in muscular young males	Possibly dopaminergic excess	Injected, then oral anticholinergics or diphenhydramine
Parkinsonism	Bradykinesia, rigidity, variable tremor; mask facies, reduced arm-swing, shuffling gait	Evolves slowly (1–4 weeks), often persists, or returns without treatment; children and elderly at high risk	Dopaminergic deficiency	Oral anticholinergics or amantadine; avoid dopaminergics
Malignant syndrome (NMS)	Stupor and delirium; usually with fever and unstable pulse, blood-pressure and respirations; variable rigidity and release of creatine kinase and myoglobin	Days to weeks; more likely with potent, injected, or high doses; can be worse in the elderly; lethal in ≥25%	Hypothalamic+extrapyramidal dysfunction proposed; not related to the malignant hyperthermia of general anesthesia	Stop antipsychotic; requires expert and intensive medical care; bromocriptine or dantrolene may help, as may ECT
"Rabbit" syndrome	Localized perioral tremor (rare and reversible)	Usually months or years	Probably a parkinsonism variant	Oral anticholinergics
Akathisia (NIA)	Motor restlessness, anxiety, agitation; can worsen agitation and increase risk of aggression or suicide	Almost immediate, and typically persists	Unknown; may have an adrenergic component	Reduce or change antipsychotic; add propranolol; benzodiazepines may help; anticholinergics usually not helpful
Tardive dyskinesias-dystonias (TD)	Oral-facial or widespread dyskinesia, choreoathetosis, variable dystonia, often slowly reversible, rarely progressive	6 to ≥12 months; worse when antipsychotic stopped; more likely and more persistent with older age	Dopaminergic excess strongly suspected	Prevention best; remission likely (more in youth), especially if drug reduced or removed

These syndromes were much more common with first-generation neuroleptic drugs (especially those of high potency) than with modern antipsychotics. They are informally considered "extrapyramidal syndromes" (EPS), although their pathophysiology for all but Parkinsonian bradykinesia is not established. Akathisia and tardive dyskinesia are easily overlooked unless considered specifically at examination, and akathisia is easily confused with, or can worsen, agitation or other symptoms of psychotic illness. With modern antipsychotics, dystonia, and parkinsonism are no longer common, but akathisia is prevalent (notably including with aripiprazole and clozapine). The malignant syndrome (also "neuroleptic malignant syndrome" [NMS]) fits into a spectrum of cerebral, drug-associated intoxications that include the so-called "serotonin syndrome" (as can occur with combining a serotonin-reuptake inhibitor with a monoamine oxidase inhibitor), the central anticholinergic syndrome associated with overdoses of older tricyclic antidepressants or other agents with high central antimuscarinic potency (including atropine, antiparkinsonism agents, and clozapine); such reactions are unusual with modern antipsychotics, but can be misdiagnosed or overlooked since their main presenting feature is *delirium* (sometimes with fever and unstable vital signs) with minimal risk of peripheral motor manifestations, and may not include increased and fluctuating muscle tension or catatonia and release of creatine kinase and myoglobin into the bloodstream that were typical with older neuroleptics of high potency intercostal muscles can lead to respiratory distress and, rarely, fatalities by asphyxiation [41]. Acute dystonias have been especially common among young males given high doses of potent older neuroleptics, especially by injection [85]. They occur infrequently with most modern antipsychotics, and are particularly likely with high doses of relatively potent antidopaminergic agents including risperidone and paliperidone. Among other modern antipsychotics, dystonias are uncommon with use of aripiprazole, and perhaps even less with asenapine, clozapine, iloperidone, and qutiapine. The typical timing of acute dystonias is to arise, not immediately, but within hours following a first-dose of neuroleptic drug, or following each injection of a long-acting antipsychotic. Usually, their risk diminishes gradually over a week or so, as parkinsonian bradykinesia emerges. The pathophysiology of the syndrome is unknown, but may involve paradoxical, initial overactivity of dopaminergic neurons in the mid- and forebrain, as an expected transient response to partial blockade of dopamine receptors. This hypothesis fits the timing of risk for acute dystonias and the gradual emergence of parkinsonism. In addition, there are anecdotal reports of dystonias having been interrupted by administration of large doses of a potent neuroleptic; such interventions are known to drive dopaminergic neurons into a state of depolarized inactivity in laboratory animals [151]. The usual treatment of acute dystonias is administration of antiparkinsonism agents, such as centrally active, antimuscarinic cholinergic agents (such as benztropine or trihexyphenidyl) or the antihistaminic-anticholinergic agent diphenhydramine, by injection initially and then orally. It remains an unresolved question whether such treatments offer prophylaxis against acute dystonias, but they are sometimes added temporarily to antipsychotic treatment in circumstances of high risk (such as with potent neuroleptics and young males). Nevertheless, it is essential to warn patients that such reactions can occur, and require emergency treatment, whether an anticholinergic agent is given prophylactically or not. Dystonias also are a component of many cases of tardive dyskinesia. Drugs with central antimuscarinic activity, including clozapine, are sometimes helpful in the clinical management of tardive forms of dystonia.

Parkinsonism is common in children as well as elderly adults. It is typically less prominent and severe with most modern antipsychotic agents than with older neuroleptics. Compared to idiopathic Parkinson's disease, antipsychotic-induced parkinsonism is marked mainly by bradykinesia or uncommonly akinesia resembling catatonia, and less prominent or more variable tremor. This reaction can be confused with the emotional flatness and withdrawal of schizophrenia and so overlooked; it can also give a potentially misleading impression of depressed mood. The pathophysiology of drug-induced parkinsonism is very likely to reflect disruption of dopaminergic neurotransmission by antipsychotic drug treatment, as a functional parallel to the anatomical loss of dopaminergic neurons in idiopathic Parkinson's disease. Drug-induced parkinsonism typically is responsive to centrally active antimuscarinic-anticholinergic agents (e.g., benztropine [Cogentin®], biperidin [Akineton®], procyclidine [Kemadrin®], or trihexyphenidyl [Artane®]). The catecholamine-potentiating agent amantadine [Symmetrel®]

is much less well evaluated in drug-induced parkinsonism than in early, mild Parkinson's disease. Efforts to remove the antiparkinsonism agents months later typically are followed by rapid return of bradykinesia, suggesting that tolerance to this effect of antipsychotic drugs is not usual [164, 209]. Dopamine agonists are not safely used to treat drug-induced parkinsonism among patients with psychotic disorders, and indeed induction of psychosis by such agents can be a problem with their use in Parkinson's disease.

The use of antipsychotic drugs to treat *psychosis in Parkinson's disease* patients is difficult to achieve without worsening the bradykinesia of the neurological disorder. Clozapine, and low doses of quetiapine have been tolerated and effective in this circumstance, but most other older and many modern antipsychotic agents tend to worsen bradykinesia in Parkinson's disease, surprisingly, including the dopamine partial-agonist aripriprazole [27, 80, 226]. ECT also may be effective and tolerated in psychosis associated with Parkinson's disease [214].

Probably related to drug-induced parkinsonism, a rare, late syndrome of localized perioral tremor (*rabbit syndrome*) has been noted after prolonged exposure to neuroleptic drugs. The usual tremor frequency (3–5 Hz) resembles that found in idiopathic Parkinson's disease, and the symptoms respond well to treatment with centrally active antimuscarinic agents, as does parkinsonism. Though arising late, this syndrome appears to be unrelated to tardive dyskinesia.

The uncommon, but potentially fatal (in approximately 15% of cases) *NMS* classically includes variable, waxing and waning catatonia, coarse tremor, stupor or delirium, fever, and autonomic instability, particularly in association with older neuroleptic agents [83, 141]. Some of these features, as well as sometimes apparently beneficial responses to a dopamine agonist, suggested a relationship to drug-induced parkinsonism when it was initially associated with first-generation neuroleptic drugs [74]. However, a syndrome primarily marked by delirium, variable fever, often without severe motor-slowing and muscle rigidity or release of creatine kinase and myoglobin from muscle into the bloodstream, as were typical of older forms of NMS, can occur with virtually any antipsychotic agent. This risk is present with modern second-generation drugs including aripiprazole and clozapine, as well as other drugs with neuroleptic-like effects, such as droperidol (Inapsine®), metoclopramide (Reglan®), prochlorperazine (Compazine®), and promethazine (Phenergan®) [7, 223]. This syndrome of cerebral intoxication arises most often within 1–2 weeks of initiation of antipsychotic treatment. It probably belongs to a group of clinically similar syndromes of acute, cerebral intoxication, and unstable vital signs, including the "serotonin syndrome" associated with certain antidepressants to be discussed below. The earlier strong association of NMS with severe muscle rigidity in association with neuroleptic treatment may be misleading, since current cases of newly-emerging delirium associated with second-generation antipsychotic drugs are unlikely to present with muscle rigidity and release of creatinine kinase or myoglobin from muscle cells. Indeed, it may be appropriate to abandon the term "NMS" in favor of the broader label, *antipsychotic-malignant syndrome* (AMS).

Safe clinical management of AMS requires rapid discontinuation of the suspected toxin and close medical supervision, ideally in an intensive care unit. The muscle relaxant dantrolene (Dantrium®) has been tried empirically to treat AMS [176, 193]. Although it can be effective in malignant hyperthermia of general anesthesia associated with excessive influx of calcium ions into muscle cells, the pathophysiology of that syndrome probably is not related to that of AMS. Moreover, the evidence that dantrium is helpful in AMS is inconsistent, and it may even worsen the outcome in some cases [174]. However, relatively high doses of a dopamine agonist such as bromocriptine (Parlodel®) appear to be helpful, and benzodiazepines may contribute some benefit [193, 210, 223]. This severe and potentially fatal neurotoxic reaction is uncommon in current psychiatric settings, perhaps due to use of newer antipsychotic agents with lower risk, more conservative antipsychotic dosing, cautious use of injected potent agents, and increased clinical vigilance.

Tardive (late) *dyskinesias* (TD) formerly occurred with a prevalence of 15–25% of adults maintained on an older neuroleptic for 6 months or more. This prevalence estimate is corrected for spontaneous and other dyskinesias, including nonspecific movement disorders of the elderly and effects of edentulism. Typical signs of TD include choreic (quick) and athetotic (writhing) movements, sometimes limited to the face or fingers, but more generalized and widespread in other patients. Dystonic and akathisia-like elements are sometimes associated with choreoathetosis. TD is painless, but sometimes disfiguring and potentially disabling if severe or widespread. Manifestations of TD are sensitive to levels of arousal and agitation, disappear in sleep, and can be brought out by distraction of the patient. TD led to medicolegal concern and risk of malpractice suits prior to introduction of modern antipsychotics. Risk of TD rises importantly (and spontaneous remission is less likely) with advancing age, probably with the presence of major affective syndromes, might be limited by use of smaller maintenance doses of antipsychotics, and is much less likely to occur with most modern antipsychotics than with first-generation neuroleptic drugs, especially those of high-potency [34, 46, 188, 200, 202]. The incidence of TD, with older neuroleptics, of ca. 3–5%/year, continues for at least 10 years, but there is also a "negative" impact of spontaneous remission, at perhaps 2–3%/year, especially in younger patients and probably with lower doses, to limit net prevalence. Moderate risk of TD is suspected with risperidone, is lower with olanzapine than with haloperidol, and is much lower with clozapine than most other antipsychotics [31, 96, 202, 232]. Risk of TD remains inadequately evaluated with most newer antipsychotics owing to the rarity of prolonged use of single, modern agents without previous exposure to first-generation antipsychotics [200, 202]. Moreover, the impact of increasingly common empirical use of multiple psychotropic agents in series or simultaneously ("polytherapy," sometimes with both older and newer agents) on risk of TD is not known. The pathophysiology of TD is likely to be related to the tendency for prolonged blockade of dopamine (and other neurotransmitter) receptors to become more abundant ("upregulate") and become supersensitive to agonists in response to prolonged blockade or following denervation [19, 199]. A proposed component of the pathophysiology is putative oxidative damage to forebrain neuronal systems, based on analogy to proposed oxidation-based neuronal damage in various neurodegenerative disorders, but this hypothesis lacks evidence in TD [53, 122, 218]. However, many patients exposed for years to older or modern antipsychotic drugs do not develop TD, suggesting that additional individual vulnerability factors other than age are involved, and might be revealed by genetic analyses [8, 122].

Chances of spontaneous remission of TD are good, especially in young patients and particularly if antipsychotic medication can be reduced or discontinued, or a very low-EPS-risk agent such as aripiprazole or clozapine can be used. TD led to medicolegal concern and risk of malpractice suits prior to introduction of modern antipsychotics, and continues to call for special vigilance for the emergence of early signs of TD. There is no specific treatment for TD, but many have been tried, usually with limited and inconclusive evidence of efficacy and safety. In severe cases, signs of TD are sometimes suppressed by administration of increased doses of the suspect agent, use of another antipsychotic agent, or giving a dopamine depleting agent such as tetrabenazine [133]. This approach has been particularly useful with the use of clozapine, especially to suppress prominent dystonic components of TD. However, the concept of giving an antidopaminergic agent to treat a condition produced by another antidopaminergic agent seems non-rational and potentially risky, although evidence of long-term worsening has not been presented. Additional treatments proposed or tried clinically include: antianxiety agents (including clonazepam [Klonopin®], lorazepam [Ativan®], buspirone [Buspar®] and others); the anticonvulsants gabapentin [Neurontin®] and levetiracetam [Keppra®]; the β-adrenergic antagonist propranolol (Inderal®), the antihypertensive α_2-adrenergic agonist clonidine (Catapres®), and the antiparkinson cerebral catecholamine enhancer amantadine (Symmetrel®); vitamin E (mainly α-tocopherol, 400–1,200 units/day) or other antioxidants; the opioid antagonist naloxone (Narcan®), branched-chain amino acids; the putative cerebral blood-flow enhancers piracetam (Noötropyl®) and the plant product *Ginkgo* (*Ginkgo biloba*). However, the effectiveness of all of these medicinal agents in TD remains in question, and some are probably ineffective, including bromocriptine (Parlodel®), selegiline (Eldepryl®), anticholinergic-antiparkinsonism agents, and fatty acid supplements [93, 189].

Cardiac Risks and Sudden-Death

Cardiac risks associated with antipsychotic agents are not common. Arrythmias can arise with acute overdoses of some agents, notably thioridazine and mesoridazine, and both myocarditis and later cardiopathies have been associated with clozapine [100]. A more predictable risk, however, is prolongation of the repolarization phase of the electrocardiac functional cycle, which has produced warnings from the FDA.

QT prolongation is recognized by prolongation (particularly >500 ms) of the QT interval corrected for heart-rate (QTc) of the electrocardiogram (usually computed by dividing observed QT by the square-root or cube-root of heart-rate). Prolongation of the QTc interval, and perhaps some aspects of the size and shape of electrocardiographic T-waves as well as a history of paroxysmal atrial tachycardia, hypokalemia, hypomagnesemia, hypertension, and advanced age appear to be risk factors for rare, but potentially fatal ventricular arrhythmias including *torsades de pointes*, and cardiac arrest [156]. The rarity of fatal cardiac reactions limits their contribution to overall mortality risk, which may well prove to be greater with prolonged metabolic abnormalities. Nevertheless, sudden cardiac deaths, though uncommon, have been associated with long-term treatment with older as well as modern antipsychotic drugs, and unanticipated cardiac arrhythmias may contribute to this risk [99, 140, 171, 172]; *decreased* mortality also has been reported in association with long-term antipsychotic treatment [206]. Prolongation of the QTc interval is especially likely with thioridazine or mesoridazine more than haloperidol among older neuroleptics, somewhat less with ziprasidone, and apparently negligibly with clozapine among second-generation antipsychotics [4, 26, 81]. Taken alone, short of severe acute overdoses, such responses to these agents are not likely, but risk increases as additional agents with inhibitory effects on cardiac repolarization are added, including tricyclic antidepressants and class Ia, sodium-channel blocking, antiarrhythmic drugs such as quinidine, procainamide (Pronestyl®), and disopyramide (Norpace®), or if other risk factors are present. In circumstances of potentially increased risk, baseline and repeated electrocardiographic studies may be warranted [62].

Clozapine has been associated with cardiac abnormalities that include acute myocarditis and later myopathies [84, 100, 147]. Carditis has been reported most often, with an uncertain incidence of approximately 0.1%, especially in young men, typically arising within the first 6 weeks of starting treatment with clozapine, and often associated with fever, malaise, chest discomfort, and dyspnea. The myocarditis is associated with elevated serum concentrations of troponin and other peptides released from damaged cardiac muscle and with variable eosinophilia. Sometimes there is associated evidence of cardiac failure, or pericarditis with effusion. The condition is supported by nonspecific electocardiographic changes, rarely is associated with QTc prolongation, may show evidence of ventricular dysfunctioning by ultrasound, and is treated supportively after discontinuing clozapine, although the fatality rate is substantial (10–50%).

There is evidence that antipsychotic drugs of all types are associated with increased risks of mortality in the elderly, and particularly patients with dementia, often in association with sudden death, strokes, and other cerebrovascular events [39, 73, 149, 154, 224]. Despite FDA warnings of such risks, the relative risks and potential benefits of antipsychotic drug treatment in elderly, demented patients, particularly those with psychotic symptoms and severe agitation, as well as the relative contributions to risk of disease vs. treatment, continue to be debated [154, 185].

Hyperprolactinemia

Many antipsychotic agents increase circulating concentrations of prolactin, presumably by blocking D_2-type dopamine receptors that inhibit release of the hormone from the anterior pituitary. Normal plasma concentrations of prolactin are below 20–25 ng/mL (approximately 440–550 milli-international-units, or mIU per liter) in women who are neither pregnant nor lactating, and below 15 ng/mL (300 mIU/L) in men. Levels observed during treatment with antipsychotic drugs can rise to 40–100 ng/mL, but concentrations over 200 ng/mL suggest the presence of a pituitary adenoma. Elevations can be expected with potent older neuroleptics such as fluphenazine or haloperidol, and particularly with risperidone and its active metabolite paliperidone among modern agents. Prolactin elevation can be associated with breast-engorgement in women and loss of libido and other sexual dysfunctions in women and men, as well as with osteopenia [94]. Hyperprolactenemia may well promote the growth of prolactin-sensitive metastases of breast carcinoma; however, its risk of causing breast cancer is not clear, but appears to be low [10, 23, 28, 30, 225].

Special Risks of Clozapine

Clozapine presents particular risks of adverse effects. Well known is its infrequent, but significant, risk of *agranulocytosis* (absolute neutrophil count [ANC] ≤500/mm³), sometimes in association with eosinophilia (eosinophil counts in peripheral blood >3,000/mm³ are particularly ominous), and with risks of sepsis and death [157]. The incidence of agranulocytosis is close to 1.0% without regular medical monitoring, and that of leukopenia is several times higher, mostly arising within the initial 6 months of exposure to clozapine [5]. This potentially lethal adverse effect led to near-abandonment of clozapine in the 1970s and 1980s, and its re-entry into clinical application in the US in 1989 as a back-up treatment with evidence of superior efficacy to many alternative antipsychotics, particularly first-generation agents. In 2002, it also became the only treatment recognized by the FDA as probably effective in reducing suicidal risk, limited to patients diagnosed with schizophrenia and based largely on a randomized comparison with olanzapine in suicidal patients diagnosed with schizophrenia [145].

The modern regulatory approval of clozapine in 1989 was based on reserving its use after failures of at least two vigorous trials of other standard antipsychotic agents, and with the need for regular monitoring of white blood cell counts\ (WBC) and absolute neurophil counts (ANC) for safety. Typically, WBC is monitored weekly for the first 6 months of treatment, biweekly for the next 6 months, and monthly thereafter. Even with regular monitoring of WBC and ANC, occasional cases of severe leukocytosis or agranulocytosis occur, sometimes emerging suddenly within a day or two. Based on FDA-approved manufacturer's (Novartis Corp) standards for monitoring the clinical use of clozapine, leukopenia is gradual as: *mild* (WBC 3,500–3,000/mm^2; ANC 2,000–1,500/mm^2), *moderate* (WBC 3,000–2,000/mm^3; ANC 1,500–1,000/mm^3), or *severe* (WBC <2,000/mm^3; ANC <1,000/mm^3). With modern blood-cell monitoring, the risk of agranulocytosis averages approximately 0.03%/month for the first 6 months, 75% less in the second 6-months, and 97% less thereafter [169].

Leukopenia has been associated with various antipsychotic agents, especially those of low-potency such as chlorpromazine, and probably as a direct chemical-toxic effect [56]. However, the severe leukopenia and agranulocytosis associated with clozapine is not well explained as a direct toxic action on the bone marrow, may involve dissimilar mechanisms, and that underlying agranulocytosis specifically may reflect unidentified, but possibly genetically determined, idiosyncracies [72, 227]. Assessment of cases of leukocytosis in association with clozapine should also consider other potential causal conditions to which the patient may have been exposed (such as cancer chemotherapy, treatment with cyclosporine or other immunosuppressants, interferons, some tetracycline and sulfa antibiotics, and rarely low-potency phenothiazines, mirtazapine, and buspirone among psychotropics). Sometimes leukopenia is partially masked or suppressed by co-treatment with lithium, which often tends to induce leukocytosis. Case reports suggest that lithium may have therapeutic effects, at least in mild drug-associated leukopenia, and there is some evidence that administration of granulocyte-colony stimulating factor may limit the severity of clozapine-induced leukocytosis [87]. Rechallenging a patient who has experienced severe leukopenia with clozapine is highly risky and usually not recommended [227].

Additional risks peculiar to clozapine include annoying sialorrhea that probably represents decreased ability to clear saliva by pharyngeal and esophageal mechanisms [168] and may increase risk of aspiration pneumonia [138]. Clozapine can also induce severe inhibition of bowel motility with risk of life-threatening ileus [64]. There is also a dose-dependent risk of epileptic seizures with clozapine, which is sometimes managed by lowering the dose or adding an anticonvulsant such as divalproex [36].

Conclusions

Introduction of effective and tolerated antipsychotic drugs with chlorpromazine in 1952 opened a revolutionary era of modern clinical and experimental psychopharmacology. With the exception of the relatively early discovery of the uniquely effective drug clozapine, a potentially lethally toxic agent, most first- and second-generation antipsychotic drugs are not clearly superior in efficacy or long-term effectiveness to the oldest phenothiazines. Most modern agents are far more expensive, but many are also better-tolerated, with lower risks of adverse neurological effects, but sometimes increased risks of potentially life-threatening metabolic abnormalities associated with obesity. Modern administrative and financial pressures encourage marked

Table 2.8 Current status of antipsychotic treatment

All antipsychotic drugs are partially effective, largely nonspecific palliatives useful in schizophrenia, but also highly effective in mania and acute psychoses, and sometimes helpful in depression

Modern antipsychotics have less risk of adverse neurological effects, but have variable metabolic and cardiovascular risks, variable efficacy, and are more costly

Most antipsychotics are not clearly distinguishable in main beneficial effects; subgroups that might better respond to or tolerate particular agents are not defined

Clozapine is exceptionally effective in schizophrenia and the only treatment with regulatory-approval as antisuicidal—all vs. high risk of clinically important and potentially lethal adverse effects, inconvenient medical monitoring, patient-reluctance, and extra costs

Development of truly innovative antipsychotic agents has slowed after some advances arising from efforts to mimic aspects of the neuropharmacology of clozapine

Innovation in the treatment of psychotic disorders awaits understanding of basic pathophysiology to define rational targets, or serendipitous discovery

shortening of time-in-hospital and use of unusually high doses and largely untested combinations of drugs in hopes of hastening recovery from acute exacerbations of psychotic illnesses. However, evidence to support such practices is lacking, and rates of clinical improvement-per-day are not increased with unusually high doses of antipsychotic drugs, or with administratively enforced shortening of hospitalization. Most antipsychotic drugs are at least partially palliative in most psychotic disorders, best evaluated in patients diagnosed with schizophrenia, and most strikingly effective in mania and acute psychoses. Long-term benefits of antipsychotic drug treatment can yield substantial symptomatic relief, improved quality-of-life, and decreased risk of long-term institutional care among chronically psychotic patients. There are also broadening applications of antipsychotic drugs to treat many other psychiatric illnesses, including mood, anxiety, and behavior disorders, and in the very young and very old. However, far greater benefits often are anticipated than are warranted, in part owing to lack of long-term adherence to recommended treatment. Moreover, long-term risks of obesity, hypertension, hyperlipidemia, and type 2 diabetes mellitus represent major limitations of some modern and many older antipsychotics, and may well represent even more severe health risks than do the adverse neurological effects of first-generation neuroleptics. The current status of antipsychotic drugs is summarized in Table 2.8.

References

1. Agid O, Kapur S, Remington G. Emerging drugs for schizophrenia. Expert Opin Emerg Drugs. 2008;13:479–95.
2. Adams CE, Fenton MK, Quraishi S, David AS. Systematic meta-review of depot antipsychotic drugs for people with schizophrenia. Br J Psychiary. 2001;179:290–9.
3. Alexander GC, Gallagher SA, Mascola A, Moloney RM, Stafford RS. Increasing off-label use of antipsychotic medications in the United States. Pharmacoepidemiol Drug Safety. 2011;20:177–84.

4. Alvarez PA, Pahissa J. QT alterations in psychopharmacology: proven candidates and suspects. Curr Drug Safety. 2010;5:97–104.
5. Alvir JM, Lieberman JA, Safferman AZ, Schwimmer JL, Schaaf JA. Clozapine-induced agranulocytosis: incidence and risk factors in the United States. N Engl J Med. 1993;329:162–7.
6. American Psychiatric Association. Diagnostic and statistical manual. 4th ed., text revision. Washington, DC: American Psychiatric Press; 2000.
7. Ananth J, Parameswaran S, Gunatilake S, Burgoyne K, Sidhom T. Neuroleptic malignant syndrome and atypical antipsychotic drugs. J Clin Psychiatry. 2004;65:464–70.
8. Arinami T, Inada T. [Genome-wide association analyses for neuroleptic-induced tardive dyskinesia (Japanese)]. Nihon Shinkei Seishin Yakurigaku Zasshi. 2011;31:155–62.
9. Avital A, Gross-Isseroff R, Stryjer R, Hermesh H, Weizman A, Shiloh R. Zolmitriptan compared to propranolol in the treatment of acute neuroleptic-induced akathisia: a comparative double-blind study. Eur Neuropsychopharmacol. 2009;19:476–82.
10. Azoulay L, Yin H, Renoux C, Suissa S. The use of atypical antipsychotics and the risk of breast cancer. Breast Cancer Res Treat. 2011;129:541–8.
11. Baldessarini RJ. Chemotherapy in Psychiatry. 2nd ed. Cambridge, MA: Harvard University Press; 1985.
12. Baldessarini RJ. Dopamine receptors and clinical medicine. In: Neve KA, Neve RL, editors. The dopamine receptors. Totowa, NJ: Humana; 1997. p. 457–98.
13. Baldessarini RJ, Cohen BM, Teicher MH. Significance of neuroleptic dose and plasma level in the pharmacological treatment of psychoses. Arch Gen Psychiatry. 1988;45:79–91.
14. Baldessarini RJ, Huston-Lyons D, Campbell A, Marsh E, Cohen BM. Do antiadrenergic actions contribute to the atypical properties of clozapine? Br J. Psychiatry. 1992;160:12–6.
15. Baldessarini RJ, Frankenburg FR. Clozapine: a novel antipsychotic agent. N Engl J Med. 1991;324:746–54.
16. Baldessarini RJ, Suppes T, Tondo L. Lithium withdrawal in bipolar disorder: implications for clinical practice and experimental therapeutics research. Am J Ther. 1996;3:492–6.
17. Baldessarini RJ, Tarazi FI. Brain dopamine receptors: a primer on their current status, basic and clinical. Harv Rev Psychiatry. 1996;3:301–25.
18. Baldessarini RJ, Tarazi FI. Pharmacotherapy of Mood Disorders, Anxiety, Psychosis and Mania (Chaps. 17 and 18). In: Brunton LL, Lazo JS, Parker KL, editors. Goodman and Gilman's the Pharmacological Basis of Therapeutics. 11th ed. New York, NY: McGraw Hill; 2005. p. 429–500.
19. Baldessarini RJ, Tarsy D. Pathophysiologic basis of neurologic side-effects of antipsychotic drugs. Ann Rev Neurobiol. 1980;3:23–41.
20. Baldessarini RJ, Tondo L, Ghiani C, Lepri B. Discontinuation rate vs. recurrence risk following long-term antidepressant treatment in major depressive disorder patients. Am J Psychiatry. 2010;167:934–41.
21. Ban TA. Clinical pharmacology of the phenothiazines. Appl Ther. 1966;8:423–7.
22. Beaulieu JM, Gainetdinov RR. Physiology, signaling, and pharmacology of dopamine receptors. Pharmacol Rev. 2011;63:182–217.
23. Berinder K, Akre O, Granath F, Hulting AL. Cancer risk in hyperprolactinemia patients: a population-based cohort study. Eur J Endocrinol. 2011;165:209–15.
24. Bhuvaneswar C, Alpert J, Harsh V, Baldessarini RJ. Adverse endocrine and metabolic effects of psychotropic drugs. CNS Drugs. 2009;23:1003–21.
25. Bleuler M, Stoll WA. Clinical use of reserpine in psychiatry: comparison with chlorpromazine. Ann N Y Acad Sci. 1955;61:167–73.
26. Blom MT, Bardai A, van Munster BC, Nieuwland MI, de Jong H, van Hoeijen DA, et al. Differential changes in QTc duration during in-hospital haloperidol use. PLoS One. 2011; 6:e23728.
27. Bloomfield K, Macdonald L, Finucane G, Snow B, Roxburgh R. The use of antipsychotic medications in patients with Parkinson's Disease at Auckland City Hospital. Intern Med J. [Epub ahead of print: 6 April 2011].

28. Bostwick JR, Guthrie SK, Ellingrod VL. Antipsychotic-induced hyperprolactinemia. Pharmacotherapy. 2009;29:64–73.
29. Buchholz S, Morrow AF, Coleman PL. Atypical antipsychotic-induced diabetes mellitus: update on epidemiology and postulated mechanisms. Intern Med J. 2008;38:602–6.
30. Bushe CJ, Bradley A, Pendlebury J. Review of hyperprolactinaemia and severe mental illness. Ann Clin Biochem. 2010;47:292–300.
31. Caligiuri MR, Jeste DV, Lacro JP. Antipsychotic-induced movement disorders in the elderly: epidemiology and treatment recommendations. Drugs Aging. 2000;17:363–84.
32. Carlsson A, Waters N, Carlsson ML. Neurotransmitter interactions in schizophrenia—therapeutic implications. Biol Psychiatry. 1999;46:1388–95.
33. Catafau AM, Tolosa E. Impact of dopamine transporter SPECT using ^{123}I-Ioflupane on diagnosis and management of patients with clinically uncertain Parkinsonian syndromes. Mov Disord. 2004;19:1175–82.
34. Cavazzoni PA, Berg PH, Kryzhanovskaya LA, Briggs SD, Roddy TE, Tohen M, et al. Comparison of treatment-emergent extrapyramidal symptoms in patients with bipolar mania or schizophrenia during olanzapine clinical trials. J Clin Psychiatry. 2006;67:107–13.
35. Centorrino F, Cincotta SL, Talamo A, Fogarty KV, Guzzetta F, Saadeh MG, et al. Hospital use of antipsychotic drugs: polytherapy. Compr Psychiatry. 2008;49:65–9.
36. Centorrino F, Price BH, Tuttle M, Bahk WM, Hennen J, Albert MJ, Baldessarini RJ. EEG abnormalities during treatment with typical and atypical antipsychotics. Am J Psychiatry. 2002;159:109–15.
37. Centorrino F, Ventriglio A, Vincenti A, Talamo A, Baldessarini RJ. Changes in medication practices for hospitalized psychiatric patients: 2009 versus 2004. Hum Psychopharmacol. 2010;25:179–86.
38. Centorrino F, Wurtman JJ, Duca KA, Fellman VH, Fogarty KV, Berry JM, et al. Weight loss in overweight patients maintained on atypical antipsychotic agents. Int J Obes (Lond). 2006;30:1011–6.
39. Chahine LM, Acar D, Chemali Z. The elderly safety imperative and antipsychotic usage. Harv Rev Psychiatry. 2010;18:158–72.
40. Chiang YL, Klainin-Yobas P, Ignacio J, Chng CM. Impact of antipsychotic side effects on attitudes towards medication in people with schizophrenia and related disorders. J Clin Nurs. 2011;20:2172–82.
41. Christodoulou C, Kalaitzi C. Antipsychotic drug-induced acute laryngeal dystonia: two case reports and a mini review. J Psychopharmacol. 2005;19:307–11.
42. Citrome L. Adjunctive lithium and anticonvulsants for the treatment of schizophrenia: what is the evidence? Expert Rev Neurother. 2009;9:55–71.
43. Clement-Cormier YC, Kebabian JW, Petzold GL, Greengard P. Dopamine-sensitive adenylate cyclase in mammalian brain: a possible site of action of antipsychotic drugs. Proc Natl Acad Sci USA. 1974;71:1113–7.
44. Cohen BM, Tsuneizumi T, Baldessarini RJ, Campbell A, Babb SM. Differences between antipsychotic drugs in persistence of brain levels and behavioral effects. Psychopharmacology (Berl). 1988;108:338–44.
45. Cooper HM, Hedges LV, Valentine J, editors. Handbook of Research Synthesis and Meta-Analysis. 2nd ed. New York, NY: Russell Sage; 2009.
46. Correll CU, Leucht S, Kane JM. Lower risk for tardive dyskinesia associated with second-generation antipsychotics: systematic review of one-year studies. Am J Psychiatry. 2004;161:414–25.
47. Csernansky JG, Mahmoud R, Brenner R. A comparison of risperidone and haloperidol for the prevention of relapse in patients with schizophrenia. N Engl J Med. 2002;346:16–22.
48. Cure S, Rathbone J, Carpenter S. Droperidol for acute psychosis. Cochrane Database Syst Rev 2004;4:CD002830.
49. Curtin F, Schulz P. Clonazepam and lorazepam in acute mania: a Bayesian meta-analysis. J Affect Disord. 2004;78:201–8.

50. Cuyun-Carter GB, Milton DR, Ascher-Svanum H, Faries DE. Sustained favorable long-term outcome in the treatment of schizophrenia: 3-year prospective observational study. BMC Psychiatry. 2011;11:143.
51. Dando S, Tohen M. Olanzapine: relapse prevention following mania. J Psychopharmacol. 2006;20 (Suppl 2):31–8.
52. Davies LM, Lewis S, Jones PB, Barnes TR, Gaughran F, Hayhurst K, et al. Cost-effectiveness of first- vs. second-generation antipsychotic drugs: results from a randomized controlled trial in schizophrenia responding poorly to previous therapy. Br J Psychiatry. 2007;191: 14–22.
53. Daya RP, Tan ML, Sookram CD, Skoblenick K, Mishra RK. Alpha-phenyl-N-tert-butylnitrone prevents oxidative stress in a haloperidol-induced animal model of tardive dyskinesia: investigating the behavioral and biochemical changes. Brain Res. 2011;1412:28–36.
54. De Fruyt J, Deschepper E, Audenaert K, Constant E, Floris M, Pitchot W, Sienaert P, Souery D, Claes S. Second generation antipsychotics in the treatment of bipolar depression: a systematic review and meta-analysis. J Psychopharmacol. 2012;26:603–17.
55. de Haan L, van Bruggen M, Lavalaye J, Booij J, Dingemans PM, Linszen D. Subjective experience and D_2 receptor occupancy in patients with recent-onset schizophrenia treated with low-dose olanzapine or haloperidol: a randomized, double-blind study. Am J Psychiatry. 2003;160:303–9.
56. Ducomb L, Baldessarini RJ. Timing and risk of bone marrow depression by psychotropic drugs. Am J Psychiatry. 1977;134:1294–5.
57. Duncan EJ, Adler LA, Stephanides M, Sanfilipo M, Angrist B. Akathisia and exacerbation of psychopathology: preliminary report. Clin Neuropharmacol. 2000;23:169–73.
58. Eapen V, John G. Weight-gain and metabolic syndrome among young patients on antipsychotic medication: what do we know and where do we go? Australas Psychiatry. 2011;19:232–5.
59. Erhart SM, Marder SR, Carpenter WT. Treatment of schizophrenia negative symptoms: future prospects. Schizophr Bull. 2006;32:234–7.
60. Essock SM, Schooler NR, Stroup TS, McEvoy JP, Rojas I, Jackson C, et al. Effectiveness of switching from antipsychotic polypharmacy to monotherapy. Am J Psychiatry. 2011;168: 702–8.
61. Falissard B, Mauri M, Shaw K, Wetterling T, Doble A, Giudicelli A, et al. METEOR study: frequency of metabolic disorders in patients with schizophrenia: focus on first and second generation and level of risk of antipsychotic drugs. Int Clin Psychopharmacol. 2011; 26:291–302.
62. FDA (US Food and Drug Administration). Guidance for industry: E14 clinical evaluation of QT/QTc interval prolongation and pro-arrhythmic potential for non-antiarrhythmic drugs; 2005. http://www.fda.gov/6922fnl.pdf.
63. Fitzgerald PB, Kapur S, Remington G, Roy P, Zipursky RB. Predicting haloperidol occupancy of central dopamine D2 receptors from plasma levels. Psychopharmacology (Berl). 2000;149:1–5.
64. Flanagan RJ, Ball RY. Gastrointestinal hypomotility: an under-recognized life-threatening adverse effect of clozapine. Forensic Sci Int. 2011;206:e31–6.
65. Frankenburg F, Baldessarini RJ. Neurosyphilis, malaria, and discovery of antipsychotic agents. Harv Rev Psychiatry. 2008;16:299–307.
66. Froimowitz M, Cody V. Incorporation of butyrophenones and related compounds into a pharmacophore for dopamine D_2 antagonists. Drug Des Discov. 1997;15:63–81.
67. Frye MA, Ketter TA, Leverich GS, Huggins T, Lantz C, Denicoff KD, Post RM. Increasing use of polypharmacotherapy for refractory mood disorders: 22 years of study. J Clin Psychiatry. 2000;1:9–13.
68. Gaebel W, Riesbeck M, Wölwer W, Klimke A, Eickhoff M, von Wilmsdorff M, et al. Relapse prevention in first-episode schizophrenia: maintenance vs. intermittent drug treatment with prodrome-based early intervention: results of a randomized controlled trial within the German Research Network on Schizophrenia. J Clin Psychiatry. 2011;72:205–18.

69. Ganguli R, Brar JS, Garbut R, Chang CC, Basu R. Changes in weight and other metabolic indicators in persons with schizophrenia following a switch to aripiprazole. Clin Schizophr Relat Psychoses. 2011;5:75–9.

70. Gardner DM, Baldessarini RJ, Waraich P. Modern antipsychotic agents: a critical overview. Can Med Assoc J. 2005;172:1703–11.

71. Gardner DM, Murphy AL, O'Donnell H, Centorrino F, Baldessarini RJ. International consensus study of antipsychotic dosing. Am J Psychiatry. 2010;167:686–93.

72. Gerson SL, Meltzer H. Mechanisms of clozapine-induced agranulocytosis. Drug Saf. 1992;7 (Suppl 1):17–25.

73. Gill SS, Bronskill SE, Normand SL, Anderson GM, Sykora K, Lam K, et al. Antipsychotic drug use and mortality in older adults with dementia. Ann Intern Med. 2007;146:775–86.

74. Gillman PK. Neuroleptic malignant syndrome: mechanisms, interactions, and causality. Mov Disord. 2010;25:1780–90.

75. Gilmer TP, Dolder CR, Lacro JP, Folsom DP, Lindamer L, Garcia P, et al. Adherence to treatment with antipsychotic medication and health care costs among Medicaid beneficiaries with schizophrenia. Am J Psychiatry. 2004;161:692–9.

76. Glass GV. Meta-analysis at 25. 2000. http://www.gvglass.info/papers/meta25.html.

77. Glass GV, McGaw B, Smith ML. Meta-analysis in social research. Beverly Hills, CA: Sage; 1981.

78. Goff DC, Hill M, Freudenreich O. Treatment adherence in schizophrenia and schizoaffective disorder. J Clin Psychiatry. 2011;72:e13.

79. Goldberg TE, Gomar JJ. Targeting cognition in schizophrenia research: from etiology to treatment. Am J Psychiatry. 2009;166:631–4.

80. Goldman JG, Vaughan CL, Goetz CG. Update expert opinion on management and research strategies in Parkinson's disease psychosis. Expert Opin Pharmacother. 2011;12:2009–24.

81. Grande I, Pons A, Baeza I, Torras A, Bernardo M. QTc prolongation: is clozapine safe? Study of 82 cases before and after clozapine treatment. Hum Psychopharmacol. [Epub ahead of print, 9 Aug 2011].

82. Gunes A, Melkersson KI, Scordo MG, Dahl ML. Association between HTR2C and HTR2A polymorphisms and metabolic abnormalities in patients treated with olanzapine or clozapine. J Clin Psychopharmacol. 2009;29:65–8.

83. Gurrera RJ, Caroff SN, Cohen A, Carroll BT, Deroos F, Francis A, et al. International consensus study of neuroleptic malignant syndrome diagnostic criteria using the Delphi method. J Clin Psychiatry. 2011;72:1222–8.

84. Haas SJ, Hill R, Krum H, Liew D, Tonkin A, Demos L, et al. Clozapine-associated myocarditis: a review of 116 cases of suspected myocarditis associated with the use of clozapine in Australia during 19932003. Drug Saf. 2007;30:47–57.

85. Haddad PM, Dursun SM. Neurological complications of psychiatric drugs: clinical features and management. Hum Psychopharmacol. 2008;23:15–26.

86. Hegarty JD, Baldessarini RJ, Tohen M, Waternaux CM, Oepen G. One hundred years of schizophrenia: a meta-analysis of the outcome literature. Am J Psychiatry. 1994;151: 1409–16.

87. Herceg M, Muzinić L, Jukić V. Can we prevent blood dyscrasia (leucopenia, thrombocytopenia) and epileptic seizures induced by clozapine. Psychiatr Danub. 2010;22:85–9.

88. Hollister LE. Clinical use of psychotherapeutic drugs: current status. Clin Pharmacol Ther. 1969;10:170–98.

89. Gatyas G. IMS Health Reports U.S. prescription sales. Norwalk, CT: IMS Health; 2010. http://www.imshealth.com/portal/site/imshealth/menuitem.a46c6d4df3db4b3d88f61101941 8c22a/?vgnextoid.

90. Gitlin MJ, Midha KK, Fogelson D, Nuechterlein K. Persistence of fluphenazine in plasma after decanoate withdrawal. J Clin Psychopharmacol. 1988;8:53–6.

91. Heres S, Davis J, Maino K, Jetzinger E, Kissling W, Leucht S. Why olanzapine beats risperidone, risperidone beats quetiapine, and quetiapine beats olanzapine: an exploratory analysis

of head-to-head comparison studies of second-generation antipsychotics. Am J Psychiatry. 2006;163:185–94.

92. Hieber R, Dellenbaugh T, Nelson LA. Role of mirtazapine in the treatment of antipsychotic-induced akathisia. Ann Pharmacother. 2008;42:841–6.

93. Howland RH. Drug therapies for tardive dyskinesia. J Psychosoc Nurs Ment Health Serv. 2011;49:13–20.

94. Inder WJ, Castle D. Antipsychotic-induced hyperprolactinaemia. Aust N Z J Psychiatry. 2011;45:830–7.

95. Janssen PA. Evolution of the butyrophenones, haloperidol and trifluperidol, from meperidine-like 4-phenylpiperidines. Int Rev Neurobiol. 1965;8:221–63.

96. Jeste DV, Rockwell E, Harris MJ, Lohr JB, Lacro J. Conventional vs. newer antipsychotics in elderly patients. Am J Geriatr Psychiatry. 1999;7:70–6.

97. Jin H, Meyer J, Mudaliar S, Henry R, Khandrika S, Glorioso DK, et al. Use of clinical markers to identify metabolic syndrome in antipsychotic-treated patients. J Clin Psychiatry. 2010;71:1273–8.

98. Jones PB, Barnes TR, Davies L, Dunn G, Lloyd H, Hayhurst KP, et al. Randomized controlled trial of the effect on Quality of Life of second- vs first-generation antipsychotic drugs in schizophrenia: Cost Utility of the Latest Antipsychotic Drugs in Schizophrenia Study (CUtLASS 1). Arch Gen Psychiatry. 2006;63:1079–87.

99. Joukamaa M, Heliövaara M, Knekt P, Aromaa A, Raitasalo R, Lehtinen V. Schizophrenia, neuroleptic medication and mortality. Br J Psychiatry. 2006;188:122–7.

100. Kamphuis H, Arends J, Timmerman L, van Marle J, Kappert J. [Myocarditis and cardiomyopathy: underestimated complications resulting from clozapine therapy (Dutch)]. Tijdschr Psychiatr. 2010;52:223–33.

101. Kane JM. Antipsychotic drug side effects: relationship to dose. J Clin Psychiatry. 1985;46: 16–21.

102. Kane JM, Fleischhacker WW, Hansen L, Perlis R, Pikalov III A, Assunção-Talbott S. Akathisia: updated review focusing on second-generation antipsychotics. J Clin Psychiatry. 2009;70:627–43.

103. Kane JM, Honigfeld G, Singer J, Meltzer HY. Clozapine for the treatment-resistant schizophrenic. Arch Gen Psychiatry. 1988;45:789–96.

104. Kane JM, Stroup TS, Marder SR. Schizophrenia: pharmacological treatment (Chap. 12.12). In: Sadock BJ, Sadock VA, Ruiz P, editors. Kaplan and Sadock's Comprehensive Textbook of Psychiatry. Philadelphia, PA: Lippincott Williams & Wilkins; 2009. p. 1546–55.

105. Karagianis J, Rosenbluth M, Tohen M, Ascher-Svanum H, Treuer T, de Lima MS, et al. Reviewing CATIE for clinicians: balancing benefit and risk using evidence-based medicine tools. Curr Med Res Opin. 2007;23:2551–77.

106. Karlsson P, Smith L, Farde L, Härnryd C, Sedvall G, Wiesel FA. Lack of apparent antipsychotic effect of the D1-dopamine receptor antagonist SCH39166 in acutely ill schizophrenic patients. Psychopharmacology (Berl). 1995;121:309–16.

107. Kebabian JW, Calne DB. Multiple receptors for dopamine. Nature. 1979;277:93–6.

108. Kelleher JP, Centorrino F, Baldessarini RJ. New formulations and new atypical antipsychotics: potential therapeutic advantages. CNS Drugs. 2002;16:249–61.

109. Kern RS, Glynn SM, Horan WP, Marder SR. Psychosocial treatments to promote functional recovery in schizophrenia. Schizophr Bull. 2009;35:347–61.

110. Ketter TA, Citrome L, Wang PW, Culver JL, Srivastava S. Treatments for bipolar disorder: can number needed to treat / harm help inform clinical decisions? Acta Psychiatr Scand. 2011;123:175–89.

111. Kim HM, Chiang C, Kales HC. After the black box warning: predictors of psychotropic treatment choices for older patients with dementia. Psychiatr Serv. 2011;62:1207–14.

112. Kishimoto T, Agarwal V, Kishi T, Leucht S, Kane JM, Correll CU. Relapse prevention in schizophrenia: a systematic review and meta-analysis of second-generation antipsychotics versus first-generation antipsychotics. Mol Psychiatry. [Epub ahead of print 29 Nov 2011].

113. Kline NS. Uses of reserpine, the newer phenothiazines, and iproniazid. Res Publ Assoc Res Nerv Ment Dis. 1959;37:218–44.

114. Knegtering H, Oolders H, Ruijsink MA, van der Moolen AE. [Depot antipsychotics in the year 2011 (Dutch)]. Tijdschr Psychiatr. 2011;53:95–105.

115. Komossa K, Depping AM, Gaudchau A, Kissling W, Leucht S. Second-generation antipsychotics for major depressive disorder and dysthymia. Cochrane Database Syst Rev. 2010;8:CD008121.

116. Kopelowicz A, Zarate R, Tripodis K, Gonzalez V, Mintz J. Differential efficacy of olanzapine for deficit and nondeficit negative symptoms in schizophrenia. Am J Psychiatry. 2000;157:987–93.

117. Krebs MO. [Development of antipsychotic drug: new directions (French)]. Therapie. 2008;63:257–62.

118. Kulkarni SK, Ninan I. Dopamine D_4 receptors and development of newer antipsychotic drugs. Fundam Clin Pharmacol. 2000;14:529–39.

119. Kumar R, Sachdev PS. Akathisia and second-generation antipsychotic drugs. Curr Opin Psychiatry. 2009;22:293–9.

120. Kurzban S, Davis L, Brekke JS. Vocational, social, and cognitive rehabilitation for individuals diagnosed with schizophrenia: a review of recent research and trends. Curr Psychiatry Rep. 2010;12:345–55.

121. Lako IM, Taxis K, Bruggeman R, Knegtering H, Burger H, Wiersma D, Slooff CJ. The course of depressive symptoms and prescribing patterns of antidepressants in schizophrenia in a one-year follow-up study. Eur Psychiatry. 2012;27:240–4.

122. Lee HJ, Kang SG. Genetics of tardive dyskinesia. Int Rev Neurobiol. 2011;98:231–64.

123. Lehmann HE, Hanrahan GE. Chlorpromazine, a new inhibiting agent for psychomotor excitement and manic states. AMA Arch Neurol Psychiatry. 1954;71:227–37.

124. Leong GB, Silva JA. Neuroleptic-induced akathisia and violence: a review. J Forensic Sci. 2003;48:187–9.

125. Lerner Y, Lwow E, Levitin A, Belmaker RH. Acute high-dose parenteral haloperidol treatment of psychosis. Am J Psychiatry. 1979;136:1061–4.

126. Leslie DL, Rosenheck RA. From conventional to atypical antipsychotics and back: dynamic processes in the diffusion of new medications. Am J Psychiatry. 2002;159:1534–40.

127. Lett TA, Wallace TJ, Chowdhury NI, Tiwari AK, Kennedy JL, Müller DJ. Pharmacogenetics of antipsychotic-induced weight gain: review and clinical implications. Mol Psychiatry. 2012;17:242–66.

128. Leucht S, Arbter D, Engel RR, Kissling W, Davis JM. How effective are second-generation antipsychotic drugs? Meta-analysis of placebo-controlled trials. Mol Psychiatry. 2009;14:429–47.

129. Leucht S, Corves C, Arbter D, Engel RR, Li C, Davis JM. Second-generation versus first-generation antipsychotic drugs for schizophrenia: meta-analysis. Lancet. 2009;373:31–41.

130. Leucht C, Heres S, Kane JM, Kissling W, Davis JM, Leucht S. Oral versus depot antipsychotic drugs for schizophrenia—critical systematic review and meta-analysis of randomized long-term trials. Schizophr Res. 2011;127:83–92.

131. Leucht S, Kissling W, Davis JM. Second-generation antipsychotics for schizophrenia: can we resolve the conflict? Psychol Med. 2009;39:1591–602.

132. Leucht S, Komossa K, Rummel-Kluge C, Corves C, Hunger H, Schmid F, et al. Meta-analysis of head-to-head comparisons of second-generation antipsychotics in the treatment of schizophrenia. Am J Psychiatry. 2009;166:152–63.

133. Leung JG, Breden EL. Tetrabenazine for the treatment of tardive dyskinesia. Ann Pharmacother. 2011;45:525–31.

134. Lieberman JA, Stroup TS, McEvoy JP, Swartz MS, Rosenheck RA, Perkins DO, et al. Clinical Antipsychotic Trials of Intervention Effectiveness (CATIE): effectiveness of antipsychotic drugs in patients with chronic schizophrenia. N Engl J Med. 2005;353:1209–923.

135. Lin CH, Lin SH, Jang FL. Adjunctive low-dose aripiprazole with standard-dose sertraline in treating fresh major depressive disorder: a randomized, double-blind, controlled study. J Clin Psychopharmacol. 2011;31:563–8.
136. Lingjaerde O. Tetrabenazine in the treatment of psychoses. Acta Psychiatr Scand. 1963;39 (Suppl 17):1–109.
137. Lipinski Jr JF, Zubenko GS, Cohen BM, Barreira PJ. Propranolol in the treatment of neuroleptic-induced akathisia. Am J Psychiatry. 1984;141:412–5.
138. Maddalena AS, Fox M, Hofmann M, Hock C. Esophageal dysfunction on psychotropic medication. Pharmacopsychiatry. 2004;37:134–8.
139. Maheux J, Vuillier L, Mahfouz M, Rouillard C, Lévesque D. Modulation of haloperidol-induced patterns of the transcription factor Nur77 and Nor-1 expression by serotonergic and adrenergic drugs in the mouse brain. Int J Neuropsychopharmacol. 2012;15:509–21.
140. Manu P, Kane JM, Correll CU. Sudden deaths in psychiatric patients. J Clin Psychiatry. 2011;72:936–41.
141. Margetić B, Aukst-Margetić B. Neuroleptic malignant syndrome and its controversies. Pharmacoepidemiol Drug Safety. 2010;19:429–35.
142. Mateos JJ, Lomeña F, Parellada E, Mireia F, Fernandez-Egea E, Pavia J, et al. Lower striatal dopamine transporter binding in neuroleptic-naive schizophrenic patients is not related to antipsychotic treatment but it suggests an illness trait. Psychopharmacology (Berl). 2007;191:805–11.
143. McEvoy JP, Lieberman JA, Stroup TS, Davis SM, Meltzer HY, Rosenheck RA, et al. Effectiveness of clozapine versus olanzapine, quetiapine, and risperidone in patients with chronic schizophrenia who did not respond to prior atypical antipsychotic treatment. Am J Psychiatry. 2006;163:600–10.
144. McIntyre RS, Yoon J, Jerrell JM, Liauw SS. Aripiprazole for the maintenance treatment of bipolar disorder: review of available evidence. Neuropsychiatr Dis Treat. 2011;7:319–23.
145. Meltzer HY, Alphs L, Green AI, Altamura AC, Anand R, Bertoldi A, et al. Clozapine treatment for suicidality in schizophrenia: International Suicide Prevention Trial (InterSePT). Arch Gen Psychiatry. 2003;60:82–91.
146. Meltzer HY, Massey BW. Role of serotonin receptors in the action of atypical antipsychotic drugs. Curr Opin Pharmacol. 2011;11:59–67.
147. Merro DB, Dec GW, Goff DC. Adverse cardiac effects associated with clozapine. J Clin Psychopharmacol. 2005;25:32–41.
148. Mishara AL, Goldberg TE. Meta-analysis and critical review of the effects of conventional neuroleptic treatment on cognition in schizophrenia. Biol Psychiatry. 2004;55:1013–22.
149. Mittal V, Kurup L, Williamson D, Muralee S, Tampi RR. Risk of cerebrovascular adverse events and death in elderly patients with dementia when treated with antipsychotic medications: literature review of evidence. Am J Alzheimers Dis Other Demen. 2011;26:10–28.
150. Mizuno T, Schmauss C, Rayport S. Distinct roles of presynaptic dopamine receptors in the differential modulation of the intrinsic synapses of medium-spiny neurons in the nucleus accumbens. BMC Neurosci. 2007;8:8.
151. Moghaddam B, Bunney BS. Depolarization inactivation of dopamine neurons: terminal release characteristics. Synapse. 1993;14:195–200.
152. Murray M. Role of CYP pharmacogenetics and drug-drug interactions in the efficacy and safety of atypical and other antipsychotic agents. J Pharm Pharmacol. 2006;58:871–85.
153. Murru A, Pacchiarotti I, Nivoli AM, Grande I, Colom F, Vieta E. What we know and what we don't know about the treatment of schizoaffective disorder. Eur Neuropsychopharmacol. 2011;21:680–90.
154. Musicco M, Palmer K, Russo A, Caltagirone C, Adorni F, Pettenati C, et al. Association between prescription of conventional or atypical antipsychotic drugs and mortality in older persons with Alzheimer's disease. Dement Geriatr Cogn Disord. 2011;31:218–24.
155. Naber D, Lambert M. The CATIE and CUtLASS studies in schizophrenia: results and implications for clinicians. CNS Drugs. 2009;23:649–59.

156. Nielsen J, Graff C, Kanters JK, Toft E, Taylor D, Meyer JM. Assessing QT interval prolongation and its associated risks with antipsychotics. CNS Drugs. 2011;25:473–90.
157. Nooijen PM, Carvalho F, Flanagan RJ. Haematological toxicity of clozapine and some other drugs used in psychiatry. Hum Psychopharmacol. 2011;26:112–9.
158. Novick D, Haro JM, Bertsch J, Haddad PM. Incidence of extrapyramidal symptoms and tardive dyskinesia in schizophrenia: thirty-six-month results from the European schizophrenia outpatient health outcomes study. J Clin Psychopharmacol. 2010;30:531–40.
159. Nunn R. Mere anecdote: evidence and stories in medicine. J Eval Clin Pract. 2011;17:920–6.
160. Nyberg S, Farde L, Halldin C. Test-retest reliability of central [^{11}C]raclopride binding at high D$_2$ receptor occupancy: PET study in haloperidol-treated patients. Psychiatry Res. 1996;67:163–71.
161. Ohaeri JU, Akanji AO. Metabolic syndrome in severe mental disorders. Metab Syndr Relat Disord. 2011;9:91–8.
162. Parkinson CN. Parkinson's law. The Economist, 19 November 1955.
163. Parsons B, Allison DB, Loebel A, Williams K, Giller E, Romano S, et al. Weight effects associated with antipsychotics: comprehensive database analysis. Schizophr Res. 2009;110:103–10.
164. Perenyi A, Gardos G, Samu I, Kallos M, Cole JO. Changes in extrapyramidal symptoms following anticholinergic drug withdrawal. Clin Neuropharmacol. 1983;6:55–61.
165. Pharoah F, Mari J, Rathbone J, Wong W. Family intervention for schizophrenia. Cochrane Database Syst Rev. 2010;12:CD000088.
166. Pierre JM. Extrapyramidal symptoms with atypical antipsychotics: incidence, prevention and management. Drug Saf. 2005;28:191–208.
167. Praharaj SK, Jana AK, Goyal N, Sinha VK. Metformin for olanzapine-induced weight gain: systematic review and meta-analysis. Br J Clin Pharmacol. 2011;71:377–82.
168. Rabinowitz T, Frankenburg FR, Centorrino F, Kando J. The effect of clozapine on saliva flow rate: pilot study. Biol Psychiatry. 1996;40:1132–4.
169. Racoosin JA, Katz R. Overview of the effect of the WBC monitoring schedule on the rate of clozxapine-associated agranulocytosis: FDA memorandum 3950B1_02_A of May 19. 2003. http://www.fda.gov/ohrms/dockets/ac/03.
170. Rado JT. Review of trials of mirtazapine for negative symptoms of schizophrenia. Psychiatr Ann. 2011;41:265–70.
171. Ray WA, Chung CP, Murray KT, Hall K, Stein CM. Atypical antipsychotic drugs and the risk of sudden cardiac death. N Engl J Med. 2009;360:225–35.
172. Ray WA, Meredith S, Thapa PB, Meador KG, Hall K, Murray KT. Antipsychotics and the risk of sudden cardiac death. Arch Gen Psychiatry. 2001;58:1161–7.
173. Reiter S, Adler L, Angrist B, Corwin J, Rotrosen J. Atenolol and propranolol in neuroleptic-induced akathisia. J Clin Psychopharmacol. 1987;7:279–80.
174. Reulbach U, Dütsch C, Biermann T, Sperling W, Thuerauf N, Kornhuber J, et al. Managing an effective treatment for neuroleptic malignant syndrome. Crit Care. 2007;11:R4.
175. Rosen SL, Baldessarini RJ. Costs of psychotropic drugs. 2011 (unpublished MS).
176. Rusyniak DE, Sprague JE. Toxin-induced hyperthermic syndromes. Med Clin North Am. 2005;89:1277–96.
177. Sackett DL, Rosenberg WM, Gray JA, Haynes RB, Richardson WS. Evidence based medicine: what it is and what it isn't. Clin Orthop Relat Res. 2007;455:3–5.
178. Sanford M. Quetiapine extended release: adjunctive treatment in major depressive disorder. CNS Drugs. 2011;25:803–13.
179. Schatzberg AF, Nemeroff CB, editors. The American Psychiatric Press Textbook of Psychopharmacology, 4th edn. Washington, DC: American Psychiatric Press; 2009.
180. Schlittler E, MacPhillamy HB, Dorfman L, Furlenmeier A, Huebner CF, Lucas R, et al. Chemistry of *Rauwolfia* alkaloids, including reserpine. Ann N Y Acad Sci. 1954;59:1–7.
181. Schneider SD, Jelinek L, Lincoln TM, Moritz S. What happened to the voices? A fine-grained analysis of how hallucinations and delusions change under psychiatric treatment. Psychiatry Res. 2011;188:13–7.

182. Seeman P. Dopamine D_2 receptors as treatment targets in schizophrenia. Clin Schizophr Relat Psychoses. 2010;4:56–73.
183. Segnitz N, Ferbert T, Schmitt A, Gass P, Gebicke-Haerter PJ, Zink M. Effects of chronic oral treatment with aripiprazole on the expression of NMDA receptor subunits and binding sites in rat brain. Psychopharmacology (Berl). 2011;217:127–42.
184. Sen G, Bose KC. *Rauwolfia serpentina*, a new Indian drug for insanity and high blood pressure. Ind Med World. 1931;2:194–201.
185. Simoni-Wastila L, Ryder PT, Qian J, Zuckerman IH, Shaffer T, Zhao L. Association of antipsychotic use with hospital events and mortality among Medicare beneficiaries residing in long-term care facilities. Am J Geriatr Psychiatry. 2009;17:417–27.
186. Simpson GM, Yadalam KG, Levinson DF, Stephanos MJ, Lo ES, Cooper TB. Single-dose pharmacokinetics of fluphenazine after fluphenazine decanoate administration. J Clin Psychopharmacol. 1990;10:417–21.
187. Singer HS. Treatment of tics and Tourette syndrome. Curr Treat Options Neurol. 2010;12:539–61.
188. Smith JM, Baldessarini RJ. Changes in prevalence, severity, and recovery in tardive dyskinesia with age. Arch Gen Psychiatry. 1980;37:1368–73.
189. Soares-Weiser K, Maayan N, McGrath J. Vitamin E for neuroleptic-induced tardive dyskinesia. Cochrane Database Syst Rev. 2011;16:CD000209.
190. Sokoloff P, Diaz J, Le Foll B, Guillin O, Leriche L, Bezard E, et al. The dopamine D_3 receptor: a therapeutic target for the treatment of neuropsychiatric disorders. CNS Neurol Disord Drug Targets. 2006;5:25–43.
191. Straus SE, Richardson WS, Glasziou P, Haynes RB. Evidence-Based Medicine: How to Practice and Teach EBM. 3rd ed. Edinburgh: Churchill-Livingstone; 2005.
192. Stroup TS, Lieberman JA, McEvoy JP, Swartz MS, Davis SM, Rosenheck RA, et al. Effectiveness of olanzapine, quetiapine, risperidone, and ziprasidone in patients with chronic schizophrenia following discontinuation of a previous atypical antipsychotic. Am J Psychiatry. 2006;163:611–22.
193. Susman VL. Clinical management of neuroleptic malignant syndrome. Psychiatr Q. 2001;72:325–36.
194. Swazey JP. Chlorpromazine in Psychiatry: A study in Therapeutic Innovation. Cambridge, MA: MIT; 1974.
195. Sylte I, Dahl SG. Molecular structure and dynamics of *cis* (Z)-and *trans* (E)-flupenthixol and clopenthixol. Pharm Res. 1991;8:462–70.
196. Tarazi FI, Baldessarini RJ. Brain dopamine D_4 receptors: basic and clinical status. Intl J Neuropsychopharmacol. 1999;2:41–58.
197. Tarazi FI, Baldessarini RJ. The dopamine D_4 receptor: significance for molecular psychiatry at the millennium. Mol Psychiatry. 2000;4:529–38.
198. Tarazi FI, Baldessarini RJ. Comparative postnatal development of dopamine D_1, D_2 and D_4 receptors in rat forebrain. Int J Dev Neurosci. 2000;18:29–37.
199. Tarsy D, Baldessarini RJ. Behavioral supersensitivity to apomorphine following chronic treatment with drugs which interfere with the synaptic function of catecholamines. Neuropharmacology. 1974;13:927–40.
200. Tarsy D, Baldessarini RJ. Epidemiology of tardive dyskinesia: is risk declining with modern antipsychotics? Mov Disord. 2006;21:589–98.
201. Tarsy D, Baldessarini RJ, Tarazi FI. Atypical antipsychotic agents: effects on extrapyramidal functions. CNS Drugs. 2002;16:23–45.
202. Tarsy D, Lungu C, Baldessarini RJ. Epidemiology of tardive dyskinesia before and during the era of modern antipsychotic drugs. Handb Clin Neurol. 2010;100:601–16.
203. Terevnikov V, Stenberg JH, Tiihonen J, Joffe M, Burkin M, Tchoukhine E, et al. Add-on mirtazapine improves depressive symptoms in schizophrenia: a double-blind randomized placebo-controlled study with an open-label extension phase. Hum Psychopharmacol. 2011;26:188–93.

204. Thuillier J. The Ten Years that Changed the Face of Mental Illness (Healy D, transl.). London: Martin Dunitz; 1999.
205. Tiihonen J, Haukka J, Taylor M, Haddad PM, Patel MX, Korhonen P. Nationwide cohort study of oral and depot antipsychotics after first hospitalization for schizophrenia. Am J Psychiatry. 2011;168:603–9.
206. Tiihonen J, Lönnqvist J, Wahlbeck K, Klaukka T, Niskanen L, Tanskanen A, et al. 11-Year follow-up of mortality in patients with schizophrenia: a population-based cohort study (FIN11 study). Lancet. 2009;374:620–7.
207. Tiihonen J, Wahlbeck K, Kiviniemi V. Efficacy of lamotrigine in clozapine-resistant schizophrenia: systematic review and meta-analysis. Schizophr Res. 2009;109:10–4.
208. Tohen M, Zarate Jr CA, Hennen J, Khalsa HMK, Strakowski SM, Gebre-Medhin P, et al. McLean-Harvard First-Episode Mania Study: prediction of recovery and first recurrence. Am J Psychiatry. 2003;160:2099–107.
209. Tonda ME, Guthrie SK. Treatment of acute neuroleptic-induced movement disorders. Pharmacotherapy. 1994;14:543–60.
210. Tural U, Onder E. Clinical and pharmacologic risk factors for neuroleptic malignant syndrome and their association with death. Psychiatry Clin Neurosci. 2010;64:79–87.
211. Turner EH, Matthews AM, Linardatos E, Tell RA, Rosenthal R. Selective publication of antidepressant trials and its influence on apparent efficacy. N Engl J Med. 2008;358:252–60.
212. Uchida H, Takeuchi H, Graff-Guerrero A, Suzuki T, Watanabe K, Mamo DC. Predicting dopamine D_2 receptor occupancy from plasma levels of antipsychotic drugs: systematic review and pooled analysis. J Clin Psychopharmacol. 2011;31:318–25.
213. Undurraga J, Baldessarini RJ. A 30-year meta-analytic review of antidepressant efficacy in acute major depression. Neuropsychopharmacology. 2012;37:851–64.
214. Usui C, Hatta K, Doi N, Kubo S, Kamigaichi R, Nakanishi A, et al. Improvements in both psychosis and motor signs in Parkinson's disease, and changes in regional cerebral blood flow after electroconvulsive therapy. Prog Neuropsychopharmacol Biol Psychiatry. 2011;35:1704–8.
215. Vázquez G, Baldessarini RJ, Yildiz A, Tamayo J, Tondo L, Salvatore P. Multi-site international collaborative clinical trials in mania: commentary. Int J Neuropsychopharmacol. 2011;14:1013–6.
216. Velligan DI, Alphs LD. Negative symptoms in schizophrenia: importance of identification and treatment. Psychiatr Times. 2008;25:1–2. http://www.psychiatrictimes.com/schizophrenia/content/article/10168/1147581?.
217. Velligan DI, Gonzalez JM. Rehabilitation and recovery in schizophrenia. Psychiatr Clin North Am. 2007;30:535–48.
218. Venkateshappa C, Harish G, Mythri RB, Mahadevan A, Srinivas Bharath MM, Shankar SK. Increased oxidative damage and decreased antioxidant function in aging human substantia nigra compared to striatum: implications for Parkinson's disease. Neurochem Res. 2012;37:358–69.
219. Ventriglio A, Vincenti A, Centorrino F, Talamo A, Fitzmaurice G, Baldessarini RJ. Use of mood-stabilizers for hospitalized adult psychotic and bipolar disorder patients. Int Clin Psychopharmacol. 2011;26:88–95.
220. Vetti HH, Molven A, Eliassen AK, Steen VM. Is pharmacogenetic CYP2D6 testing useful? Tidsskr Nor Laegeforen. 2010;130:2224–8.
221. Viguera AC, Baldessarini RJ, Hegarty JM, Van Kammen D, Tohen M. Risk of discontinuing maintenance medication in schizophrenia. Arch Gen Psychiatry. 1997;54:49–55.
222. Wahlbeck K, Cheine M, Essali A, Adams C. Evidence of clozapine's effectiveness in schizophrenia: a systematic review and meta-analysis of randomized trials. Am J Psychiatry. 1999;156:990–9.
223. Waldorf S. Neuroleptic malignant syndrome. Am Assoc Nurse Anesth J. 2003;71:389–94.
224. Wang PS, Schneeweiss S, Avorn J, Fischer MA, Mogun H, Solomon DH, et al. Risk of death in elderly users of conventional vs. atypical antipsychotic medications. N Engl J Med. 2005;353:2335–41.

225. Wang PS, Walker AM, Tsuang MT, Orav EJ, Glynn RJ, Levin R, et al. Dopamine antagonists and the development of breast cancer. Arch Gen Psychiatry. 2002;59:1147–54.
226. Weintraub D, Chen P, Ignacio RV, Mamikonyan E, Kales HC. Patterns and trends in antipsychotic prescribing for Parkinson disease psychosis. Arch Neurol. 2011;68:899–904.
227. Whiskey E, Taylor D. Restarting clozapine after neutropenia: evaluating the possibilities and practicalities. CNS Drugs. 2007;21:25–35.
228. Wiesel FA, Farde L, Nordström AL, Sedvall G. Central D_1- and D_2-receptor occupancy during antipsychotic drug treatment. Prog Neuropsychopharmacol Biol Psychiatry. 1990;14:759–67.
229. Winkleman Jr NW. Chlorpromazine in the treatment of neuropsychiatric disorders. JAMA. 1954;155:18–21.
230. Williams JM, Galli A. The dopamine transporter: a vigilant border control for psychostimulant action. Handb Exp Pharmacol. 2006;175:215–32.
231. Wirshing WC. Movement disorders associated with neuroleptic treatment. J Clin Psychiatry. 2001;62 (Suppl 21):15–8.
232. Woerner MG, Correll CU, Alvir JM, Greenwald B, Delman H, Kane JM. Incidence of tardive dyskinesia with risperidone or olanzapine in the elderly: results from a 2-year, prospective study in antipsychotic-naïve patients. Neuropsychopharmacology. 2011;36:1738–46.
233. Wong DF. In vivo imaging of D_2 dopamine receptors in schizophrenia: the ups and downs of neuroimaging research. Arch Gen Psychiatry. 2002;59:31–4.
234. Xia J, Merinder LB, Belgamwar MR. Psychoeducation for schizophrenia. Cochrane Database Syst Rev. 2011;6:CD002831.
235. Yildiz A, Vieta E, Leucht S, Baldessarini RJ. Efficacy of antimanic treatments: meta-analysis of randomized, controlled trials. Neuropsychopharmacology. 2011;36:375–89.
236. Yildiz A, Vieta E, Tohen M, Baldessarini RJ. Factors modifying drug and placebo responses in randomized trials for bipolar mania. Int J Neuropsychopharmacol. 2011;14:863–75.
237. Zervas IM, Theleritis C, Soldatos CR. Using ECT in schizophrenia: review from a clinical perspective. World J Biol Psychiatry. 2012;13:96–105.
238. Zink M, Englisch S, Meyer-Lindenberg A. [Polypharmacy in schizophrenia]. Nervenarzt. 2011;82:853–88.
239. Zyss T, Banach M, Zieba A. [Akathisia—diagnosis, pathophysiology and therapy (Polish)]. Psychiatr Pol. 2009;43:387–402.

Chapter 3
Mood-Stabilizing Agents

Where there is much light, the shadow is deep.

Johann von Goethe

Introduction

Mood-stabilizing treatments include a variety of types of medicinal agents with important applications in the treatment of *bipolar* (manic-depressive) disorder patients. Recent estimates indicate that the prevalence of these illnesses exceeds that of the psychotic disorders, with lifetime estimates ranging from approximately 1% for type I bipolar disorder (with mania) to several percent if all forms of bipolar disorder (including type II with hypomania and mainly recurrent depression). Possibly related conditions include recurrent depression with mild hypomania-like features proposed to lie within a "bipolar spectrum" of disorders, poorly studied cyclothymic disorder, and hyperthymic or cyclothymic personalities, and conditions that do not meet standard diagnostic criteria for severity or duration (not-otherwise-specified [NOS] bipolar disorder), and are reviewed extensively elsewhere [138, 369, 383]. Also, some conditions diagnosed as schizoaffective disorders, particularly of the bipolar type, share many features of bipolar disorders and may be at least partially benefited by similar treatments.

The general, and broad, concepts of mania and melancholia, and clinical observations indicating the possibility of their occurrence within the same person at different times or even the same time ("mixed," rapidly alternating, or complex states of mood and behavior) were recognized from ancient times, notably by Soranus of Ephesus and Aretaeus of Cappadocia (both in modern-day Turkey and Alexandria, Egypt). These ancient concepts, involving various types of excited and depressed or withdrawn states, may well have included common neuropsychiatric disorders as well as conditions that would be recognized now as primary mood disorders [1, 4]. The view that both types of abnormal states of mood and behavior could be found at different times in the same person persisted through

R.J. Baldessarini, *Chemotherapy in Psychiatry: Pharmacologic Basis of Treatments for Major Mental Illness*, DOI 10.1007/978-1-4614-3710-9_3,
© Springer Science+Business Media New York 2013

medieval times into the Enlightenment and the nineteenth century [271, 299, 301, 340]. In the nineteenth century, milder forms of recurrent mood disturbance including cyclothymia were included with the major syndromes of mania and melancholia by Karl Kahlbaum and others [21]. This tendency to extend the range of mood disorders evolved into the very broad concept of "manic-depressive" illness introduced by Emil Kraepelin in the late 1890s [340]. Bipolar manic-depressive illness was not again clearly distinguished from "unipolar" melancholia or depression (with single episodes or recurrences) until the mid-twentieth century by European investigators including Karl Leonhard, Carlo Perris, and Jules Angst [4, 373, 275, 276]. Bipolar and unipolar (major depressive) disorders were not recognized separately in the standard American psychiatric nomenclature until 1980 [11]. This history makes bipolar disorder one of the oldest and also one of the most recently distinguished of the major mental illnesses. It is also important to note that classification of the bipolar disorders as "mood" disorders can be misleading in failing to emphasize that they involve broad disturbances of thinking and behavior as well as of emotional regulation [139, 340].

In the 1970s the type-II subtype of bipolar disorder, with recurrent major depression and hypomania, was proposed to distinguish it from type I bipolar disorder with mania, and eventually accepted into the standard nomenclature two decades later [11, 113]. Originally type II bipolar disorder was described among patients who tended to respond unfavorably to treatment with lithium, probably owing to the excess of depression in the syndrome and the limited beneficial effects of most mood-stabilizing medicines on the depressive phase of bipolar disorders. More recent proposals would extend the bipolar disorder concept even more broadly to include juvenile as well as adult forms [107], and recurrent disorders that are mainly depressive, with some features suggestive of hypomania ("bipolar spectrum" [5]). Such broadening of the concept suggests a return to the older, very broad "manic-depressive" group of syndromes of Kraepelin [340]. However, type I bipolar disorder is relatively stereotyped in its presentations, at least in adult forms, has a high level of heritability, and appears to be an unusually stable diagnostic entity over prolonged follow-up, based on diagnostic criteria of the diagnostic systems of both the American Psychiatric Association (DSM) [11] and the World Health Organization (ICD) [293, 294, 383]. For these reasons, bipolar I disorder is especially likely to be amenable to genetic and other biological investigations [23]. Moreover, the therapeutics of type I bipolar disorder is far more extensively investigated than is the treatment of the other proposed types.

Despite the relatively stereotyped and stable presentations of acute recurrent episodes of adult bipolar I disorder, a high proportion of cases of apparently unipolar depression or other disorders of adults and juveniles are not readily recognized clinically as probable cases of bipolar disorder. Indeed, the international average latency from illness-onset to the diagnosis of any form of bipolar disorder is 5–10 years, and probably longer with juvenile onset [283, 338]. Such extraordinary delays in diagnosis appear often to be based on clinical recognition of the depressive phase of bipolar disorders but not hypomania or even mania, or mixed states with both manic and depressive features present either simultaneously or in rapid alternation [292].

Table 3.1 Treatments for bipolar disorder

Lithium

The standard "mood stabilizer" with benefits against all phases of the illness, especially long term (slow to act in acute mania). Its popularity has eroded owing to risks of toxic effects, need for regular medical monitoring, patient fears, and competitive marketing of alternatives

Anticonvulsants

Carbamazepine is approved only for acute mania or mixed states and appears to be less effective than lithium in all phases of bipolar illness

Divalproex is approved only for treating mania and may have beneficial effects in acute bipolar depression, but is widely used empirically for long-term prophylaxis

Lamotrigine is not antimanic and is moderately effective against recurrences of bipolar depression and less against recurrences of mania

Other anticonvulsants lack evidence of efficacy in any phase of bipolar illness

Antidepressants

Overused, with strong patient demand but lack explicit regulatory approval for bipolar depression; short-term efficacy is controversial and long-term prophylactic effects appear to be limited. Probably are safer in type-II bipolar disorder than in type-I, owing to risks of mood switching into mania that may or may not be reduced by co-treatment with a mood stabilizer

Antipsychotics

All appear to be antimanic, although only chlorpromazine and most second-generation agents have regulatory approval for acute mania

Sedative–hypnotics

High-potency agents (e.g., clonazepam, lorazepam) are employed adjunctively in acute mania and may have useful later effects vs. anxiety or for emerging irritability

It also follows that the relative ease of recognizing depression clinically and the high levels of subjective distress associated with bipolar depression have encouraged use of antidepressant drugs, particularly relatively safe modern antidepressants, as the most common, and among the most sustained, forms of treatment for bipolar disorder patients in the United States. This practice occurs despite limited and conflicting evidence concerning the efficacy and safety of this class of medicines for bipolar disorder patients and lack of explicit regulatory approval for their use to treat or prevent depressive phases of bipolar disorder [25, 27, 34, 47, 344].

A growing list of pharmacologically dissimilar medicines has been used clinically, with variable regulatory approval, for the treatment of acute mania or recurrences of bipolar I disorder (Table 3.1) [34, 88, 141, 192, 235]. No treatment has explicit regulatory approval for the treatment of bipolar II disorder, although antidepressants, with or without mood-stabilizers, are widely used for this syndrome as well as other forms of recurrent depression believed to lie within a proposed bipolar spectrum. *Lithium carbonate* is the oldest and most extensively studied mood-stabilizing agent. The antidementia drug memantine also may have mood-stabilizing effects [200]. The definition of "mood stabilizer" has been debated, but essentially involves long-term stabilization of at least some aspects of mood and behavioral disturbances in bipolar disorder patients without worsening of other phases of the disease. Lithium remains a standard treatment internationally, although this inexpensive and unpatentable mineral (a "natural product") has largely been displaced by faster acting and safer agents for mania, and more aggressively marketed, though not necessarily more effective, alternatives for long-term prophylaxis. Most antipsychotic drugs (see Chap. 2) are highly effective and rapidly acting as antimanic treatments. In addition, several drugs developed as anticonvulsant or antiepileptic agents (particularly *carbamazepine* [Tegretol® and others] and *divalproex* [Depakote®]) have antimanic effects that are recognized by regulatory agencies, and the anticonvulsant *lamotrigine* (Lamictal®) has long-lasting mood-stabilizing effects that are selective for bipolar depression, but virtually lacks antimanic efficacy. Except for lithium salts, the long-term, prophylactic, efficacy of such mood-stabilizing medicines, particularly against recurrences of bipolar depression, remains the least secure aspect of their clinical pharmacology. In addition, high-potency sedative–anxiolytic drugs including clonazepam (Klonopin®) and lorazepam (Ativan®) are useful adjunctively in the short-term clinical management of mania and are sometimes used during long-term maintenance treatment with standard mood stabilizers to address commonly comorbid anxiety or emerging irritability, even though these "off-label" indications lack explicit regulatory approval. Salient characteristics of commonly employed mood-stabilizing agents are summarized in Table 3.2 [34, 235].

Lithium Salts

Development of Lithium

Lithium is a light alkali metal of atomic weight 6.94, with a high natural abundance of the ^{7}Li stable isotope and far less ^{6}Li, found in a variety of salt forms, notably in pegmatites of granitic or other igneous rocks. Lithium bromide was used to treat mania in the mid-nineteenth century by neuropsychiatrist William Hammond of New York (Union Surgeon General during the American Civil War), who later viewed the sedative actions of various bromide salts as more or less similar [150, 370]. The less hygroscopic carbonate salt was used in Denmark at the end of the nineteenth century by physicians Carl and Fritz Lange to treat depression [309]. Lithium carbonate also was used to treat gout in the same era, based on the laboratory observation of very high water solubility of lithium urate, though without proof that this treatment controlled the hyperuricemia or arthritic manifestations of gout. The possibility that some mood disorders may be due to "cerebral gout" was proposed by Alfred Garrod in England in the late 1800s [123]. Use of lithium carbonate continued to the mid-twentieth century as an accepted, but inadequately tested, treatment for gout. The salt was introduced into modern psychiatry by John Cade in Melbourne, Australia in 1949, who recognized both its short-term beneficial effects in acute mania and potential for protecting against recurrences of bipolar disorder [68]. His clinical use of the salt arose from its already accepted use for gout as well as his experience with its effects in laboratory animals in which he was studying the metabolism of purines, of which uric acid is a principal metabolic product.

Table 3.2 Properties of FDA-approved mood-stabilizing anticonvulsants

Property	Carbamazepine	Divalproex	Lamotrigine
Brand names	Equetro®, Tegretol®	Depakote®	Lamictal®
Indication-approvals			
Mania/mixed-state	2004	1994	No
Bipolar depression	No	No	No
Maintenance	No	No	2003
Doses (mg/day)			
Typical	400–1,200	750–1,500	200–400
Extreme	200–1,600	250–4,200	25–600
Serum concentrations (μg/mL)	6–12	50–120	5–10
Elimination half-life (h)	25–65 (later 12–17)	9–16	14–34
Metabolism	CYP-3A4	Mitochondrial oxidation, glucuronidation	Glucuronidation
Actions	Blocks voltage-gated Na⁺ channels; can potentiate GABA	Potentiates GABA, inhibits GABA-transaminase; blocks voltage-gated Na⁺ channels and T-type Ca²⁺ channels; inhibits histone deactylase to alter DNA function	Blocks voltage-gated Na⁺ channels; inhibits glutamate release
Adverse effects	Sedation, headache, leukopenia, SIADH, rashes (rare Stevens–Johnson); withdrawal seizures	Weight gain, hepatic toxicity (especially in children given other anticonvulsants), hyper-ammonemia CNS intoxication blood dyscrasias; withdrawal seizures	Common rashes; rare dermonecrolysis (Stevens–Johnson); CNS intoxication; some leukopenia; withdrawal seizures
Teratogenicity	Multiple cardiac and other anomalies, spina bifida	Multiple cardiac and other anomalies, severe risk of spina bifida; cognitive effects (?); autism (?)	Possible midline clefts
Drug interactions	Induces metabolism of many drugs and itself; grapefruit inhibits 3A4	Increases lamotrigine, carbamazepine, tricyclic antidepressants; blocks folate absorption; salicylates can increase valproate concentrations	Valproate increases; estrogen increases
Other applications	Epilepsy, migraine, neuropathic pain; trigeminal neuralgia	Epilepsy, migraine, neuropathic pain; possibly effective in bipolar depression; experimental in cancers and dementia prevention; probably not effective vs. HIV	Epilepsy, migraine, neuropathic pain; adjunct for unipolar depression (?)

Valproate can be given in "loading" doses to 10–30 mg/kg of body weight in mania [188, 189]

Indication approvals (US-FDA) are stated by year for first formulation approved [192]

Oxcarbamazepine (Trileptal®) is the 10-keto chemical analog of carbamazepine and has been employed empirically to treat bipolar disorder, typically at doses of 600–1,200 (extreme doses: 300–2,400) mg/day (with a short half-life of 1–5 h), but it remains unapproved and has limited research support for efficacy in bipolar disorder

[?] indicates reported findings of uncertain significance

Lithium carbonate was soon developed further and established for safe clinical application, largely through the efforts of internist Mogens Schou in Denmark [51, 139, 146, 305].

Lithium treatment for mania, and eventually for long-term prevention of recurrences of acute episodes of mania, mixed states, or depression in bipolar disorder patients was widely accepted internationally by the 1960s. Nevertheless, full acceptance of lithium carbonate as a treatment for mania or bipolar disorder, particularly in the United States, was delayed owing largely to concern about the potential toxicity of lithium. In the same year as Cade's initial, favorable report concerning use of lithium carbonate to treat mania and to prevent recurrences in 1949, a series of American case reports appeared concerning severe toxic reactions and several fatalities involving poorly monitored use of lithium salts as substitutes for sodium chloride, including by patients in renal or cardiac failure [153]. Most of such risks can be avoided by selection of lithium for treatment of patients without major medical abnormalities of fluid and electrolyte balance or of renal function, by use of moderate, controlled doses, and by regular monitoring of serum concentrations at the approximate daily nadir of rising and falling serum levels following each dose [378]. Medical reluctance to accept the potential risks of lithium treatment, and its status as an unpatentable, orphan substance with little commercial backing, delayed its acceptance and regulatory approval in the United States [32, 51]. It was finally approved for mania in 1970 and for unspecified prevention of "recurrences" (presumably of mania) in long-term maintenance treatment in 1974 [129, 192].

Actions of Lithium

Pharmacodynamic actions of lithium are complex and elude a simple, coherent, and comprehensive theory of its stabilizing effects on mood and behavior, although several of its actions on neurotransmission appear to be consistent with antimanic actions, at least. Reported effects include alterations of the distribution of cations including Na^+, K^+, Ca^{2+}, and Mg^{2+}, as well as subtle changes in neuronal membrane polarization. Lithium also interferes with the production, storage, and release of norepinephrine and dopamine and can increase their reuptake at cerebral nerve terminals, but may increase release of serotonin in forebrain [26, 142, 186, 298, 306]. In addition, lithium exerts complex effects on monoamine receptor functions, usually tending to reduce the functioning of adrenergic and other cerebral mechanisms implicated in mood elevation and behavioral activation, including preventing up-regulation and supersensitivity of monoamine receptors in association with their long-term blockade, as well as exerting possible decreases in the sensitivity of excitatory glutamate receptors [89, 98, 261, 269, 284]. Lithium also exerts complex effects on neuronal cellular, molecular second messenger, and effector mechanisms that mediate synaptic neurotransmission [53, 67, 304]. These effects include inhibition of monoamine-stimulated production of cyclic-adenosine monophosphate (cyclic-AMP), reduced dissociation of the trimeric ($\alpha\beta\gamma$) subunit configuration GTP (the purine nucleoside, guanosine triphosphate)-associated proteins (G-proteins), impaired production of phosphatidylinositol bisphosphae (PIP_2) by inositol-1-phosphatase, and decreased activity of phospholipase-C and its conversion of PIP_2 to phosphoinositol trisphosphate (IP_3) and diacylglycerol (DAG), which liberates free intracytoplasmic Ca^{2+} ions required for many physiological processes. There also is some evidence that the antimanic anticonvulsants carbamazepine and valproate can exert effects on the metabolism of membrane phospholipids to parallel those of lithium [307, 379, 385]. In addition, inhibition of glycogen synthase kinase (GSK-3) is a selective effect of lithium and probably also valproate [143, 179, 223, 379]. Lithium also can inhibit other protein kinases (PK), especially types PKC-α and PKC-β [98]. Inhibition of PKC is an effect shared by the effective antimanic agents valproate and tamoxifen [372, 387].

Lithium also interferes with the actions of the pituitary peptides, antidiuretic hormone (ADH; arginine-vasopressin) at its renal tubule receptors located mainly in the distal convoluted tubules and collection ducts, as well as thyroid-stimulating hormone (TSH) at receptors in the thyroid gland. Both actions are mediated by phospholiapse-C (PLC-3) through G-protein coupled receptor mechanisms, and are, again, related to the effects of lithium on membrane phospholipid metabolism. Clinical effects of interfering with the actions of ADH include spontaneous excess excretion of dilute urine (polyuria), sometimes with elevated circulating concentrations of vasopressin, and loss of renal concentrating responses to exogenous synthetic analogues of vasopressin such as desmopressin (DDAVP). Effects on TSH can lead to impairment of thyroid function with reduced output of thyroxin, and such effects are often anticipated by increased circulating concentrations of TSH before chemical or clinical evidence of hypothyroidism is found [58].

Additional cerebral effects of lithium and possibly also the antimanic anticonvulsant valproate include growing evidence of neuroprotective actions that spare cerebral neuronal cell death. Such effects are of great interest currently as potential leads to effective treatments for a variety of neurodegenerative disorders, including Parkinson's disease, amyotrophic lateral sclerosis, Huntington's disease, and various dementias [72, 78, 117, 210, 211, 218, 265].

Dosing and Monitoring of Lithium Treatment

Several lithium salts have been used to treat bipolar disorder (e.g., carbonate, citrate, and nitrate), but the carbonate has been most often accepted as standard. It is available as generic tablets or capsules, usually at 150, 300, 450, or 600 mg of lithium carbonate (Li_2CO_3; MW = 73.9 g/mol, or 2 mEq of lithium per 73.9 mg) per capsule, or approximately 8 mEq of lithium per 300 mg of the carbonate. Liquid preparations of lithium citrate ($Li_3CH_2COHCH_2[COO]_3 \cdot [H_2O]_4$; MW = 282 g/mol, or 3 mEq of lithium per 282 mg) also are commercially available for clinical use with flavoring agents, usually provided at 8 mEq/5 mL (teaspoonful), or the equivalent of 300 mg of the carbonate.

A major limitation to the safe use of lithium is that its *therapeutic index* (margin or safety), or the ratio of toxic to clinically effective median doses or plasma concentrations, is perhaps the lowest of any widely employed psychotropic agent. That is, peak plasma concentrations, which are reached at 2–4 h after ingestion of lithium carbonate, can exceed 3 mEq/L, only 3–5 times above clinically useful, safe levels, and can be acutely intoxicating, especially to elderly or infirm patients. Assays of serum concentrations are based on various laboratory methods, including atomic-absorption spectroscopy, ion-selective electrodes, flame photometry, dye binding, or commercially marketed porphyrin-colorimetric or phosphatase enzymatic methods suitable for office or clinic use [6, 100, 136].

Serum samples for assays of lithium usually are obtained in the approximate daily nadir of rising and falling circulating levels of lithium ion, typically 10–12 h after a last dose, when changes over time are at their lowest, and measurements are most stable reliable, and appropriate for safe dosing and predictive purposes. However, it is important to note that peak plasma concentrations 2–3 times above nominally appropriate daily trough concentrations are encountered routinely, and may prove to be toxic for some patients. Target trough concentrations that usually are well tolerated and clinically effective typically range from 0.50 to 0.90 mEq/L, and a standard therapeutic range can be considered to be at 0.60–0.75 mEq/L, particularly for long-term maintenance treatment [174, 378]. Concentrations as high as 1.25 mEq/L had been used in acute mania before safer and faster acting alternatives became widely accepted. However, sustained serum concentrations above 1.0 mEq/L probably provide limited additional benefit over that provided by more moderate concentrations, increase adverse effects, and are best reserved for acute mania, rapid cycling, or treatment-resistant cases, especially in young, vigorous patients, and when other treatment options have failed. Trough circulating concentrations below 0.50 mEq/L (as low as 0.30–0.40 mEq/L) may prove useful when standard doses are not well tolerated, as in elderly patients, and especially if lithium is combined with another mood-stabilizing agent such as an anticonvulsant. Given the limited margin of safety of lithium, its clinical administration is best guided by serum concentrations rather than mg doses. However, typical oral doses range from 600 to 1,800 mg/day (300 mg three times a day being a particularly common dose), and occasionally lower or higher but rarely greater than 2,400 mg/day, in divided doses.

In general, dose–response relationships for lithium and other mood-stabilizing agents have not been investigated extensively. In a landmark study, lithium carbonate was randomly assigned for maintenance treatment up to 2 years at several ranges of mean serum concentrations that covered the usual dosing spectrum. Beneficial effects were scored as percentage of weeks ill less than a hypothetical maximum of 100%; adverse effects were rated with a checklist, with no theoretical limit but for eventual coma and death at very high concentrations. Approximately maximum benefit was associated with serum concentrations of 0.61–0.75 mEq/L, with little or no gain at higher levels, in contrast to continuously rising risks of adverse effects (Fig. 3.1) [222]. Other efforts to quantify relationships between serum concentrations of lithium and beneficial or adverse clinical effects have been confounded by study designs involving adjustments of ongoing doses or post-hoc analyses of

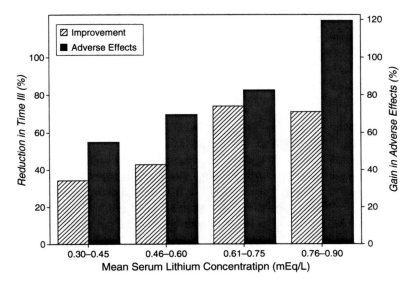

Fig. 3.1 Dose–response study of long-term beneficial and adverse effects of lithium carbonate in bipolar disorder. Data are based on a 2-year, prospective study of DSM-III bipolar disorder patients randomized to four ranges of average daily trough serum concentrations of lithium. Improvement (*striped bars*) is percentage of weeks ill below a theoretical maximum of 100 %; adverse effects are based on a side-effect checklist. Note that approximately maximum benefit was found at serum concentrations of 0.61–0.75 mEq/L, with little or no gain at higher concentrations, in contrast to continuously rising risks of adverse effects. Adapted from Maj et al. [222]

nonrandomly treated patients who required or tolerated different doses or serum concentrations. Secondary adjustment of circulating lithium concentrations risk producing stressful effects of removing lithium, especially from relatively high to low serum concentrations, and particularly if the change is made abruptly [39, 110, 127, 272].

Slow-release preparations of lithium carbonate (e.g., Eskalith® and Lithobid®) have been used with the hope that dosing could safely be limited to once or twice a day, especially for convenient long-term use. Sustained-release lithium carbonate may limit central nervous system (CNS) intoxication risks slightly but may yield somewhat less stable, day-to-day serum concentrations than do ordinary preparations of the salt, and can increase lower intestinal distress and complicate management of acute overdoses owing to delayed intestinal absorption. Efforts to develop a true once-daily preparation of lithium carbonate have not been successful. Once-daily dosing also has been tried with ordinary preparations of lithium carbonate and can be safe if doses are moderate and patients are young and in good health [62, 349]. However, despite some exaggerated claims, once-daily dosing appears not to have major beneficial effects on renal safety in humans (specifically, reducing 24-h urine volume). Once-daily dosing is limited in safety and tolerability, and is attractive mainly for convenience. In an unpublished review of nine studies in human subjects reported between 1981 and 1998, we found that 24-h urine volume was reduced by an average of only 14% (2.61 ± 0.88 vs. 2.25 ± 4.24 L/day) with once- vs. divided- (two or three times) daily dosing of lithium carbonate, and without differences in serum creatinine concentration or clearance. In view of the preceding considerations, once-daily dosing is *not* recommended as a routine practice, despite its convenience and potential for enhancing long-term treatment adherence.

Elimination of Lithium

Lithium is eliminated almost entirely by the kidneys (much less in sweat, saliva, or tears). The plasma elimination half-life of lithium ion averages approximately 20–24 h and increases moderately with advancing age. Elimination is also sensitive to circulating and urine concentrations of sodium ion, with which lithium competes for resorption in the proximal nephron, such that lower availability of Na^+ enhances retention of Li^+, with potentially toxic consequences. Cerebral concentrations of lithium are lower than those in plasma by 20–40% and attain peak and nadir levels somewhat more slowly than blood. Brain/blood ratios tend to increase and to become less closely correlated at advanced ages, adding to the need for caution when using lithium with elderly patients [244]. Studies of the cerebral pharmacokinetics of lithium are facilitated by use of nuclear magnetic-resonance spectroscopy (NMS) technology, which can detect ^7Li in brain tissue in living patients [244, 315].

Anticonvulsants

Several anticonvulsant drugs developed to treat epilepsy are effective in various phases of bipolar disorder [8, 148, 188, 289]. Three such agents are FDA approved (Tables 3.2 and 3.3; Fig. 3.2). They include *carbamazepine* ([Carbatrol®, Equetro®, Tegretol®, and others] for acute mania and mixed states), *valproate* (usually as the sodium divalproate salt [Depakote®]) for mania, and *lamotrigine* (Lamictal®) for long-term maintenance treatment, mainly to reduce risk of recurrences of bipolar depression.

Carbamazepine

The iminostilbene or dibenzazepine, carbamazepine (5H-dibenz-[b,f]-azepine-5-carboxamide; Fig. 3.2; Table 3.2) was synthesized in 1960 by Walter Schindler at Geigy Corporation in Basle, Switzerland, in a series of tricylic compounds that had led to the discovery of the antidepressant imipramine in the late 1950s. This anticonvulsant, specifically, as a slow-release formulation (Equetro®), only recently received FDA approval in 2004 for the treatment of acute mania and mixed bipolar states

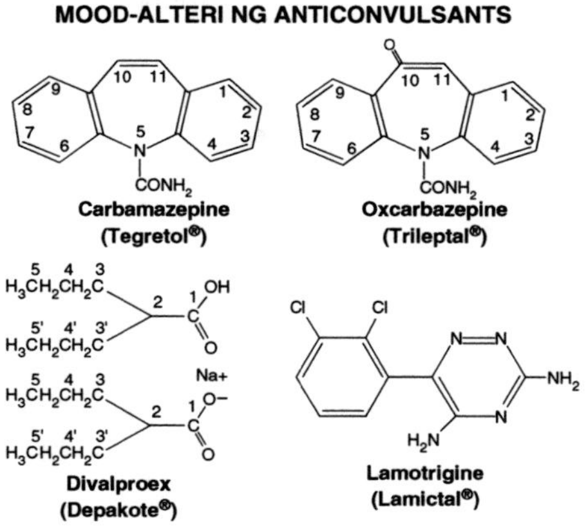

MOOD-ALTERING ANTICONVULSANTS

Fig. 3.2 Mood-altering anticonvulsants. Carbamazepine ([Equetro®, Tegretol®, and others]) and valproate (usually a mixture of the sodium valproate salt and valproic acid, divalproex [Valproex®]) are FDA approved only for treatment of acute mania (carbamazepine also for acute bipolar mixed states), and lamotrigine [Lamictal®] is approved for long-term maintenance treatment, aimed mainly at the depressive phases of bipolar disorder. Oxcarbazepine is simpler to use clinically and is used empirically in bipolar disorder, off-label. Other anticonvulsants (including gabapentin [Neurontin®]) have been used, but lack evidence of short-term or sustained antimanic, antidepressant, or mood-stabilizing effects in bipolar disorder patients

Table 3.3 Metabolism of mood-stabilizing anticonvulsants

Drug	Preferred	Induced	Inhibited
Carbamazepine	CYPs: 1A2, 2C8, 2C9, 3A4	CYP: 2C9, 3A	None
		UGT	
Lamotrigine	UGT	UGT	None
Oxcarbazepine	UGT	3A4/5	CYP: 2C19 (moderate)
		UGT	UGT (weak)
Valproate	CYP: 2C9, 2C19	None	None
	UGT		

Adapted from McNamara [231]
CYP: hepatic microsomal cytochrome oxidase isozymes
Induced, inhibited: pathways increased or decreased by each drug
Preferred: usual preferred pathway of metabolism for each drug
UGT uridine diphosphate glucuronosyl transferase (for glucuronide conjugation)

despite prolonged clinical use of the agent to treat bipolar disorder patients [48]. Different formulations of carbamazepine appear to be similar in efficacy [105]. The anticonvulsant was introduced in Japan to treat mania in the early 1970s [256, 257, 317]. However, evidence pertaining to possible prophylactic, mood-stabilizing effects of carbamazepine remains limited, despite decades of international use of the drug for long term as well as antimanic treatment for bipolar disorder patients, and it is not FDA approved for long-term mood stabilization. Some evidence from controlled trials suggests that carbamazepine and lithium may have similar effects in acute mania and perhaps for long-term treatment of bipolar disorder, but other studies indicate that the anticonvulsant may be somewhat less effective [81, 95, 96, 282]. In addition to its main anticonvulsant effects and usefulness in bipolar disorder, carbamazepine also is effective in migraine (which is prevalent in bipolar disorder patients), neuropathic pain, and trigeminal neuralgia [254].

Carbamazepine typically is given at doses of 400–1,200 mg/day, seeking serum concentrations of 6–12 µg/mL. Such levels have been found to be effective in epilepsy, although relationships of efficacy and tolerability to doses or circulating drug concentrations are little studied in bipolar disorder. Elimination half-life is initially 15–65 h in adults, but the ability of carbamazepine to induce hepatic cytochrome-P450 (CYP) microsomal oxidases, including its preferred pathway through isozyme CYP-3A4, leads to self-induction of its own clearance, with falling serum concentrations, later half-life estimates of only 12–17 h, and often a need for upward adjustment of doses. Carbamazepine increases clearance and lowers plasma concentrations of many other drugs, including psychotropic agents such as antidepressants and antipsychotics as well as other anticonvulsants including valproate. Adverse effects of carbamazepine include excessive sedation, headache, leukopenia, the syndrome of inappropriate secretion of antidiuretic hormone (SIADH) with hyponatremia, benign rashes as well as rare but potentially fatal Stevens–Johnson dermatomuconecrolysis. Carbamazepine also exerts important teratogenic effects on fetal development. Most anticonvulsants risk inducing withdrawal seizures if discontinued abruptly in patients with epilepsy, although specific risks of either seizures or increased rates of major recurrences in bipolar disorder patients following discontinuation are not well documented [122].

Valproic Acid

Valproic acid (2-propylpentanoic or di-*n*-propylacetic acid [Fig. 3.2]) is a relatively old compound, its synthesis by B.S. Burton and use in industrial solvents dating from 1882 [52]. It is the propyl derivative of valeric (propylacetic) acid; this precursor is a component of the plant valerian (*Valeriana officinalis*), whose roots have been an herbal remedy from ancient times, with purported sedative–hypnotic and anxiolytic effects. Valproic acid was initially discovered serendipitously by French chemist Pierre Eymard in 1962 to have anticonvulsant activity, when it was used as a supposedly inert, short-chain fatty-acid component of solvents being employed to dissolve novel agents being tested as potential new anticonvulsants [161, 237]. Its potential psychotropic properties, including for mania, were first noted by Lambert and his colleagues in France 1966 [208], and extended to support a GABA hypothesis for the treatment of mania by Emrich and his colleagues [106]. It is available as the free acid (Depakene®), the sodium salt (Epilim®), and the carboxamide (valpromide [Depamid®]), but is commonly used as the sodium complex of two molecules of valproic acid, divalproex (Depakote®). Valproate typically is administered in doses of 750–1,500 mg/day to achieve serum concentrations in the range of

50–120 µg/mL, again, established for the treatment of epilepsy and not well tested in bipolar disorder.

Valproate not only is a broadly effective anticonvulsant, but like carbamazepine, is useful for migraine and neuropathic pain. Valproate and other mood stabilizers may also have useful adjunctive actions in apparently unipolar major depression, although differentiation of unipolar from bipolar depression is often challenging, and the available data on the use of anticonvulsants in unipolar depression are very limited [154, 266, 308, 361]. Valproate has also been considered for a range of other potential clinical applications, including adjunctive use in the treatment of several forms of cancer, perhaps through its actions as an inhibitor of histone deactylase to alter DNA function [93, 260, 368]. It may also slow development of dementia of the Alzheimer type and perhaps other neurodegenerative disorders, although the evidence of such effects remains preliminary and inconsistent [78, 115, 196, 324].

Valproate is efficiently metabolized primarily by mitochrondrial β-oxidation, resulting in hydroxy or double-bonded metabolites, followed by glucuronide conjugation and excretion. The oxidized, 2-*en*-metabolite is active neuropharmacologically and relatively long acting, but the elimination half-life of valproic acid, the principal active compound in plasma, is short (9–16 h in adults; Table 3.3). Some oxidized metabolites are further elaborated by hepatic microsomal oxidases (notably, 4-*en*, and 2,4-*en*) to provide reactive, unsaturated species that may contribute to potential hepatic toxicity, before being converted to less active glucuronide conjugates for renal excretion [188]. Such potentially toxic by-products are more likely to be produced by vigorous oxidation in infant or juvenile livers and by induction of microsomal oxidases by co-administration of other anticonvulsants such as carbamazepine or phenytoin.

Divalproex probably has long-term prophylactic benefits against recurrences of mania, bipolar mixed states, and bipolar depression, but the evidence supporting this largely clinical impression remains remarkably limited [61], and such prophylactic indications are not FDA-approved. Moreover, further study of the matter is constrained by end of patent protection for divalproex, as well as by its effective domination of the mood-stabilizer market in recent years, at least in the United States.

Lamotrigine

The phenyltriazine anticonvulsant, lamotrigine ([Lamictal®], 3,5-diamino-6-[2,3-dichlorophenyl]-1,2,3-triazine; Fig. 3.2) was the first agent given FDA approval (in 2003) primarily for long-term mood-stabilizing effects (only for adults aged 18 or older) since lithium carbonate in the 1970s. It is unusual in lacking antimanic effects. It is one of few treatments specifically FDA approved for the treatment of bipolar depression, at least long term, with much more limited effects in acute major depressive episodes in bipolar disorder or unipolar depression [327]. Lamotrigine is moderately effective, long term against bipolar depression, but it both lacks antimanic efficacy and shows only marginal prophylactic benefits against recurrences of mania. Its short-term applications are further limited by the need to increase its doses very slowly to limit risk of potentially life-threatening dermatologic-mucosal toxicity (Stevens-Johnson syndrome), often over several weeks.

Other Anticonvulsants

In addition to the anticonvulsants already described, the 10-keto-analog (not a metabolite) of carbamazepine, *oxcarbazepine* (Trileptal®), is also of interest (Table 3.2). Compared to carbamazepine, it has less tendency to induce the metabolic clearance of other drugs or itself or to induce blood dyscrasias. Accordingly, it is somewhat simpler to use clinically than carbamazepine, except for a relatively high risk of the syndrome of inappropriate secretion of antidiuretic hormone [SIADH] with potentially dangerous hyponatremia [347]. However, oxcarbazepine is not FDA approved for use in any phase of bipolar disorder and, despite years of off-label empirical application, evidence of its efficacy in mania or as a proposed mood stabilizer is remarkably limited [35, 74, 182, 228, 357, 373].

Other anticonvulsants have been considered as potentially antimanic or mood stabilizing, but either lack supportive evidence (e.g., felbamate, gabapentin, levetiracetam, phenytoin, pregabalin, topiramate, valnoctamide, vigabatrin, and zonisamide) or have failed to show efficacy in controlled trials. Topiramate ([Topamax®], 2,3:4,5-bis-*O*-(1-methylethylidene)-β-D-fructopyranose sulfamate) is sometimes used to limit weight gain associated with other psychotropic drugs but lacks antimanic or mood-stabilizing activity [206, 350, 373]. Gabapentin ([Neurontin®], 2-[1-(aminomethyl)-cyclohexyl]acetic acid) also has been used off-label, largely for nonspecific purposes related to its anxiolytic and sedative actions in bipolar disorder patients, despite its lack of convincing evidence of having appreciable antimanic or mood-stabilizing effects [73, 220, 373]. In addition, despite their prevalent empirical use in the treatment of psychotic disorders, the evidence that mood-stabilizing agents have major beneficial effects in such illnesses is very limited [354].

Common Actions of Mood-Stabilizing Anticonvulsants

Not all anticonvulsants are effective in treating acute manic or mixed episodes of bipolar disorder; some may or may not have beneficial effects in acute bipolar depression, and the long-term mood-stabilizing effects of most anticonvulsants remain incompletely defined and are recognized by FDA approval only for lamotrigine. Best established as antimanic agents are valproate and carbamazepine, whereas lamotrigine lacks antimanic effects but has long-term benefits, especially to protect against recurrences of bipolar depression, topiramate lacks beneficial effects in bipolar disorder, and the status of oxcarbazepine remains uncertain

[79, 289, 350, 351, 373]. These clinical findings of apparent differential efficacy in epilepsy, with or without efficacy in bipolar disorder, encourage consideration of action mechanisms that may differentiate the apparent subtypes based on their spectrum of clinical effects. In addition, there might be mechanisms shared by anticonvulsants and other drugs found to be effective in the treatment of bipolar disorder, particularly lithium, that may support hypotheses of drug action as a lead to rational development of additional, novel treatments.

Many chemically and pharmacologically dissimilar anticonvulsants share an ability to slow the post-action-potential recovery phase repolarization of neuronal cell membranes by voltage-dependent sodium channels, especially during stimulation at relatively high frequencies. Such effects, notably, are associated with carbamazepine, lamotrigine, and valproate but also with oxcarbazepine and topiramate [231]. Since these are agents with and without well-established clinical beneficial effects in bipolar disorder, it seems implausible that these actions represent specific mechanisms underlying mood-stabilizing effects. It may, nevertheless, be possible that actions of anticonvulsants at sodium channels are necessary but insufficient for beneficial effects on mood, and that additional effects may be required for effects found in bipolar disorder patients.

Among well-established molecular actions of both lithium and some anticonvulsants in clinically plausible concentrations or doses are effects on second messenger and intracellular signaling systems that involve phosphorylated inositol [307, 385]. This hexahydroxylated-cyclohexane, or hexol, carbohydrate is produced from glucose and is a common component of lipids found in the CNS, particularly in phosphatidylinositol (PI) and phosphatidylinositolphosphate (PIP). Phosphorylated inositol compounds play critical roles in maintaining neuronal membrane potentials and controlling concentrations of free intracellular Ca^{2+} ion [172, 205, 286]. Carbamazepine, lithium, and valproate all have evidence of ability to reduce cerebral tissue concentrations of inositol, though by dissimilar mechanisms [227]. Lithium inhibits inositol monophosphatase, whereas valproate inhibits myo-inositol-1-phosphate synthase which produces myo-inositol-1-phosphate from glucose-6-phosphate; the mechanism for carbamazepine is not clear [235]. In turn, loss of inositol can limit synthesis of phosphoinositides, including phosphatidylinositol-4,5-bisphosphate (PIP_2) and its availability as a substrate of phospholipase-C (PLC) to produce DAG and inositol-1,4,5-trisphosphate (IP_3). In turn, losses of IP_3 limit activation of phosphokinase-C (PKC), the availability of free intracellular Ca^{2+}, and may interfere with the functioning of glutamate transmission through its N-methyl-D-aspartate (NMDA) receptors. In contrast to these apparently shared mechanisms among clinically effective mood-stabilizing agents, topiramate, which lacks efficacy in bipolar disorder, can *enhance* the actions of IP_3 and the release of intracellular Ca^{2+} [255].

Another candidate of molecular target for mood-altering effects is inhibition of protein kinase C (PKC). This enzyme, or family of approximately ten similar isozymes, contributes to regulating the function of other proteins through the phosphorylation of hydroxyl groups of their serine and threonine amino acids. PKC enzymes, in turn, are activated by increased concentrations of DAG formed from phosphoinositides, by Ca^{2+}, and phospholipids such as phosphatidylserine, and so are involved in several signal-transduction molecular cascades important for neuronal functioning [232]. Inhibition of PKC isozymes is a property shared, at least, by lithium and valproate but not by gabapentin or topiramate [90, 149], suggesting this effect as an additional candidate target for novel mood-stabilizing agents. This prediction was tested in clinical trials of the estrogen receptor inhibitor tamoxifen, used in the treatment of carcinoma of the breast, and one of few clinically employed drugs that also exerts inhibitory effects on PKC in brain tissue [386]. In small, controlled trials, tamoxifen has shown potent antimanic actions [372, 373, 388].

A third potential target is the metabolic cascade involving the 20-carbon, tetra-double-bonded eicosatetraenoic compound arachidonic acid derived from phospholipids [103, 284, 285, 287, 288, 297]. This fatty acid and docosahexaenoic acid (DHA) account for approximately 20% of cerebral lipids. Arachidonic acid (AA) can be produced from phospholipids by the actions of lipases C (PLC) and A_2 (PLA$_2$), and from DAG by diacylglycerol-lipase. Arachidonic acid is itself involved in second-messenger signaling and it has regulatory actions on phospholipases and protein kinases, among other physiological functions. Carbamazepine, lamotrigine, lithium, and valproate all have evidence of decreasing transcription of cytosolic phospholipase A_2 and of reducing levels of its transcription factors (AP-2 and NF-κB) so as to reduce production of arachidonic acid from phospholipids, as well as reducing transcription of cyclooxygenase-2 (COX-2), which produces prostaglandins and other inflammatory factors. Synthesis of [^{11}C]-labeled fatty acids as radiotracers allows molecules involved in brain arachidonic acid signaling to be localized and measured in human brain tissue with position emission tomography (PET) [103, 384]. Pathways involving arachidonic acid might contribute to development of novel drugs, potentially including innovative treatments for mood disorders.

Additional mechanisms that might yield novel mood-altering agents include antagonistic actions at T-type Ca^{2+} channels, which appear to be characteristic of lamotrigine and valproate, but not carbamazepine. Also, anticonvulsants with and without activity in bipolar disorder can exert complex effects on neurotransmission mediated by the inhibitory amino acid γ-aminobutyric acid (GABA), and some anticonvulsants as well as lithium can inhibit actions of the excitatory amino acid L-glutamate, often tending toward more influence of GABA on cerebral function and less of glutamate [145, 183, 184]. Another interesting potential target is the enzyme, myristolated alanine-rich C-kinase (MARKS), which is a substrate for PKCs and a mediator of synaptic plasticity. Cerebral concentrations of MARKS are decreased by lithium and valproate but not by carbamazepine, antipsychotic or antidepressant drugs [231]. Another potential target molecule is glycogen synthase kinase (GSK-3), which exerts regulatory, and often inhibitory, effects on many enzymes. It is selectively inhibited by lithium and valproate [379]. Both mood stabilizers also affect the production of various factors involved in regulating gene expression (e.g., the activator protein AP-1 and transcription factor PEBP-2β), and they can increase cerebral levels of the protein Bcl-2, which is implicated in neuroprotective actions.

Additional potential molecular targets for development of novel treatments for bipolar disorder and major depression also have been considered, including several kinases, the corticotrophin system, nerve growth factors, and gene-expression modulators [77, 187, 219, 227, 295]. Agents with effects on the excitatory amino acid L-glutamate have shown some promise [200, 386]. This growing list of increasingly complex molecular effects of mood-stabilizing agents may lead to useful theories concerning either the pathophysiology of bipolar disorder or to innovative treatments, but few have yielded such practical results to date. The rational prediction of antimanic effects of tamoxifen as an inhibitor of cerebral PKC is a notable and encouraging exception.

Antipsychotic–Antimanic Drugs

In addition to their effects in psychotic disorders discussed in Chapter 2 with their chemistry and neuropharmacology, virtually all antipsychotic agents have powerful and relatively rapid antimanic actions. Most modern antipsychotics have FDA approval for use in acute mania (Table 3.4). Also, the combination of olanzapine with the serotonin reuptake inhibitor (SRI) antidepressant fluoxetine (Symbyax®) is effective in acute bipolar depression and possibly also long term [65, 66, 101, 326]. Several modern antipsychotics (including aripiprazole, olanzapine, risperidone, and possibly quetiapine and ziprasidone) may have long-term, mood-stabilizing effects [229, 358]. These are most likely for recurrences of mania and hypomania, and perhaps mixed states, but less certain regarding recurrences of highly prevalent and relatively long-lasting bipolar depressive and dysthymic phases.

Evidence supporting long-term beneficial effects of most antipsychotic agents in bipolar disorder patients is limited and potentially confounded by the adverse (apparent relapse-inducing) effects of discontinuing ongoing treatment relatively rapidly in long-term trials with patients who typically are only recently recovering from an acute episode of mania [33, 39, 141, 362]. In addition, many trials involve "enrichment" by continuing vs. discontinuing antipsychotic drug treatment among

Table 3.4 Indication approval status for antipsychotics in bipolar disorder

Drugs		Indications			
Agent	Brand	Mania	Mixed	Depressed	Maintenance
Aripiprazole	Abilify®	2004	2004	No	2006
Asenapine	Saphris®	2009	2009	No	No
Chlorpromazine	Thorazine®	1973	No	No	No
Iloperidone	Fanapt®	No	No	No	No
Olanzapine	Zyprexa®	2000	2000	2003 (+fluoxetine)	2009
Paliperidone	Invega®	No	No	No	No
Quetiapine	Seroquel®	2004	2004	2006	2008 (+Li or VPA)
Risperidone	Risperdal®	2003	2003	No	2009 (LAI)
Ziprasidone	Geodon®	2004	2004	No	2009 (+Li or VPA)

Adapted from Ketter et al. [192].
Indications (with FDA approval year) include acute mania, mixed states, or bipolar depression.
LAI long-acting injectable; *Li* lithium carbonate; *VPA* valproate; [+] indicates agents approved for use adjunctive use only.

patients who have initially responded to a test drug of interest, typically in acute mania, with a high risk of relapses into mania as the primary outcome when ongoing treatment is discontinued [141]. Such designs risk both the confounding effects of treatment discontinuation and may be limited in generalization compared to random or alternative clinical starting points, including from depression or euthymia. They also raise potential ethical questions about the nature of informed consent and of clinical care during experimental therapeutic trials.

Treatment of Bipolar Disorder

Short-Term Treatment: Mania

Treatment of acute episodes of mania was the earliest indication for treatment of bipolar disorder, starting with bromides and other sedatives in the nineteenth century, and introduction of lithium carbonate in 1949 and chlorpromazine in the early 1950s. Indeed, most medicinal treatments for bipolar disorders have been introduced by testing for short-term efficacy in acute mania. Lamotrigine is a notable exception, in being a drug with substantial long-term benefits, especially for depressive phases of bipolar disorder, but little antimanic effect as well as impractically slow initial dosing schedules for rapid treatment. Although lithium carbonate is effective and approved for acute mania, onset of its beneficial actions is relatively slow and its risks of adverse effects relatively high, especially if dosed aggressively. For lithium, anticonvulsants, and antipsychotics, there is a substantial literature arising from reports of randomized, controlled trials in acute mania (Fig. 3.3, Table 3.5) [79, 193, 325, 373].

Lithium salts are relatively specific for the treatment of mania but their latency to useful clinical effects is relatively long (5–10 days), and potential toxicity limits the feasibility of rapid dose increases. As most acutely manic patients are treated in hospitals, such slowness of action is incompatible with current economics-driven pressures to limit length of hospitalization. For this reason and due to the limited margin of safety (therapeutic index or ratio of toxic-to-effective median doses and blood concentrations) of lithium in acutely disturbed, poorly cooperative, and metabolically compromised patients, lithium is best used following initial control of mania by other effective and more rapid means. Short-term efficacy of lithium in mania is best established in young adults. In addition, lithium's efficacy in mania almost certainly extends to adolescents, but is poorly studied in children, and lithium may be less effective and is less well tolerated in the elderly. Moreover, most antimanic or mood-stabilizing treatments probably are less effective in patients with mixed manic-depressive or rapidly cycling bipolar illnesses [335], although not necessarily with psychotic features [374]. Currently, most manic patients are treated initially with an antipsychotic agent (most of which are more and more quickly effective in mania than in schizophrenia) or an antimanic anticonvulsant such as

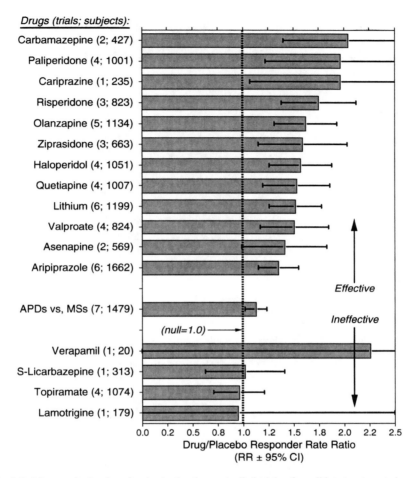

Fig. 3.3 Meta-analysis of randomized, placebo-controlled trials of candidate treatments in acute mania. Plot of drug/placebo responder-rate (usually ≥50% improvement in mania ratings) ratio (RR) with 95% confidence intervals (CI) for 12 agents that were statistically superior to placebo, and 4 that were not. With each drug is a summary of trials and total numbers of subjects tested. The *vertical dotted line* is the null value (no difference between drug and placebo) of RR = 1.0. *APDs* antipsychotic drugs; *MSs* mood-stabilizing anticonvulsants or lithium. Adapted from Yildiz et al. [373, 374]

divalproex in doses up to 10–20 mg/kg/day, which probably is more effective than other anticonvulsants [189]. In addition, temporary use of a sedative can be helpful initially; commonly employed are high-potency benzodiazepines, such as loraze-pam [Ativan®] or clonazepam [Klonopin®] [241]. Lithium salts can then be added more safely in gradually increasing doses, in preparation for aftercare, in which mood-stabilizing anticonvulsants also are useful.

Table 3.5 Response rates and numbers-needed-to-treat in acute mania and acute bipolar depression

Drugs	Subjects (*n*)	Responders (%)	Response difference	NNT
Acute Mania				
Risperidone	273	58	0.28	3.6
Olanzapine	124	56	0.26	3.8
Carbamazepine	223	52	0.22	4.5
Lithium	134	52	0.22	4.5
Quetiapine	209	49	0.19	5.3
Ziprasidone	268	49	0.19	5.3
Valproate	261	48	0.18	5.6
Aripiprazole	260	46	0.16	6.2
Asenapine	357	46	0.16	6.2
Placebo	1,639	30	0.00	–
Acute Bipolar Depression				
Quetiapine	648	59	0.23	4.3
Olanzapine + fluoxetine	82	56	0.20	5.0
Lamotrigine	541	47	0.11	9.1
Placebo	1,215	36	0.00	–

Data are adapted from Ketter et al. [192, 193]

Subject counts are pooled across reported placebo-controlled trials

Response difference: proportion based on drug-minus-placebo response

NNT number-needed-to-treat reciprocal of response difference

Response rates for lack of statistical differences between specific agents and classes, although second-generation antipsychotics may be somewhat more effective or more rapidly acting in these short-term trials [373, 374], and lamotrigine is only marginally superior to placebo for bipolar depression [69]

Research findings from controlled trials, especially the limited number involving head-to-head comparisons, suggest that antipsychotic drugs may be superior in apparent efficacy in acute mania to anticonvulsants or lithium [79, 325]. Modest differences (ca. 10%) in the apparent efficacy of antipsychotic drugs vs. lithium or mood-stabilizing anticonvulsants can be seen in a summary of direct comparisons of such treatments in a recent, comprehensive, quantitative review and meta-analysis of placebo-controlled trials of various treatments in acute mania (Fig. 3.3) [373]. Of 16 agents tested (anticonvulsants, antipsychotics, lithium carbonate), four were not more effective than placebo (lamotrigine, *S*-licarbazepine [the hydroxy-metabolite of oxcarbazepine], topiramate, and the calcium-channel blocker verapamil). Although one can rank the computed differences in effects of the active drugs vs. placebo, an important observation is that the 95% confidence intervals of these measures all overlap, suggesting that apparent differences in efficacy are not substantial or robust, particularly in view of the small numbers of controlled trials involved (ranging from 1 to 6).

A possibly surprising finding in most modern trials of treatment for mania is that placebo-associated response rates are remarkably high, typically in the range associated with treatment of acute episodes of major depression, now often above 30%. Moreover, placebo-associated response rates in manic patients have been rising consistently over the past several decades, whereas rates associated with active treatments have remained relatively stable over the years [373, 374]. Placebo-associated responses in mania or other illnesses involve many factors that are often considered "nonspecific" but can be important clinically. In mania, being treated in a quiet, orderly, nonprovocative, protective, and structured environment, as well as use of sedative–hypnotic agents to restore sleep, and having regular meals, all can have a major beneficial impact. However, to account for the evident secular effect of selectively rising placebo-associated rates in controlled trials in mania, other factors probably are involved. Since the effect correlates strongly with rising numbers of subjects and of collaborating sites per trial, it is likely that less reliable diagnostic and clinical assessments involved with larger and more complex trials contribute to rising variance in ratings and a tendency to regress to average ratings. Such trends are especially likely given the recent tendency to rely on multiple, dissimilar sites and cultures in order to generate large numbers of subjects to assure statistically significant outcomes in the face of declining effect sizes (drug–placebo differences). Such regression-to-the-mean effects are more likely to occur during treatment with an inactive placebo [352, 353, 374].

Another important observation is that most modern controlled trials for the treatment of mania are very brief, averaging only 3 weeks in duration. For this reason, it remains unclear whether the apparent advantage of antipsychotic drugs represents greater efficacy or instead, greater speed of action. The distinction may not matter greatly from a practical perspective, particularly in view of current pressures to limit length of hospitalization. Nevertheless, it is important to emphasize that, although it is often possible to reach substantial reductions of ratings of manic symptoms within 3 weeks, actual clinical remission and recovery (sustained remission) require many weeks or even months. Indeed, full functional recovery, even after a first lifetime episode of mania, may not be achieved even after years of treatment and follow-up [331]. It is important to emphasize, as was noted about the treatment of acute episodes of psychotic illness (Chap. 2), that contemporary, economics-driven emphasis on limiting length of stay in hospitals means that many manic and other patients are discharged improved, but far from recovered, and require particularly close monitoring and support during the early months of aftercare.

Short-Term Treatment: Bipolar Depression and Mixed States

Treatment of depressive phases of bipolar disorder remains highly unsatisfactory, treatment trials for the prominent depressive phases of type II bipolar disorder are uncommon, and studies specific to bipolar mixed states remain rare [46, 356]. Antidepressants are widely used in the treatment of bipolar disorder but lack specific regulatory approval for use in bipolar depression [25, 27, 34]. They are associated with some increased risk of inducing pathologically elevated or complex states, including mania, hypomania, mixed states, psychosis, or sustained rapid cycling or emotional instability [7, 339]. Moreover, the ability of mood-stabilizing treatments to limit this risk, though widely assumed, remains unproved [339]. Additional risk of inducing pathologically elevated mood states in association with antidepressant treatment of bipolar disorder patients is surprisingly moderate. Average rates here increased from approximately 13% without to only 15% with antidepressants, presumably reflecting the relatively high base rate of spontaneous mood switching without antidepressants [339]. In general, clinicians tend to avoid use of antidepressants in type I more than type II bipolar disorder patients and appear to be more willing to accept risks of inducing hypomania among type II patients [12, 13, 217, 263, 333, 343]. Presumably, such avoidance of antidepressants in type I patients arises from concerns about inducing potentially disruptive or dangerous states of mania, agitation, or psychosis. On the other hand, states of elevated or complex mood and behavior, even at the limited intensity of hypomania, among bipolar II disorder patients can have significant, adverse clinical consequences. It remains uncertain whether antidepressant-associated pathological increases in mood and behavior, particularly among patients considered to have unipolar depressive disorders, are an indication of previously undiagnosed bipolar disorder or an adverse pharmacological effect [102, 107]. Regardless of the interpretation, a capacity of switching into mania-like states during antidepressant treatment should be noted and considered in future treatment trials, even among patients considered to have a unipolar depressive disorder.

A more difficult question is whether antidepressants are as effective in acute bipolar depression as in unipolar depressive disorders. Evidence concerning such relative efficacy remains inconsistent and conflicted [134, 262, 263, 310, 333, 353]. Some of the inconsistencies in this research literature may reflect the probable negative impact of adding an antidepressant to the treatment of patients whose depressive or dysphoric condition is marked by anxiety, anguish, agitation, suicidal preoccupations, or psychosis [333]. In addition, current definitions of mixed bipolar states are remarkably narrow and may discourage clinical consideration of prevalent but sometimes subtle mixtures of agitation, anguish, and some excitement with dysphoria [92, 290, 333]. As further research considers both definitions of mixed states and their optimal treatment, a prudent position is to avoid the use of antidepressants in already agitated patients, regardless of diagnosis, and to prefer agents with relatively short duration of action for use in bipolar depression, in moderate and cautiously increased doses, ideally combined with a mood-stabilizing agent, and to reduce doses when mood and behavior begin to respond.

Evidence from randomized, controlled trials concerning potential treatments for acute bipolar depression remains limited (Table 3.6). Encouraging findings have emerged concerning treatment with valproate, quetiapine, and possibly other second-generation antipsychotic agents (Table 3.4) as well as the combination of olanzapine with fluoxetine [60, 88, 94, 314, 326]. The efficacy of lamotrigine in acute bipolar depression appears to be marginal, and its requirement of slow, gradual dose increases to avoid dermatological reactions limits its rapid application in acute depression [69]. Quetiapine is also supported in controlled studies in type II bipolar depression, with less compelling evidence for lithium, antidepressants, and the dopamine agonist pramipexole [322]. Experimentally, cautious administration of the glutamate-NMDA receptor antagonist ketamine, addition of the antinarcolepsy

Table 3.6 Meta-analyses of randomized, placebo-controlled trials in acute bipolar depression

Treatment	Trials	RR (95% CI)[a]	p-Value
Valproate	1	3.59 (1.11–11.6)	0.03
Imipramine	1	2.49 (1.06–5.82)	0.04
Phenelzine	1	2.29 (0.97–5.39)	0.06[b]
Olanzapine + fluoxetine	1	1.74 (1.37–2.22)	<0.0001
Quetiapine	5	1.36 (1.22–1.50)	<0.0001
Lamotrigine	5	1.23 (0.98–1.43)	0.09[b]
Olanzapine	1	1.22 (1.00–1.48)	0.09[b]
Lithium	1	1.12 (0.92–1.37)	0.27[b]
Aripiprazole	2	1.05 (0.88–1.25)	0.57[b]
Paroxetine	1	1.04 (0.82–1.32)	0.73[b]

Adapted from Vieta et al. [358, 359, 360] and Kemp et al. [190]
[a] Pooled response rate ratio (RR for drug/placebo) for 7.95 ± 1.54 weeks of treatment with 95% confidence intervals; data are ranked by RR. Treatments include lamotrigine at 50 and 200 mg/day and quetiapine at 300 and 600 mg/day, both pooled
[b] Not statistically different from placebo

stimulant armodafinil (the active *R*-isomer of modafinil), or of omega-3 fatty acids to standard treatment also may provide additional benefits in bipolar depression [70, 99, 297, 316, 368].

Evolving physical methods of treating severe, and especially otherwise treatment-unresponsive, major depression, include repeated transcranial magnetic stimulation (rTMS), vagal nerve stimulation (VNS), and deep brain stimulation (BDS), as well as continued use of electroconvulsive treatment (ECT). These methods are not adequately evaluated in the treatment of bipolar depression or for comparative efficacy between unipolar and bipolar depression. They are reviewed elsewhere [124, 128, 167, 168, 191, 207, 225, 242].

Mixed states are among the least well-studied components of bipolar disorders, although some encouraging research findings are emerging. Treatment involving antidepressants probably risks inducing worsening of agitation in such states, but mood stabilizing and antipsychotic agents appear to have a useful place [264, 356]. Although mixed states have been recognized since the conceptualization of bipolar disorder and its broader antecedent, manic-depressive illness, in the previous century [292], the current concept has become excessively narrowly defined. In the current diagnostic system of the American Psychiatric Association, mixed states are limited to fully expressed, simultaneous mania and major depression [11]. Such severe mixed states account for only a minority of complex mood states with admixtures of depression or dysphoria and increased energy, excitement, or agitation that can occur in both types I and II bipolar disorder [92, 290, 333, 356]. Options for the treatment of such states usually are considered to include primary reliance on mood-stabilizers or modern antipsychotic agents, and perhaps sedatives, as well as electroconvulstive therapy in severe cases [346].

Research on the treatment of mixed states has been hampered by widespread consideration of them as variants of mania, and by including patients in mixed states with others with more or less pure mania in therapeutic trials, often without separate analysis of the subgroups. This kind of evidence has led to regulatory approval of several modern antipsychotic agents for both mania and mixed states (aripiprazole, asenapine, olanzapine, quetiapine, risperidone, ziprasidone; Table 3.4). However, further studies are needed to clarify the relative efficacy of particular treatments in each type of bipolar disorder state, as well as in relatively subtle forms of complex or mixed states (Table 3.4), and to clarify the apparently closer relationship of many mixed states to bipolar depression than to mania, including a high risk of depression and of suicide attempts over time [31, 46, 199].

Initiating Long-Term Prophylactic Treatment

Most patients diagnosed with bipolar disorder have multiple recurrences over many years, sometimes starting in preadult years, and persisting into senescence. The numbers, timing, and polarity of episodes vary greatly among individuals. There had been an impression dating to the work of Kraepelin that manic-depressive illnesses tend to progress toward shortening intervals between acute episodes of mania and depression. However, evidence pertaining to this hypothesis has been highly inconsistent, and it is more likely that longitudinal patterns are largely unpredictable and possibly somewhat chaotic, with a progressive course in only a minority of patients, at least as modified by current treatments [30]. Nevertheless, the high probability of recurrences that can be clinically dangerous implies that most patients diagnosed with bipolar disorder, especially type I with mania, are likely to become candidates for long-term, often indefinitely continued, treatment with a mood-stabilizing treatment regimen of one or more medicines plus long-term support and rehabilitation aimed at prophylaxis or prevention of recurrences and disability.

Despite the general unpredictability of the long-term course of bipolar disorders, approximately half of bipolar I disorder patients follow a course that is dominated by either manic or hypomanic, or by depressive recurrences in an excess of 2:1 or more. There is a nearly 50% greater risk of predominant mania than predominant depression in the approximately half of bipolar disorder patients with predominant polarity, or about 25% overall [29, 30, 46, 85]. In part, an excess of the number of manic episodes reflects their shorter average duration and greater frequency than depressive recurrences. However, depression and dysthymia represent the greatly dominant long-term morbidity based on the proportion of weeks ill rather than episode recurrence counts. Indeed, depressive components of bipolar disorder account for fully three-quarters of time in residual illness even with sustained treatment [29]. Perhaps even more important for prognosis and treatment planning are striking associations between the polarity of first lifetime episodes and later patterns of morbidity, such that initial mania predicts later mania or psychosis and initial depression or mixed states predict future illnesses of similar type [29, 31, 46]. There is also substantial evidence that bipolar disorder patients who tend to shift from mania into depression, rather than the obverse, tend to respond more favorably to long-term maintenance treatment with lithium and perhaps other mood-stabilizing treatments [108, 204].

The decision to recommend sustained long-term treatment of patients with bipolar disorder is a complex clinical undertaking that is likely to require substantial longitudinal experience with individual patients and collaboration with them, due deliberation, and a high level of flexibility. Many experienced clinicians avoid making long-term prognoses or treatment plans soon after clinical improvement has been achieved following a first-lifetime acute manic or mixed episode. Many, especially young, patients are very reluctant to contemplate life-long treatment with medicines that may be costly and are likely to produce nontrivial or even dangerous adverse effects. Despite typically great reluctance to accept long-term treatment after a single major episode of illness in a bipolar disorder, it is wise to insist on continued medication and close clinical follow-up for at least several months after initial response to treatment, as a contribution to achieving full remission and sustained recovery. Moreover, rates of relapse are high within the months following an acute episode or mania or bipolar depression, and particularly high if medication is discontinued abruptly or rapidly, as some patients may be inclined to do, often without discussion with a treating clinician [28, 39, 40]. After an intermediate phase of treatment continued for perhaps a year (commonly considered "continuation therapy"), more prolonged maintenance treatment can be considered, based on a broad assessment of strengths and vulnerabilities of individual patients, and the nature of their previous experience with illness.

Long-term treatment is increasingly likely as the number of recurrences rises, especially with two or more episodes within several years, and particularly if they have been severe and associated with hospitalization or with suicide attempts [389]. Some characteristics of bipolar disorder patients tend to predict less favorable responses to long-term mood-stabilizing treatments. These include recent rapid cycling (usually defined as four or more episodes within a 12-month period), a tendency to shift from depression into mania rather than the opposite, comorbid substance abuse, and probably a relatively high prevalence of complex or mixed states [30, 31, 41, 85, 108, 151, 204], in addition to the major factor of poor or erratic adherence to treatment. Contrary to expectation, prolonged delay from illness onset to the start of sustained mood-stabilizing treatment or a higher total count of previous recurrences appear not to predict an inferior treatment response, at least to lithium maintenance treatment, despite often devastating clinical and developmental impacts of untreated illness [18, 37, 39, 42, 63]. Some bipolar disorder patients are very reluctant to accept a diagnosis, prognosis, or treatment, and only after having some experience of adverse outcomes associated with discontinuation of recommended treatment may agree to accept prolonged treatment and clinical supervision [18, 28, 63].

Lithium salts and antimanic anticonvulsants may be of value for some conditions other than bipolar disorder. These include cases of "secondary" mania associated with neuromedical disorders (particularly in the elderly), responses to mood-elevating agents (e.g., antidepressants, stimulants, or corticosteroids), as well as episodes of aggressive outbursts [180, 203]. Other potential medical applications of lithium, as in hyperthyroidism and leukopenia, have limited support [116, 120, 162].

The possibility that mood-stabilizing agents may have some value in the treatment of either acute or recurrent nonbipolar depression also has been considered, with most of the limited evidence available pertaining to lithium, including use to supplement the effects of antidepressants [49, 50, 80, 87]. There is very limited support for lamotrigine or anticonvulsant mood-stabilizing agents [9, 173, 359]. Some of the apparently beneficial effects of mood-stabilizers in depression may reflect under-diagnosis of bipolar disorder or the presence of mainly depressive conditions that may share some characteristics of bipolar disorder, such as a strong family history, relatively early age at onset and high recurrence rate, prominent psychomotor retardation during depression, or possible mixed features, subtle hypomania, or psychosis [171, 234]. It is also important to note that mania can arise as a complication of general medical or neuromedical disorders and in response to mood-elevating drugs [203].

Choice of Long-Term Treatment

There is no ideal long-term treatment for bipolar disorders, and medications alone are unlikely to be sufficient for comprehensive care and overall clinical management. Given the far from perfect control of ongoing symptoms and major recurrences in most cases of bipolar disorder, it has become increasingly common to employ multiple treatments in hopes of obtaining additive or even synergistic beneficial effects [75, 119, 247, 279]. However, very few specific combinations have undergone systematic, long-term assessment of their added benefits or risks. Examples of the few combinations that have been studied include some evidence that lithium plus carbamazepine is more effective in stabilizing mood than carbamazepine alone [20, 96], and that lithium with valproate is more effective than valproate alone, though little more effective than lithium alone [24, 126]. Adding an antipsychotic agent to either lithium or valproate may also add long-term beneficial effects, particularly for protection against recurrences of mania [359, 360]. The combination of olanzapine with fluoxetine, given long term, may yield superior benefits to lamotrigine alone [65], whereas adding lamotrigine to lithium may also add benefit [348]. Despite evidently wide assumption of effective protection, it remains uncertain whether including a mood-stabilizer with an antidepressant can avoid risk of pathologically increased moods or complex behavioral states sometimes associated with antidepressant treatment of bipolar disorder patients ("switching" [263, 264, 339]). In general, most combination treatments aimed at mood stabilization in bipolar disorders are neither adequately evaluated for their ability to provide additional benefits, nor for their safety or risk of unanticipated drug–drug interactions.

Among the growing number of options for long-term, prevention-oriented treatments for bipolar disorder patients, it is virtually impossible to rank specific treatments by either relative efficacy or tolerability, similar to findings regarding most treatments for acute mania [188, 373, 374] (Fig. 3.3). Lithium is the longest established agent with mood-stabilizing effects. In a review of trials that varied greatly in the quality of their design, lithium was superior to either a placebo or to

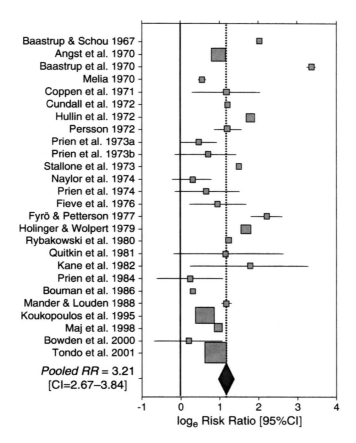

Fig. 3.4 Meta-analysis of 26 trials of long-term lithium treatment (1967–2001), including randomized placebo-controlled trials and clinical comparisons to before vs. during treatment. The pooled relative risk (RR) of relapse with vs. without lithium was 3.21 [95% CI: 2.67–3.84]. Data are plotted as natural logarithms of observed RR values with their 95% confidence intervals when available. The *vertical solid line* is the null value ($\log_e 1 = 0$), and the *vertical dotted line* is the pooled RR, also indicated by the *black diamond* as pooled RR with CI. Across the years, there was a nonsignificant tendency toward falling efficacy of lithium (Spearman nonparametric $r_s = -0.366$ [$p = 0.07$]). Adapted from Baldessarini et al. [44]

Table 3.7 Summary of randomized, placebo-controlled, long-term trials of lithium carbonate in bipolar disorder patients

Outcome	Trials	Relapse risk (%)		Odds ratio (95% CI)	p-Value
		Lithium	Placebo		
Any illness	5	147/369 (39.8%)	243/401 (60.6%)	0.43 (0.32–0.58)	0.0001
Mania	4	37/268 (13.8%)	70/297 (23.6%)	0.52 (0.32–0.82)	0.0003
Depression	4	67/268 (25.0%)	88/297 (29.6%)	0.79 (0.54–1.16)	0.22

Adapted from Geddes et al. [125]

nontreatment (typically in "mirror-image" comparisons of morbidity before vs. during treatment), by a ratio of 3.2-fold [44] (Fig. 3.4). In a more selective meta-analytic review of randomized, controlled, long-term trials, the overall superiority of lithium over a placebo averaged 2.3-fold, with clearer superiority against recurrences of mania than against bipolar depression [125] (Table 3.7). In maintenance treatment, lithium and probably also other mood-stabilizers except lamotrigine appear to be more effective against hypomanic > manic > mixed > depressive phases in type I bipolar disorder [125, 332], and is at least as effective as alternative treatments in rapid-cycling bipolar disorder [248, 335]. Lithium also appears to be effective in type II bipolar disorder in protecting against recurrences of major depression, and especially of hypomania, although controlled, long-term treatment trials in bipolar disorders other than type I are uncommon [249, 322, 332, 382]. Lithium also may have beneficial effects when used adjunctively in nonbipolar depression, although the place of routine lithium treatment in unipolar major depression without even suggestive evidence of bipolarity is neither secure nor FDA approved [80, 140, 213, 234]. Such effects of lithium are especially poorly evaluated for long-term prophylactic treatment, but appear to be substantial in short-term supplementation when antidepressants alone provide inadequate benefits. This effect is best demonstrated with tricyclic antidepressants but also may occur less consistently with some serotonin-reuptake inhibitors [49, 50, 112, 159, 240].

Several recent meta-analytic reviews considered controlled trials of medicines commonly employed for long-term treatment of bipolar disorder patients [57, 277, 278, 358]. In one analysis, only lithium and the antipsychotic agents aripiprazole, olanzapine, and risperidone showed evidence of protective effects against recurrences of mania, whereas lamotrigine, quetiapine, and valproate were not significantly different from placebo [358]. In the very few head-to-head trials available for review, lithium was more effective than carbamazepine, but other long-term trials involving olanzapine, valproate, lamotrigine, and lithium in various comparisons showed little difference in outcomes [20, 57]. Other reviews of aripiprazole suggest that evidence supporting its long-term beneficial effects in bipolar disorder is quite limited [342]. Against recurrences of bipolar depression, divalproex, quetiapine, and possibly also risperidone and olanzapine appeared to be effective, whereas aripiprazole, lithium, and even lamotrigine were not significantly superior to placebo [358]. Evidence that antidepressants have long-term benefits in limiting risk of recurrences of bipolar depression is very limited and largely negative [130, 131, 291]. In other reviews, overall superiority to placebo treatment also was found for olanzapine and quetiapine, probably for aripiprazole and lithium, and possibly lamotrigine and valproate [277, 278] (Table 3.9). Based on numbers-needed-to-treat (NNT, or the number of cases required to show superiority to placebo treatment, computed as the reciprocal of the difference in responses with a test agent and placebo [193]), most drugs tested were superior against recurrences of mania, with NNT estimates of 2–5 (aripiprazole, lithium, olanzapine, quetiapine), whereas valproate had an NNT above 20, and lamotrigine again appeared to be essentially ineffective (Table 3.9). Against depression, only lamotrigine, lithium, and quetiapine had NNT values below 10, whereas

Table 3.8 Meta-analyses of placebo-controlled, long-term treatment trials in bipolar disorder: relative drug/placebo rates and their 95% confidence intervals[a]

Treatment	Trials	Mania-mixed	p-Value
Olanzapine	1	2.50 (1.75–3.57)	<0.001
Aripiprazole	1	2.00 (1.12–3.85)	0.02
Lithium	3	1.59 (0.99–2.56)	0.05
Risperidone	1	1.54 (1.19–1.96)	<0.001
Valproate	1	1.27 (0.78–2.04)	0.34[b]
Quetiapine	4	1.12 (0.62–2.04)	0.70[b]
Lamotrigine	2	1.04 (0.75–1.47)	0.80[b]
Treatment	Trials	Depression	p-Value
Valproate	1	2.50 (1.22–5.00)	0.01
Quetiapine	4	2.00 (1.27–3.12)	0.003
Lamotrigine	2	1.43 (0.74–2.78)	0.29[b]
Risperidone	1	1.43 (1.18–1.72)	<0.001
Olanzapine	1	1.28 (0.96–1.72)	0.08[b]
Lithium	3	1.14 (0.87–1.49)	0.35[b]
Aripiprazole	1	1.10 (0.52–2.33)	0.80[b]

[a]Adapted from Vieta et al. [358]
Meta-analytically pooled rate-ratio (RR for drug/placebo) with 95% confidence intervals for preventing recurrences over 78 ± 64 weeks of prophylactic treatments that include quetiapine at 300 and 600 mg/day pooled, and risperidone as a long-acting injectable preparation. Data are ranked by RR values for prophylactic effects vs. mania or mixed states and against depression
[b]Not statistically significant vs. placebo

Table 3.9 Mood stabilizers: prophylactic effects in bipolar-I disorder

Agents	RCTs	Overall efficacy (RR)	Number needed to treat (NNT)		Efficacy ratio (M/D)
			Mania (M)	Depression (D)	
Olanzapine	2	3.11 (2.42–4.02)	4.49 (3.44–6.44)	11.8 (6.56–57.1)	4.16
Quetiapine	1	2.63 (2.14–3.24)	2.38 (2.07–6.79)	3.34 (2.78–4.18)	1.40
Aripiprazole	1	2.07 (1.02–4.22)	5.05 (3.06–25.2)	15.8 (5.25–31.6)	3.13
Lithium	6	1.96 (1.70–2.27)	3.84 (3.35–4.48)	7.34 (5.70–10.3)	1.91
Valproate	1	1.88 (1.00–3.18)	21.3 (6.79–35.8)	10.5 (5.64–74.3)	1/2.03
Lamotrigine	3	1.16 (0.93–1.43)	49.5 (36.0–63.0)	5.01 (3.71–7.75)	1/9.88

Adapted from Popovic et al. [277]
Values pooled from meta-analyses, with (95% CI), and number of ≥6-month, randomized, placebo-controlled trials (RCTs)
Ranked by overall efficacy (pooled rate-ratio [RR] for drug/placebo for all recurrence types)
Efficacy ratio: number-needed-to-treat (NNT = 1/rate-difference; lower = more effective) to prevent mania (M) or depression (D) (inverted efficacy ratio means effects are greater [lower NNT] for depression than mania)
Note: olanzapine was most effective vs. mania, lamotrigine vs. depression, and lithium was balanced but favored mania

for aripiprazole, olanzapine, and valproate, NNT exceeded 10 (Table 3.9). However, most of these comparisons involved as few as one or two trials, so as to preclude firm conclusions (Tables 3.8 and 3.9). Long-term trials involving specific combinations of medicines are too few to sustain systematic pooling of findings or meta-analysis.

Overall Conclusions Concerning Long-Term Treatment in Bipolar Disorders

In general, the body of data arising from controlled, long-term therapeutic trials just summarized indicates that several antipsychotic drugs and lithium are effective in reducing rates of recurrence or delaying latency to first recurrences or relapses of mania, and that quetiapine and lamotrigine appear to be particularly effective against recurrences of bipolar depression, with less compelling support for such effects of lithium and of the combination of olanzapine with fluoxetine.

From a methodological perspective, many modern trials have employed survival analysis methods to compare times from the end of an index episode of illness to the start of the next episode under dissimilar treatment conditions, sometimes including a placebo. This statistical method was developed for analysis of data involving death as the primary outcome measure and not for assessing therapeutic benefits. Nevertheless, survival analysis is attractive for making use of the contributions of all subjects while they remain in a treatment trial. The method even tolerates missing data or early dropouts, both of which are very common, especially with longer trials as are required for studies of prophylactic effects in an illness that may recur at approximately yearly intervals [86, 185, 312]. Although the method is technically appealing, it has rarely been tested for the validity of time-to-a-next-episode recurrence as an implicit surrogate measure for prolonged stability or wellness. For such a test, we compared measures of the proportion of weeks well over several years of clinical follow-up to a first interval from the end of an index episode of illness to the start of a next recurrence in 459 bipolar I, II, and recurrent unipolar depressive disorder patients treated with lithium. The interval between two episodes was considered equivalent to recurrence latency as provided by survival analysis. We found a moderate, but limited, correlation (overall nonparametric $r_s = -0.61$) indicating lesser long-term morbidity following a longer delay to a second episode of illness [45].

Reductions of delays of recurrent major episodes of illness in bipolar disorders with use of sustained treatment with drugs proposed to provide prophylactic, mood-stabilizing effects are only partial, largely technical indications of clinical benefit. Such measures, while statistically convenient as discrete and quantitative outcomes in experimental therapeutic trials, may not provide sufficient information of interest to clinicians and patients. More clinically relevant are long-term estimates of sustained wellness, freedom from subsyndromal as well as major depressive as well as manic phases or episodes of illness, and improved functioning. Such clinically relevant outcomes would include minimizing the pervasive tendency for many bipolar disorder patients to experience sustained subsyndromal morbidity of moderate

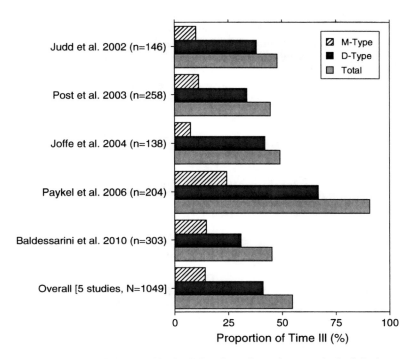

Fig. 3.5 Residual morbidity among bipolar-I disorder patients given standard, clinical treatments in five modern, long-term studies. Percent of time in follow-up is shown for *M*-type (mania and hypomania) and *D*-type (depression, dysthymia, and major or minor dysphoric-mixed states) morbidity. Note that the total proportion of time ill averaged 54% of weeks despite clinical treatment, and that three-quarters of residual illness was accounted for by *D*-type morbidity, with an overall ratio of 3.0-fold more time in *D* vs. *M* illness-states. Data are adapted from Judd et al. [181], Post et al. [280], Joffe et al. [178], Paykel et al. [268], Baldessarini et al. [29]

intensity, particularly dysthymia or mood instability, and their prevalent difficulties of returning to antemortem levels of occupational and interpersonal functioning and independence [331]. Indeed, several recent, large, long-term studies have indicated that bipolar I and II disorder patients given clinical treatments selected by current community standards were unwell as much as half of weeks of follow-up, though not necessarily in major episodes of mania, depression, or mixed states. Moreover, nearly three-quarters of this morbidity has been depressive, involving major depression, dysthymia, or mixed states of variable severity (Fig. 3.5) [29]. Such findings support the proposal that bipolar depression and related states remain a particularly urgent, and only partly met, therapeutic target [47]. Its importance is particularly urgent since continued depressive illness and sometimes under-appreciated anxiety, which is common in bipolar disorders, are strongly associated with disability and dysfunction, substance abuse, and excess mortality, either due to suicide and other violence in younger years, or to intercurrent medical illnesses in older patients [55, 259, 337].

As for patients with other major psychiatric disorders, the pharmacological treatments considered in this Chapter are best applied clinically in a systematic program of comprehensive care that includes supportive psychosocial interventions and rehabilitative efforts. These can be combined with effective medication and carried out cost-effectively, often in group settings [157, 238, 296, 300, 302].

Adherence to Long-Term Treatment

Generally, treatment adherence is a major challenge in the treatment of any psychiatric disorder over prolonged times. It is almost certainly a major reason for treatment failures and supposed "breakthrough" recurrences of acute illness despite presumed treatment. In addition, nonadherence and its consequences can contribute to the impression that relatively simple treatment regiments are inadequate and so encourage increased prescribing of complex combinations of medicines with largely untested additive value and safety, and with added risk of intolerable side effects. It is difficult to evaluate treatment adherence to long-term treatment in any psychiatric illness. It is particularly difficult to obtain secure, quantitative data concerning adherence rates among outpatients who may be seen only occasionally and briefly and are left largely on their own to self-medicate. Nevertheless, based on self-reporting, at least a third of bipolar disorder patients acknowledge some level of incomplete adherence to recommended treatment, and that rate is almost surely a severe underestimate [28].

Factors contributing to erratic treatment adherence can arise from the nature of bipolar disorder as a usually intermittent, rather than chronic disorder, in which long-term treatment may be viewed by patients as more of a sustained burden, expense, and inconvenience than a help. Indeed, the concept that prophylaxis against future illness is important even when one is feeling relatively well is not easy for many patients to accept, and eventual acceptance may await some experience with consequences of discontinuing treatment. There also is some research-based information about characteristics that may be associated with poor treatment adherence among bipolar disorder patients. Such unfavorable factors include reluctance or inability to accept a diagnosis and treatment, youth, past evidence of not taking medicine including lapses of prescriptions, relatively severe illness with disability and psychotic features, obsessive-compulsive features, recent mania or hypomania, and substance abuse, particularly of alcohol or cannabis [28, 137].

An essential consideration in selecting treatments for long-term use with bipolar disorder patients is the tolerability of the regimen. Adverse effects, whether seemingly trivial or clinically important, are a very common reason for failure of patients to adhere to recommended long-term treatment. Moreover, relief of subjective distress usually follows discontinuation, and such responses can contribute to patient skepticism about the cost:benefit ratio of continuing treatment and encourage further nonadherence to recommended treatment. Ironically, treatment discontinuation is especially likely as symptoms of bipolar illness re-emerge, contributing to greater risk of major recurrences just when medication is most needed, and of setting up a cycle of treatment discontinuation-associated recurrences.

Particularly common adverse effects of specific types of medicines given with mood-stabilizing intent are reviewed later. However, risks of adverse effects vary widely among individuals, and often they are discovered only when the drugs are tried, calling for flexible and repeated negotiations with each patient to seek tolerable long-term regimens that are likely to be accepted and followed. In addition to objective side effects, there may be important and sometimes limiting subjective and social reactions to particular medicines. Lithium has acquired a reputation for being complicated and potentially toxic; antipsychotic drugs may carry a stigmatizing effect from their initial use in psychotic disorders; and anticonvulsants are sometimes viewed as being excessively risky or intolerable, arising in part from the potentially severe dermatological effects of lamotrigine and carbamazepine, as well as the high risk of teratogentic effects associated with use of valproate or carbamazepine during early pregnancy. More generally, bipolar disorder patients, especially in their younger years, often are particularly sensitive to being identified as chronically mentally ill and in need of long-term psychiatric supervision and treatment.

Treatment Discontinuation

It has become increasingly clear that discontinuation of ongoing, long-term medication with most psychotropic drugs (mood-stabilizers, antipsychotics, antidepressants, anxiolytics) can lead to important adverse consequences that are not always recognized as related to treatment discontinuation [33, 40, 362]. Such effects are to be distinguished from more classical, neuromedical reactions to agents associated with physiological dependence, including alcohol, sedatives, opiates, most antianxiety agents, and some short-acting antidepressants [33, 39, 40]. Such withdrawal reactions typically emerge within days of stopping treatment. In contrast, recurrences of primary illnesses for which treatment is being given often emerge over weeks or even several months. They have been identified by the much higher rates of recurrence (new episode) or relapse (return of a most-recent illness) following treatment discontinuation than has been observed in the same patients prior to treatment [319], and by emerging much earlier following abrupt or rapid than slow and gradual drug discontinuation [110].

In bipolar disorder, the phenomenon of excess and early risk of illness recurrence following rapid discontinuation of treatment has best been demonstrated with lithium, but should be assumed to be possible with virtually any long-administered psychotropic agent (Fig. 3.6) [27, 33, 37, 39, 110], including reports of mania following rapid discontinuation of antidepressants [239, 250]. In our studies involving discontinuation of long-term lithium treatment either abruptly or rapidly (1–14 days) vs. gradually (>14 days), there were marked differences in recurrence risks and time to recurrence in the following weeks, the relative latencies differed by 4.6-fold, and the median time to recurrence was only 18 weeks after rapid discontinuation

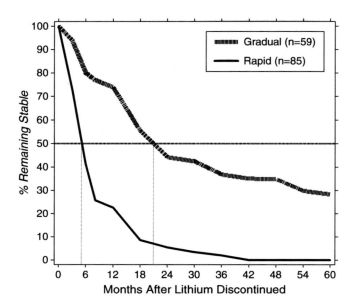

Fig. 3.6 Survival analysis of time to recurrence vs. rate of lithium discontinuation. Time to recurrence (50% risk: *horizontal dotted line*) is compared among 144 bipolar I disorder patients following abrupt or rapid vs. gradual (with dose reduction over ≥2 weeks) discontinuation of long-term lithium maintenance treatment, computed by Kaplan–Meier survival analysis. Median times to recurrence (*vertical dotted lines*) were 4.5 vs. 20.5 months following rapid vs. gradual discontinuation (4.6-fold difference). Based on findings reported by Faedda et al. [110] and Baldessarini et al. [33, 39]

vs. nearly 2 years after gradual discontinuation (Fig. 3.6). The later recurrences are closer to the natural history of recurrence cycles in untreated bipolar disorders [46]. Such observations, though striking, are based on clinical observations, in which decisions about when and how rapidly to discontinue treatment were neither controlled nor randomized. On the other hand, designing and conducting prospective protocols to test the hypothesis that rapid drug discontinuation is more risky than gradual discontinuation would encounter very challenging ethical and practical constraints.

Risks of excessive and early recurrences or relapses following discontinuation of treatments for bipolar and other disorders not only have important clinical implications, and may involve ethical dilemmas and complications of the research consenting process, but may also contribute to artifactual outcomes and misinterpretations in many controlled therapeutic trials [33]. Misleading interpretations are especially likely when continued or long-term treatments are being compared to supposed "nontreatment," making it important to emphasize that discontinuation is not the same as no treatment. Typically the comparison condition involves removal of an effective treatment to continue on a placebo, often carried out over only a few days, and often among patients who are barely starting to recover from an index episode of bipolar mania or depression. Comparisons of continued vs. discontinued treatment in vulnerable patients do not provide adequate evidence of sustained, prophylactic effectiveness of a treatment, even though an initial treatment response followed by relapse when treatment is discontinued may support the conclusion that a given treatment was effective short term.

Such treatment-discontinuation protocols are increasingly favored in combination with so-called enrichment trial designs that aim at saving time and expense by carrying out an initial trial in acute illness, followed by later discontinuation of treatment following evidence of short-term efficacy usually prior to full clinical recovery. Enriched trial designs risk of generating findings that may not generalize to other circumstances [141]. That is, the "enrichment" process seeks to study patients who respond initially to a particular treatment, usually for the sake of economy, but likely to have marketing expectations. Moreover, secure evidence is lacking to support the implicit hypothesis that short- (weeks or a few months) and long-term treatment responses (many months or years) are synonymous. Full clinical recovery from acute episodes in bipolar disorders may require many weeks or even months [331], whereas discontinuation designs typically involve quite time-limited "stabilization" phases before discontinuing to a placebo. Even if recovery is adequate and sustained, the impact of treatment discontinuation, especially if rapid, can lead to the impression that continued treatment was not only effective but necessary.

A specific clinical circumstance in which discontinuation of any treatment is likely is during and sometimes following pregnancy. It is particularly likely, especially in areas with high rates of litigation concerning fetal and neonatal abnormalities that may or may not be related to maternal drug exposures in pregnancy, so that a woman's physicians will advise avoiding fetal exposure to any nonessential drug, especially early in pregnancy [121]. Such practices are so pervasive that many well-informed women refuse to continue even medically important and low-risk treatments during pregnancy. Such practices appear to be driven, at least nominally, by concern for fetal and developmental health of offspring, although a strong element of liability management by clinicians also may be involved. Only recently, gradually, and in some areas, particularly in academic medical centers, have there been efforts to balance concerns for both maternal and fetal health in pregnancy [84, 363].

In our recent collaborative studies of women with bipolar disorders during pregnancy and the postpartum, there have been several striking findings relevant to maternal mental health [366, 367] (Fig. 3.7). Among pregnant and nonpregnant, age-matched women diagnosed with bipolar disorder who elected to discontinue long-term lithium treatment, there were major risks of illness episode recurrences within the following 9 months, leading to approximately 55% recurrence risks in both groups, at indistinguishable rates of recurrence per month (Fig. 3.7a). This finding might suggest that pregnancy is more or less "risk neutral" with respect to illness recurrences; alternatively, it may reflect a dominant impact of treatment discontinuation that is much greater than any positive or negative impact of pregnancy itself. In addition, the same comparison limited to women who remained stable for the first 9 months, considered the next 6 months. This comparison indicated a very high, and nearly fivefold greater, recurrence risk among postpartum than nonpregnant

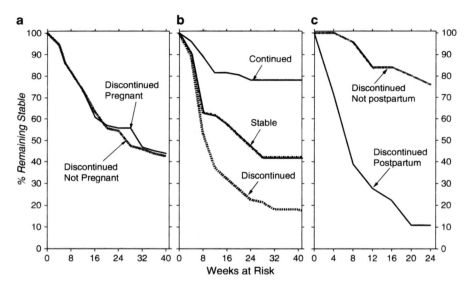

Fig. 3.7 Effects of discontinuing mood-stabilizing treatment and pregnancy. (**a**) Comparison of age-matched pregnant ($n=42$) vs. nonpregnant ($n=59$) women with bipolar I disorder after discontinuing long-term lithium treatment: proportion remaining stable over time, based on Kaplan–Meier survival analysis. Recurrence risk was approximately 55% in both groups. Adapted from Viguera et al. [366]. (**b**) Comparison of women with bipolar disorder during pregnancy with mood stabilizers continued ($n=27$) or discontinued ($n=62$) at the start of pregnancy, compared with a middle group of women who had been stable for at least 12 months before pregnancy without mood-stabilizer treatment ($n=19$). Recurrence risk was mainly depressive and in the first trimester, and was nearly three times greater after discontinuing vs. continuing mood-stabilizer treatment. Adapted from Viguera et al. [367]. (**c**) Comparison of women who had just given birth ($n=20$) or had not been pregnant ($n=25$) and had remained stable without lithium treatment for the duration of pregnancy or an equivalent time. Exposure time is the first 6 months postpartum or the equivalent time among nonpregnant women. The latency to 25% recurrence risk was only 5 weeks postpartum vs. nearly 25 weeks among nonpregnancy, age-matched, women. Adapted from Viguera et al. (2001)

women, with most of the risk appearing soon after childbirth (Fig. 3.7c). This finding supports other indications that the days and weeks immediately following childbirth are a period of particularly high risk for recurrences in bipolar disorders [330, 375].

During pregnancy itself, only 22% of another, prospectively followed, sample of women with bipolar disorders who continued to receive mood-stabilizing treatments experienced a recurrence, whereas approximately 82% of those who discontinued treatment became ill again within 9 months (Fig. 3.7b). In these studies, there was a strong tendency for greater risks following rapid discontinuation of treatment [366, 367]. A particularly interesting finding is that approximately 58% of pregnant women who had been stable for at least 12 months before pregnancy without mood-stabilizing treatment became ill during pregnancy without treatment (Fig. 3.7b). This finding suggests that pregnancy itself may represent a stressor to increase recurrence risks, although not as great as stressor as the early postpartum period.

It is also noteworthy that most of the illnesses that emerged during pregnancy were either depressive or dysphoric mixed states, and that risk was greatest during the first trimester, and fell in later months of pregnancy. This pattern of high risks of depression early in pregnancy had been reported more than a century ago and largely forgotten [330]. The high risk of depressive and dysphoric illness during pregnancy contrasts to postpartum risks, which include depression but also many cases of mania, mixed states, or psychosis [367]. Its implications include the probable impracticality of avoiding risks to the fetus with little added risk to the mother by discontinuing maintenance treatment early in pregnancy, and restarting it later.

In summary, discontinuing ongoing maintenance treatment, especially abruptly or rapidly, can lead to marked increases in recurrence risks and much shorter times to recurrence. Such findings are common to most psychotropic drugs and have important clinical, therapeutic, ethical, and scientific implications, as well as major challenges for developing sound policies concerning clinical management of pregnant women with bipolar disorders before, during, and following pregnancy, especially in the first trimester and postpartum. Finally, remarkably little is known about potential risks to fetal and neonatal development of having a severely mentally ill and distressed mother. However, models based on effects of stress in pregnancy laboratory animals suggest that adverse neurodevelopmental responses are likely, and preliminary clinical evidence indicates stress responses in newborns of mothers who had been acutely mentally ill during pregnancy [64, 82, 170, 226, 323].

Antisuicidal Effects of Long-Term Treatment

A particular concern for the care of patients with bipolar disorders is their extraordinarily high risks of premature mortality from both violent and natural causes. Their suicide rates are among the highest in any psychiatric disorder, and violent deaths ascribed to accidents also are not uncommon among patients with bipolar disorders [195, 336, 337]. Suicide rates are perhaps 20 times above those in the age-matched general population, especially early in the course of the illness in adolescent and early adult years. Excess mortality is also associated with mortality due to complications of substance abuse and to intercurrent medical illnesses, particularly with stress-sensitive cardiovascular and pulmonary diseases [19, 259, 231]. Even though mortality rates are increased by only two- to threefold among bipolar disorder patients compared to the older general population, due to the very large numbers of medically ill older persons, the total numbers of deaths involved are similar to those associated with suicide and other forms of violence [259]. It is notable that risk of suicidal behavior is remarkably similar among patients with both types I and II bipolar disorders, differing by less than 10% [253, 337]. There is growing evidence

Table 3.10 Summary of meta-analyses of effects of lithium maintenance treatment against suicidal risks

Conditions	Studies (n)	Risk ratio (95%CI)	p-Value
All studies with bipolar cases[a] [35, 38]	34	4.91 (3.86–6.23)	<0.001
Suicides[a] [35, 38]	25	4.89 (3.46–6.91)	<0.001
Attempts[a] [35, 38]	20	4.86 (3.58–6.59)	<0.001
Bipolar disorder[a] [35, 38]	15	5.53 (3.72–8.20)	<0.001
Major affective disorders[a] [35, 38]	19	4.58 (3.38–6.21)	<0.001
Unipolar depression[b] [150]	8	2.01 (1.81–2.22)	<0.001
Randomized trials[a] [35, 38]	8	4.29 (1.46–12.6)	0.008
Open trials [35, 38]	26	5.00 (3.83–6.53)	<0.001
Lithium vs. anticonvulsants[c] [36]	6	2.86 (2.29–3.57)	<0.001
Anticonvulsants vs. lithium [133]	1	1.21 (ca. 0.30–1.45)	>0.05

Based on meta-analyses of 34 studies of bipolar disorder and mixed, recurrent major affective disorders.
[a]Involving 3,200 patients, a meta-analysis of 8 studies of lithium in 329 patients with recurrent major depressive disorder.
[b]Another on comparisons of lithium vs. anticonvulsants (mainly valproate and some carbamazepine).
[c]An insurance database study of 48,000 patients, comparing anticonvulsants (carbamazepine, lamotrigine, oxcarbazepine, valproate) to lithium.

that long-term treatment of bipolar disorder patients, generally, can reduce mortality from all causes, and not only from suicide [14, 15].

Efforts to define treatment effects that may reduce suicide risk generally are surprisingly recent and still limited. As discussed in Chapter 2, the only treatment with FDA approval for reduction of suicidal risk is clozapine in the treatment of schizophrenia patients, although its ability to reduce mortality from suicide remains less clear than reductions in attempts or of emergency interventions to protect against impending suicidal risk [160, 233]. Other forms of commonly employed interventions also lack regulatory approval for suicide prevention, including treatment with anticonvulsants, antidepressants, electroconvulsive treatment (ECT), rapid hospitalization, "contracts for safety," or psychotherapy [35, 38, 246].

In contrast, there is a large body of evidence to indicate marked reductions in the risks of suicide attempts and fatalities among bipolar disorder patients in association with long-term treatment with lithium [35, 38, 201, 245, 334] (Table 3.10), with some exceptions [224, 258]. Risk of both suicides and attempts appears to be about 70–80% lower during long-term lithium treatment of bipolar disorder patients; the ratio of attempts/suicides is about double (lower lethality), and these benefits are sustained for many years. Death by acute overdose of lithium is possible, but the rate of overdosing during long-term lithium treatment is low: vomiting typically limits acute intoxication, and dialysis is highly effective in removing excesses of lithium [35]. Indeed, the reported lethality of lithium overdoses is surprisingly low, and similar to that of divalproex, modern antidepressants, or antipsychotics [35]. If lithium is discontinued the risk of suicidal behavior in the subsequent months rises sharply, and much more following abrupt or rapid discontinuation [43]. The apparent antisuicide effect of lithium may be larger than is associated with long-term use of mood-stabilizing anticonvulsants, particularly divalproex, to treat bipolar disorder patients [36, 138]. An antisuicidal effect of lithium may also extend to the treatment of some patients with recurrent, apparently unipolar major depressive disorders [150].

Mechanisms that may mediate the effect of lithium treatment against suicidal behavior may include beneficial effects of the treatment on depressive and mixed states that are particularly associated with suicide [335, 336]. Alternative possibilities are more direct effects on such states as anger and impulsivity as well as dysphoria, possibly without close correspondence to beneficial effects on mania and bipolar depression specifically [3, 51, 246]. A plausible additional possibility is that the extra care and clinical attention typically provided to patients treated long-term with either lithium or clozapine, itself, may have a protective effect against suicide. This possibility, however, is made less likely since the principal controlled trial that compared clozapine to olanzapine to prevent suicide in previously suicidal schizophrenia patients matched the treatment groups for clinical contact time, and still found superior benefits with clozapine, suggesting a pharmacological contribution [233].

Finally, there have been recent suggestions that suicidal risk may be increased with the use of some anticonvulsants, based mainly on retrospective analyses of incidental findings of adverse outcomes in controlled trials, mainly among patients with epilepsy [16, 22, 267]. However, evidence that such an effect may extend to the use of anticonvulsants for psychiatric indications including bipolar disorder is not established, and there is evidence to the contrary [133, 276].

Adverse Effects

Mood-stabilizing agents have a wide range of adverse or toxic effects as well as interactions with other drugs that may be encountered among psychiatric patients [51, 58, 215]. Prevalent adverse effects of mood-stabilizing drugs are summarized in Table 3.11.

Lithium

Lithium has a predictable association with acute intoxication resembling that of ethanol and closely correlated with circulating serum concentrations of lithium ion. Early signs include tremor, ataxia, slurred speech, and confusion, advancing to coma and death. For lithium salts, a recently reported annual rate of clinically significant intoxication was well below 1% of treated cases, and involved serum concentrations averaging only 2 mEq/L, and involvement of other drugs that interfere with lithium clearance in approximately half of the cases [97]. Acute overdoses of lithium salts resulting in serum concentrations ≥3 mEq/L typically lead rapidly to delirium followed by coma, sometimes with residual motor and cognitive deficits. Concentrations of lithium of ≥5 mEq/L can be fatal without treatment, which

Table 3.11 Adverse effects of mood-altering medicines used for bipolar disorder

Adverse Effects	Lithium	Anticonvulsants	Antipsychotics	Antidepressants
Acute efficacy poor	No	LTG in mania	No	Debated
Prophylaxis uncertain	No	CBZ, VPA	Yes	Yes
Sedation or lethargy	Variable	Variable	Yes	Uncommon
Neurological	Mild cognitive, rare EPS	Cognitive	Variable EPS	No
Headache	No	Yes	No	Rare
Gastrointestinal distress	Early	Variable	No	SRIs
Weight gain	Yes	Yes	CLZ, ONZ, QTP	Variable
Hypertension	No	No	No	Variable
Type II diabetes	No	No	CLZ, ONZ, QTP	No
Hypothyroidism	Uncommon	No	No	No
Renal effects	Predictable	No	No	No
SIADH/hyponatremia	Rare	oxCBZ, CBZ	Variable	SRIs
Dermatologic	Yes	LTG, CBZ	Rare	No
Hepatic damage	no	VPA, CBZ	Rare	No
Pancreatitis	No	VPA	Rare	No
Bone marrow dysfunction	No	CBZ, VPA	CBZ	Rare
Masculinization (PCOS)	No	VPA	No	No
Drug interactions	NSAIDs, diuretics, CCBs	CBZ; VPA with LTG	Rare	SRIs
Teratogenic	Uncommon	VPA>>CBZ>LTG	Not proved	Not certain

CBZ carbamazepine; *CCBs* calcium-channel blockers; *CLZ* clozapine; *EPS* extrapyramidal syndromes; *LTG* lamotrigine; *NSAIDs* nonsteroidal anti-inflammatory drugs; *ONZ* olanzapine; *oxCBZ* oxcarbazepine; *PCOS* polycystic ovary syndrome with increased testosterone secretion; *QTP* quetiapine; *SIADH* syndrome of inappropriate secretion of antidiuretic hormone with water retention and risk of hyponatremia; *SRIs* serotonin-reuptake inhibitors; *VPA* valproate. Among drug interactions, carbamazepine induces metabolism of itself and other agents; some SRIs block metabolism of other agents includes intensive medical care and usually immediate hemodialysis [10, 97, 174]. If slow-release preparations of lithium salts are involved, there is a risk of delayed intestinal absorption with secondary increases of serum concentrations that may require more prolonged or repeated dialysis, often over several days, even after the patient initially regains full consciousness. In general, serum concentrations of lithium may or may not correspond closely to clinically observed signs and symptoms during recovery from lithium intoxication. An experimental method of removal of excesses of lithium is by use of an orally or rectally administered cation-exchange resin such as sodium polystyrene sulfonate [132]. Several types of drugs can increase renal retention and raise serum concentrations of lithium to toxic levels, as is discussed later regarding drug–drug interactions [174, 311].

In addition to cerebral intoxication, lithium has a series of prevalent and characteristic adverse effects at therapeutic doses. These include frequent, usually temporary, initial polyuria and thirst. Polyuria may be more likely when lithium is combined with an SRI antidepressant [381]. Perhaps 10–20% of lithium-treated patients develop later, persistent and clinically significant, but usually reversible, *diabetes insipidus*. This reaction is associated with elevated circulating levels of ADH and unresponsiveness to exogenous synthetic antidiuretic peptides such as desmopressin [147, 311]. Lithium-induced diabetes insipidus can be reversed by the actions of diuretic drugs, and amiloride has been considered particularly safe for this approach by having limited lithium-retaining effects and sparing losses of potassium [54, 198]. However, many clinicians prefer more conservative approaches, such as lowering the dose of lithium and use of alternative mood-stabilizing treatments. High 24-hour urine volumes (≥ 5 L) have been associated with elevated risk of histopathological changes in renal biopsies suggestive of chronic granulomatous, nephron-distorting, inflammation [163, 164]. Nevertheless, irreversible azotemic renal failure due to lithium treatment is uncommon (long-term risk may be approximately 1%) and usually is sensitively and simply detected by quarterly or semiannual monitoring of serum creatinine concentrations, watching for gradually rising trends rather than any particular value, but with closer monitoring at levels above 1.3 or 1.4 mg/dL [56, 174].

Weight-gain and dermatological disorders including severe acne, worsening of psoriasis, and mild alopecia all have been associated with lithium treatment. However, subtle neurological or cognitive effects, as well as weight-gain and fear of renal damage, are leading reasons to resist or discontinue long-term lithium therapy [135]. Neurological effects include a high risk of mild resting tremor and impaired handwriting or other fine-motor skills, which may be responsive to reduced doses of lithium carbonate or adding a beta-adrenergic antagonist such as propranolol or nadolol [91]. There may also be varying degrees of psychomotor and subjective cognitive impairment, mild mental confusion, and even low-grade delirium, which can occur even at trough plasma concentrations in a nominally therapeutic range (with much-higher intervening peak concentrations), particularly in elderly or neurologically impaired patients, as well as adding to the variable cognitive deficits that have been documented in bipolar disorder [158, 380].

Table 3.12 Drug effects on serum lithium concentrations

Increase levels	Minor effects	Decrease levels
Thiazide diuretics	Salicylates	Mannitol
NSAIDs	Acetaminophen	Acetazolamide
ACE-inhibitors	Lamotrigine	Theophylline
Furosamide	Valproate	Caffeine
Other loop diuretics	Antipsychotics	
Amiloride (weak)		
Alprazolam (weak)		

Adapted from Grandjean and Aubry [144].
ACE angiotensin converting enzyme; *NSAIDs* nonsteroidal anti-inflammatory agents. Effects are on circulating concentrations of lithium: agents that decrease clearance of lithium and increase levels can have potentially dangerous effects. Agents are listed in approximate order of risks.

Clinical hypothyroidism or myxedema is unusual during lithium therapy, but diffuse, nontoxic, and nonprecancerous goiter can occur in a substantial minority of patients long exposed to lithium [58, 212]. Sometimes serum thyroid hormone indices decline into the low normal or subnormal range, usually following earlier elevations of TSH, secondary to blocking of its actions at the thyroid by lithium. Some clinicians believe that adding supplemental thyroid hormone not only can avoid or reduce goiter, but may also contribute to minimizing depression or slowing of rapid cycling. Thyroxin (T4) may be more physiological, has a longer half-life, and has been better studied for prophylaxis in bipolar disorder, but triiodothyronine (T3) appears to have more psychotropic impact, at least in unipolar depression [71, 209, 281, 377]. Lithium also has been associated with hyperparathyroidism, although evidence of its possible association with risk of bone fractures and on rates of bone healing remains largely negative [58, 230, 355].

Drug–drug interactions can be clinically important and even lethal with lithium treatment. Several types of drugs interfere with the renal clearance of lithium (Table 3.12). Important interactions of this type have been associated with sodium-losing diuretics, including thiazides, furosamide (Lasix®), and other diuretics acting at the loop of Henle ("loop-diuretics"), with lesser effects of the renal distal convoluted tubule-active diuretic amiloride (Midamor® [114]). Several nonsteroidal anti-inflammatory agents (NSAIDs) also have important effects on lithium clearance and can increase circulating concentrations of lithium to potentially toxic levels. This interaction is well established for ibuprofen, is much less likely with some COX-2 inhibitors, and may be limited also with acetaminophen and naproxen [214, 320]. Another class of agents with potent ability to increase circulating concentrations of lithium are the angiotensin-converting enzyme (ACE) inhibitor antihypertensive drugs such as lisinopril (Zestril® [114, 236]). Interactions of lithium with other psychotropic drugs, notably antipsychotic drugs (including haloperidol), lamotrigine, and valproate appear to be limited [144].

Anticonvulsant Mood Stabilizers

Carbamazepine and valproic acid can induce excessive sedative and nonspecific central depressant effects, confusion, mild cognitive impairment, headaches, and gastrointestinal distress, varying with dose and individual susceptibilities. Oxcarbazepine, more than carbamazepine, has been associated with the SIADH, which can induce potentially severely neurotoxic hyponatremia [347]. Carbamazepine also is associated with rashes and, very rarely, with Stevens–Johnson dermatomuconecrolysis, as well as with potential hepatic toxicity and suppression of white blood cell production.

Rashes are particularly strongly associated with lamotrigine, especially when combined with valproate, which can interfere with the clearance of lamotrigine, whereas carbamazepine has the opposite interactive effect [215]. Lamotrigine is also associated with much more common, nonspecific rashes; although they usually are clinically benign, they can induce fear on the part of patients as well as clinicians, and to reluctance to continue the treatment. The incidence of clinically significant dermatological reactions to lamotrigine among bipolar disorder patients averages approximately 1/100 treated cases, most appearing within the first 8 weeks of treatment [303]. Much of the dermatological risk can be avoided by very slow increases in doses of lamotrigine, typically starting at 25–50 mg/day unless valproate is also used, in which case, starting doses of 12.5–25 mg/day and maximum doses well below usual target doses of 200–300 mg/day are safer, and typically attained gradually over 2 or 3 weeks [165, 194, 303].

Hepatic toxicity has been associated with carbamazepine and particularly with valproate, which is also associated with pancreatitis [344]. Hepatic toxicity in association with valproate is rare and has most often been encountered among very young children or infants treated with multiple anticonvulsants, but can be severe. This association may reflect higher rates of conversion of valproic acid to unsaturated metabolites that are potentially hepatotoxic, as was discussed earlier in the basic pharmacology of this agent. Additional effects are increased circulating concentrations of ammonia, sometimes with toxic encephalopathy and without other evidence of liver failure [111, 376]. There are preliminary suggestions that treatment with the quaternary ammonium metabolite L-carnitine (L-3-hydroxy-4-[trimethylazaniumyl] butanoate) might limit the hepatotoxic effects of valproate, at least in acute overdoses [274].

Drug–drug interactions with anticonvulsant agents are common. These include the striking ability of carbamazepine to induce the activity and abundance of some hepatic cytochrome-P450 microsomal oxidases, with enhanced clearance and reduced circulating concentrations of many types of drugs, including carbamazepine itself. Similar interactions occur with other anticonvulsants, antipsychotics, and antidepressants, sometimes requiring upward adjustment of doses to overcome the effect, but with risks of intoxication if metabolities are CNS-active or if the drug metabolism-inducing agent is removed [176]. Oxcarbazepine is much less likely to have such effects [74]. Valproic acid has some ability to exert opposite effects of inhibiting drug-metabolizing pathways in the liver, including glucuronidation, with increased plasma concentrations of some other agents, notably lamotrigine. Some potentiating effects of valproate can also occur in interactions with carbamazepine and benzodiazepines [221].

Valproate also has been associated with development of polycystic ovaries with the syndrome (PCOS) of obesity, infertility, variable diabetes, and masculinizing effects associated with increases in circulating concentrations of androgenic steroids in young women with bipolar disorder or epilepsy [177]. This adverse syndrome occurs about twice as often among women with epilepsy treated with valproate compared to other anticonvulsants [169].

Pregnancy

Pregnancy presents complex challenges for safe clinical management of women with major psychiatric disorders. As noted earlier, risks of recurrences of bipolar disorder during and especially soon following pregnancy are high, especially if mood-stabilizing treatment is discontinued (Fig. 3.7). Pregnancy and childbirth also are associated with variable and sometimes substantial changes in the metabolism and clearance of psychotropic and other drugs that can differ among trimesters and following delivery [197, 270, 328].

Mood-stabilizing drugs, as a group, also carry especially high risks of structural anomalies during early fetal development. Assessment of sometimes severe teratogenic or other adverse developmental effects associated with drug exposures during pregnancy is complicated by their generally low incidence, coupled with a substantial risk of spontaneous or unexplained fetal malformations, on the order of 3–5% among children born of mothers with no known exposures to potential teratogenic foreign (xenobiotic) substances [252]. Until recently, there appears to have been an unbalanced focus on risk of fetal malformations in association with use of psychotropic medicines during pregnancy, and marked regional differences in attitudes about such risks continue. Much of this focus appears to be driven by concerns about potential legal liability and suits in addition to clinical concern about fetal and neonatal health [363]. In general, there is a high level of avoidance of all foreign substances during pregnancy. The costs of such a conservative strategy include increased risk of re-emerging maternal illness without treatment, and even greater risk when a maintenance treatment is interrupted, especially abruptly, as is likely when an unanticipated diagnosis of pregnancy is made [84, 364]. An emerging challenge is to balance concern for fetal risks with the impact of an acute psychiatric illness in a pregnant women on the patient herself, her family, and even on fetal development.

Most drugs that can pass the blood–brain diffusion barrier to exert effects in the CNS also pass the placental barrier to reach the developing fetus; many are also secreted in maternal milk in variable, but often nontrivial concentrations to exert

Table 3.13 Perinatal risks of psychotropic drugs

Drug types	Spontaneous abortion	Major fetal abnormalities	Neonatal syndromes	Neurocognitive development
Antidepressants				
TCAs	Moderate	Low (≤1%)	Low	Low
SRIs	Moderate	Pulmonary hypertension (6×)	Moderate withdrawal	Low
Atypical	Moderate	Low (≤3%)	Low	Low
MAOIs	Uncertain	Uncertain	Uncertain	Uncertain
Anticonvulsants and lithium				
Carbamazepine	Low	Spina bifida (2–3×) and others	Low	Low
Gabapentin	Low	Low	Low	Low
Lamotrigine	Low	Clefts (6%)	Low	Low
Oxcarbazepine	Low	Low	Low	Low
Valproate	Low	Spina bifida (3–5×) and others (7×)	Low	Low
Lithium	Low	Cardiac (5–10×)	Low muscle tone, hypothyroid, polyuria	Low
Antipsychotics				
Butyrophenones	Low	Low	Extrapyramidal	Low
Phenothiazines	Low	Low	Extrapyramidal	Low
Second generation	Low	Low	Probably low	Probably low
Sedative–anxiolytics				
Benzodiazepines	Low	Clefts (moderate)	Moderate withdrawal	Low

Adapted from Einarson [104]. x = times increase.
Approximate base risks *without* drug exposure: all major fetal anomalies (ca. 2.5%); spina bifida (ca. 0.1%, with some reduction given folic acid ca. 4 mg/day for ≥90 days pre- and postconception); Ebstein's cardiac malformation (mitral valve malformation and variable septal defects; ca. 0.005%) associated selectively with prenatal exposure to lithium (risk to ca. 0.1%). Maternal depression or epilepsy per se probably do not increase risks of major malformations but may be associated with impaired neurocognitive development.

effects on the newborn. It is unfortunate that most of the commonly employed mood-stabilizing drugs, including carbamazepine, divalproex, lamotrigine, and lithium have adverse fetal or later developmental effects associated with their use, particularly early in pregnancy [104, 121, 329, 345]. It is particularly unfortunate that recurrence risks of bipolar disorder appear to be highest in the first trimester, challenging the former hope that interrupting treatment temporarily would be safe for both mother and fetus [367]. Moreover, the diagnosis of pregnancy is sometimes delayed for at least one menstrual cycle, whereas some fetal malformations arise very early in gestation, particularly spina bifida—sometimes before pregnancy is recognized [371]. A summary of proposed perinatal effects of various psychotropic drugs, including mood-stabilizing agents, is provided in Table 3.13. It is important to emphasize that knowledge of these risks and strategies for optimal clinical management remain strikingly underdeveloped and inadequately studied.

Maternal exposure to lithium in the first trimester of pregnancy has long been associated with increased risk of inducing major cardiac malformations, most commonly Ebstein's tricuspid value and variable septal malformations. Recent estimates place the risk at 5–10 times above the spontaneous base rate for Ebstein's anomalies of approximately 1/20,000 live births [83]. It is important to emphasize that the risk with lithium-exposure(<1/1,000) is far below the rates and severity of complex teratogenic effects associated with some anticonvulsant mood-stabilizers, especially valproate and carbamazepine (several per 100). Most cases of Ebstein's malformations can be detected during pregnancy by ultrasonography. If the pregnancy is continued to term, most of them, particularly malformed tricuspid valves, can be repaired surgically postnatally [313].

Lithium also can induce transient neonatal hypotonia ("floppy-baby" syndrome) and it is occasionally associated with thyroid hypofunction in the newborn [118, 202]. Soon after childbirth, maternal elimination and tolerance of lithium typically decrease sharply as its renal clearance decreases, calling for caution to avoid overdosing [174]. Lithium and some anticonvulsants pass into human breast milk (for lithium, at about half of maternal circulating concentrations) to yield serum concentrations about one-quarter those of lithium-treated mothers, with largely uncertain consequences for the newborn [76, 365].

Mood-stabilizing anticonvulsants also cannot be considered safe during pregnancy. As a group, they carry elevated risks of a range of fetal malformations [252, 329, 345]. Valproate is especially dangerous, with risks of malformations of derivatives of the fetal neural crest, including spina bifida, at rates of several percent of exposed women. There are also risks of cardiac and various other malformations with valproate, occurring at several times higher rates than with carbamazepine and other anticonvulsants likely to be used to treat bipolar disorder patients. It remains uncertain whether such risks are dependent on the daily maternal dose or serum concentrations, but use of multiple agents may increase teratogenic risks [329]. Exposure to valproate during pregnancy has been associated with a complex "fetal valproate syndrome," which includes variable admixtures of neural tube defects, aortic coarctation, Ebstein anomaly, cardiac atrial septal defects, craniosynostosis, and abnormalities of the limbs, eyes, respiratory system, and abdominal wall, with a characteristic facial appearance, and developmental delays in as many as 50% of newborns from exposed pregnancies [17]. Supplements of folic acid are recommended routinely to limit risks of spontaneous or idiopathic spina bifida and are also given to women taking anticonvulsants for epilepsy or bipolar disorder. Although the value of folic acid supplementation in preventing neural tube defects associated with early pregnancy is well established, there remains some uncertainty as to whether such supplementation can successfully counteract the effects of teratogenic substances such as valproate [175]. To be optimally effective, it is usual to start supplementation with folate well before pregnancy to saturate tissues, particularly to limit risk of spina bifida, which begins to develop within the first weeks of pregnancy. Treatment options for spina bifida are very limited, but there are experimental efforts to develop intrauterine surgical interventions to limit damage to the exposed fetal spinal cord [2]. Following delivery, some neonates exposed to valproic acid during pregnancy have been reported to have decreased blood coagulation and bleeding tendencies, possibly related to deficiencies of vitamin K and modifiable by giving supplements of it [155, 156]. There may also be neurodevelopmental delays in early childhood to follow exposure to valproate in utero, but such effects typically are very difficult to ascribe to particular causal factors [252].

Exposure to other mood-stabilizing anticonvulsants during pregnancy appears to carry substantially lower risks than those associated with maternal use of valproate. Initially, it had been hoped that lamotrigine would be free of congenital anomalies, but growing case series have detected increased risks of cleft lip or palate at rates of approximately 0.4% of those exposed [166, 251, 252]. Carbamazepine is reported to have similar qualitative developmental effects as valproate, but at substantially lower rates, that average 2–3-times below those of valproate [252, 329]. Oxcarbazepine remains insufficiently studied regarding potential teratogenic or other adverse developmental effects. Possible long-lasting effects of exposure to mood-stabilizing agents in utero or in breast milk, including effects of withdrawal, in infancy remain poorly documented [243, 318, 341].

Conclusions

Mood-stabilizing agents, and particularly lithium alone or with an anticonvulsant, are the cornerstone of rational long-term medical management of bipolar disorder patients. Available mood-stabilizing agents include lithium and several anticonvulsant drugs. Carbamazepine and valproate are effective in acute mania and have limited long-term research support for long-term prophylactic effects for which they are, nevertheless, often employed. Lamotrigine lacks antimanic efficacy but has moderate ability to limit recurrences of bipolar depressive episodes. Most antipsychotic drugs are rapidly antimanic and may have long-term beneficial effects, although their testing for prophylactic benefits in bipolar disorder have been limited and sometimes confounded by possible discontinuation-associated worsening that can lead to misleading interpretations. Other drugs remain much less well evaluated (e.g., oxcarbazepine) or are ineffective (e.g., topiramate, calcium-channel blockers). The current status of the therapeutics of bipolar disorder includes substantial reductions of recurrences of episodes of mania or psychosis, but morbidity continues to be detected in perhaps 40–50% of weeks of follow-up, much of it at subsyndromal, but still clinically significant. Perhaps three-quarters of this residual morbidity is accounted for by depressive-dysthymic or mixed states, making improved treatment of depressive phases of bipolar disorder an urgent priority.

Current expert opinion in the United States favors use of lithium or other mood stabilizers or some antipsychotics (especially quetiapine and the combination of olanzapine with fluoxetine), rather than antidepressants, to treat bipolar depression and protect against recurrences. ECT can be a useful temporary alternative in difficult cases, with beneficial effects in both depression and mania, and there are suggestive reports of beneficial effects in bipolar depression for direct dopamine agonists used to treat Parkinson's disease. Overly aggressive and long-term use of antidepressants in bipolar disorder is very common, evidently in response to the prevalent but difficult-to-treat depressive-dysthymic components of the disorder. This practice lacks vigorous research support in that antidepressant drugs show inconsistent short-term effects, and quite limited effectiveness above that of mood stabilizers alone, and possibly unfavorable risk:benefit relationships in long-term use, with risks of destabilizing effects in perhaps 10–15% of cases of bipolar I disorder. These effects include at least moderate increases of mood cycling and rapid "switching" from depression into diagnostically confusing and potentially dangerous, complex, mixed-dysphoric manic-depressive states, as well as manic excitement or psychosis and behavioral dyscontrol. It is widely supposed, but not proved, that lithium or anticonvulsants can protect against such "switching."

Bipolar depression and mixed states carry very high risks of suicide and other violence in younger years, and of excess mortality associated with medical disorders in later years. Lithium may have protective effects against suicidal behavior and appears to be more effective than anticonvulsant mood stabilizers or other alternative treatments. Psychoeducational, group, and other psychotherapeutic interventions, as well as rehabilitative efforts, can also enhance stability, functional status, and quality of life as well as treatment adherence in bipolar disorder patients in a cost-effective manner.

References

1. Adams F, editor. The Extant Works of Aretaeus, the Cappadocian. Boston: Milford House; 1972/1856.
2. Adzick NS, Thom EA, Spong CY, Brock III JW, Burrows PK, Johnson MP, et al. Randomized trial of prenatal versus postnatal repair of myelomeningocele. N Engl J Med. 2011;364:993–1004.
3. Ahrens B, Müller-Oerlinghausen B. Does lithium exert an independent antisuicidal effect? Pharmacopsychiatry. 2001;34:132–6.
4. Akiskal HS. Mood disorders: historical introduction and conceptual overview, Chap. 13.1. In: Sadock BJ, Sadock VA, editors. Mood Disorders, Kaplan & Sadock's Comprehensive Textbook of Psychiatry, vol. 1. 9th ed. Philadelphia: Wolters Kluver/Lippincott Williams & Wilkins; 2009. p. 1629–45.
5. Akiskal HS, Benazzi F. DSM-IV and ICD-10 categories of recurrent [major] depressive and bipolar II disorders: evidence that they lie on a dimensional spectrum. J Affect Disord. 2006;92:45–54.
6. Aliasgharpour M, Hagani H. Evaluation of lithium determination in three analyzers: flame emission, flame atomic absorption spectroscopy and ion selective electrode. N Am J Med Sci. 2009;1:244–6.
7. Altshuler LL, Post RM, Leverich GS, Mikalauskas K, Rosoff A, Ackerman L. Antidepressant-induced mania and cycle acceleration: a controversy revisited. Am J Psychiatry. 1995;152:1130–8.
8. Altamura AC, Lietti L, Dobrea C, Benatti B, Arici C, Dell'Osso B. Mood stabilizers for patients with bipolar disorder: state of the art. Expert Rev Neurother. 2011;11:85–99.

9. Amann B, Born C, Crespo JM, Pomarol-Clotet E, McKenna P. Lamotrigine: when and where does it act in affective disorders? systematic review. J Psychopharmacol. 2011;25:1289–94.
10. Amdisen A. Clinical features and management of lithium poisoning. Med Toxicol Adverse Drug Exp. 1988;3:18–32.
11. American Psychiatric Association (APA). Diagnostic and Statistical Manual, third and fourth revisions (DSM-III, -IV, and IV-text-revision). Washington, DC: American Psychiatric Press; 1980, 1994, 2000.
12. Amsterdam JD, Shults J. Efficacy and mood conversion rate of short-term fluoxetine monotherapy of bipolar II major depressive episode. J Clin Psychopharmacol. 2010;30:306–11.
13. Amsterdam JD, Wang G, Shults J. Venlafaxine monotherapy in bipolar type II depressed patients unresponsive to prior lithium monotherapy. Acta Psychiatr Scand. 2010;121:201–8.
14. Angst J, Angst F, Gerber-Werder R, Gamma A. Suicide in 406 mood-disorder patients with and without long-term medication: a 40 to 44 years' follow-up. Arch Suicide Res. 2005;9: 279–300.
15. Angst F, Stassen HH, Clayton PJ, Angst J. Mortality of patients with mood disorders: follow-up over 34-38 years. J Affect Disord. 2002;68:167–81.
16. Arana A, Wentworth CE, Ayuso-Mateos JL, Arellano FM. Suicide-related events in patients treated with antiepileptic drugs. N Engl J Med. 2010;363:542–51.
17. Ardinger HH, Atkin JF, Blackston RD, Elsas LJ, Clarren SK, Livingstone S, et al. Verification of the fetal valproate syndrome phenotype. Am J Med Genet. 1988;29:171–85.
18. Baethge C, Baldessarini RJ, Bratti IM, Tondo L. Prophylaxis-latency and outcome in bipolar disorders. Can J Psychiatry. 2003;48:449–57.
19. Baethge C, Baldessarini RJ, Khalsa HMK, Hennen J, Salvatore P, Tohen M. Substance abuse in first-episode bipolar I disorder: indications for early intervention. Am J Psychiatry. 2005;162:1008–10.
20. Baethge C, Baldessarini RJ, Mathiske-Schmidt K, Hennen J, Berghofer A, Muller-Oerlinghausen B, et al. Long-term combination therapy versus monotherapy with lithium and carbamazepine in 46 bipolar I patients. J Clin Psychiatry. 2005;66:174–82.
21. Baethge C, Salvatore P, Baldessarini RJ. On cyclic insanity (1882) by Karl Ludwig Kahlbaum, MD: translation and commentary. Harv Rev Psychiatry. 2003;11:78–90.
22. Bagary M. Epilepsy, antiepileptic drugs and suicidality. Curr Opin Neurol. 2011;24:177–82.
23. Baldessarini RJ. Plea for integrity of the bipolar disorder concept. Bipolar Disord. 2000;2:3–7.
24. Baldessarini RJ. Commentary: the bipolar affective disorder: lithium/anticonvulsant evaluation (BALANCE) study. Bipolar Disord. 2010;12:669–72.
25. Baldessarini RJ, Henk HJ, Sklar AR, Chang J, Leahy LF. Psychotropic medications for bipolar disorder patients in the United States: polytherapy and adherence. Psychiatr Serv. 2008;59:1175–83.
26. Baldessarini RJ, Kopin IJ. Effect of drugs on release of [³H]-norepinephrine from central nervous system tissues by electrical stimulation in vitro. J Pharmacol Exp Ther. 1967;156:31–8.
27. Baldessarini RJ, Leahy LF, Arcona S, Gause D, Zhang W, Hennen J. Prescribing patterns of psychotropic medicines in the United States for patients diagnosed with bipolar disorders. Psychiatr Serv. 2007;58:85–91.
28. Baldessarini RJ, Perry R, Pike J. Factors associated with treatment nonadherence among US bipolar disorder patients. Hum Psychopharmacol. 2008;23:95–105.
29. Baldessarini RJ, Salvatore P, Khalsa H-MK, Gebre-Medhin P, Imaz H, González-Pinto A, et al. Morbidity in 303 first-episode bipolar I disorder patients. Bipolar Disord. 2010;12:264–70.
30. Baldessarini RJ, Salvatore P, Khalsa H-MK, Imaz-Etxeberria H, Gonzalez-Pinto A, Tohen M. Episode cycles with increasing recurrences in first-episode bipolar-I disorder patients. J Affect Disord. 2011;136:149–54.
31. Baldessarini RJ, Salvatore P, Khalsa H-MK, Tohen M. Dissimilar morbidity following initial mania vs. mixed-states in type-I bipolar disorder. J Affect Disord. 2010;126:299–302.
32. Baldessarini RJ, Stephens J. Lithium carbonate for affective disorder: clinical pharmacology and toxicology. Arch Gen Psychiatry. 1970;22:72–77.

33. Baldessarini RJ, Suppes T, Tondo L. Lithium withdrawal in bipolar disorder: implications for clinical practice and experimental therapeutics research. Am J Ther. 1996;3:492–6.
34. Baldessarini RJ, Tarazi FI. Pharmacotherapy of psychosis and mania. Chap. 18. In: Brunton LL, Lazo JS, Parker KL, editors. Goodman and Gilman's Goodman and Gilman's the Pharmacological Basis of Therapeutics. 11th ed. New York: McGraw-Hill; 2005. p. 461–500.
35. Baldessarini RJ, Tondo L. Antisuicide effects of lithium and anticonvulsants: update. Bipolar Disord. 2008;10:114–5.
36. Baldessarini RJ, Tondo L. Meta-analytic comparison of antisuicidal effects of lithium versus anticonvulsants. Pharmacopsychiatry. 2009;42:72–5.
37. Baldessarini RJ, Tondo L, Baethge C, Lepri B, Bratti IM. Effects of treatment-latency on response to maintenance treatment in manic-depressive disorders. Bipolar Disord. 2007;9:386–93.
38. Baldessarini RJ, Tondo L, Davis P, Pompili M, Goodwin FK, Hennen J. Decreased risk of suicides and attempts during long-term lithium treatment: meta-analytic review. Bipolar Disord. 2006;8:625–39.
39. Baldessarini RJ, Tondo L, Faedda GL, Viguera AC, Baethge C, Salvatore P, et al. Latency, discontinuation, and re-use of lithium treatment. In: Bauer M, Grof P, Müller-Oerlinghausen B, editors. Lithium in Neuropsychiatry: The Comprehensive Guide. London: Taylor & Francis; 2007. p. 465–81.
40. Baldessarini RJ, Tondo L, Ghiani C, Lepri B. Illness risk following rapid versus gradual discontinuation of antidepressants. Am J Psychiatry. 2010;167:934–41.
41. Baldessarini RJ, Tondo L, Hennen J. Effects of rapid cycling on response to lithium maintenance treatment in 360 bipolar I and II disorder patients. J Affect Disord. 2000;61:13–22.
42. Baldessarini RJ, Tondo L, Hennen J. Latency and episodes before treatment: effects on pre-treatment morbidity but not response to maintenance treatment in bipolar I and II disorders. Bipolar Disord. 2003;5:169–79.
43. Baldessarini RJ, Tondo L, Hennen J. Lithium treatment and suicide risk in major affective disorders: update and new findings. J Clin Psychiatry. 2003;64 (Suppl 5):44–52.
44. Baldessarini RJ, Tondo L, Hennen J, Viguera AC. Is lithium still worth using? An update of selected recent research. Harv Rev Psychiatry. 2002;10:59–75.
45. Baldessarini RJ, Tondo L, Lepri B. Empirical test of a basic assumption underlying survival analysis as applied to experimental therapeutics in psychopharmacology. J Clin Psychopharmacol. 2010;30:72–5.
46. Baldessarini RJ, Undurraga J, Vázquez GH, Tondo L, Salvatore P, Ha K, et al. Predominant recurrence polarity among 928 adult international bipolar-I disorder patients. Acta Psychiatr Scand. 2012;125:293–302.
47. Baldessarini RJ, Vieta E, Calabrese JR, Tohen M, Bowden C. Bipolar depression: overview and commentary. Harv Rev Psychiatry. 2010;18:143–57.
48. Ballenger JC, Post RM. Carbamazepine in manic-depressive illness: a new treatment. Am J Psychiatry. 1980;137:782–90.
49. Bauer M, Adli M, Bschor T, Pilhatsch M, Pfennig A, Sasse J, et al. Lithium's emerging role in the treatment of refractory major depressive episodes: augmentation of antidepressants. Neuropsychobiology. 2010;62:36–42.
50. Bauer M, Döpfmer S. Lithium augmentation in treatment-resistant depression: meta-analysis of placebo-controlled studies. J Clin Psychopharmacol. 1999;19:427–34.
51. Bauer M, Grof P, Müller-Oerlinghausen B, editors. Lithium in Neuropsychiatry: The Comprehensive Guide. London: Taylor & Francis; 2007.
52. Beatriz S, Fagundes R. Valproic acid: review. Rev Neurosci. 2008;16:130–6.
53. Beaulieu JM, Caron MG. Looking at lithium: molecular moods and complex behavior. Mol Interv. 2008;8:230–41.
54. Bedford JJ, Weggery S, Ellis G, McDonald FJ, Joyce PR, Leader JP, et al. Lithium-induced nephrogenic diabetes insipidus: renal effects of amiloride. Clin J Am Soc Nephrol. 2008;3: 1324–31.
55. Bellivier F, Yon L, Luquiens A, Azorin JM, Bertsch J, Gerard S, et al. Suicidal attempts in bipolar disorder: results from an observational study (EMBLEM). Bipolar Disord. 2011;13:377–86.

56. Bendz H, Schön S, Attman PO, Aurell M. Renal failure occurs in chronic lithium treatment but is uncommon. Kidney Int. 2010;77:219–24.
57. Beynon S, Soares-Weiser K, Woolacott N, Duffy S, Geddes JR. Pharmacological interventions for the prevention of relapse in bipolar disorder: systematic review of controlled trials. J Psychopharmacol. 2009;23:574–91.
58. Bhuvaneswar C, Alpert J, Harsh V, Baldessarini RJ. Adverse endocrine and metabolic effects of psychotropic drugs. CNS Drugs. 2009;23:1003–21.
59. Biel MG, Peselow E, Mulcare L, Case BG, Fieve R. Continuation versus discontinuation of lithium in recurrent bipolar illness: a naturalistic study. Bipolar Disord. 2007;9:435–42.
60. Bond DJ, Lam RW, Yatham LN. Divalproex sodium vs. placebo in the treatment of acute bipolar depression: systematic review and meta-analysis. J Affect Disord. 2010;124:228–34.
61. Bowden CL, Calabrese JR, McElroy SL, Gyulai L, Wassef A, Petty F, Pope HG Jr, Chou JC, Keck PE Jr, Rhodes LJ, Swann AC, Hirschfeld RM, Wozniak PJ. Randomized, placebo-controlled 12-month trial of divalproex and lithium in treatment of outpatients with bipolar I disorder. Arch Gen Psychiatry. 2000;57:481–9.
62. Bowen RC, Grof P, Grof E. Less frequent lithium administration and lower urine volume. Am J Psychiatry. 1991;148:189–92.
63. Bratti IM, Baldessarini RJ, Baethge C, Tondo L. Pretreatment episode count and response to lithium treatment in manic-depressive illness. Harv Rev Psychiatry. 2003;11:245–56.
64. Brennan PA, Pargas R, Walker EF, Green P, Newport DJ, Stowe Z. Maternal depression and infant cortisol: influences of timing, comorbidity and treatment. J Child Psychol Psychiatry. 2008;49:1099–107.
65. Brown E, Dunner DL, McElroy SL, Keck PE, Adams DH, Degenhardt E, et al. Olanzapine/fluoxetine combination vs. lamotrigine in the 6-month treatment of bipolar I depression. Int J Neuropsychopharmacol. 2009;12:773–82.
66. Brown EB, McElroy SL, Keck Jr PE, Deldar A, Adams DH, Tohen M, et al. Seven-week, randomized, double-blind trial of olanzapine/fluoxetine combination versus lamotrigine in the treatment of bipolar I depression. J Clin Psychiatry. 2006;67:1025–33.
67. Brunello N, Tascedda F. Cellular mechanisms and second messengers: relevance to the psychopharmacology of bipolar disorders. Int J Neuropsychopharmacol. 2003;6:181–9.
68. Cade JF. Lithium salts in the treatment of psychotic excitement. Med J Aust. 1949;2:349–52.
69. Calabrese JR, Huffman RF, White RL, Edwards S, Thompson TR, Ascher JA, et al. Lamotrigine in the acute treatment of bipolar depression: results of five double-blind, placebo-controlled clinical trials. Bipolar Disord. 2008;10:323–33.
70. Calabrese JR, Ketter TA, Youakim JM, Tiller JM, Yang R, Frye MA. Adjunctive armodafinil for major depressive episodes associated with bipolar I disorder: a randomized, multicenter, double-blind, placebo-controlled, proof-of-concept study. J Clin Psychiatry. 2010;71:1363–70.
71. Calabrese JR, Shelton MD, Rapport DJ, Kujawa M, Kimmel SE, Caban S. Current research on rapid cycling bipolar disorder and its treatment. J Affect Disord. 2001;67:241–55.
72. Calderó J, Brunet N, Tarabal O, Piedrafita L, Hereu M, Ayala V, et al. Lithium prevents excitotoxic cell death of motoneurons in organotypic slice culture of spinal cord. Neuroscience. 2010;165:1353–69.
73. Carey TS, Williams Jr JW, Oldham JM, Goodman F, Ranney LM, Whitener L, et al. Gabapentin in the treatment of mental illness: the echo chamber of the case series. J Psychiatr Pract. 2008;14 (Suppl 1):15–27.
74. Centorrino F, Albert MJ, Berry JM, Kelleher JP, Fellman V, Line G, et al. Oxcarbazepine: clinical experience with hospitalized psychiatric patients. Bipolar Disord. 2003;5:370–4.
75. Centorrino F, Ventriglio A, Vincenti A, Talamo A, Baldessarini RJ. Changes in medication practices for hospitalized psychiatric patients: 2009 vs. 2004. Hum Psychopharmacol. 2010;25:179–86.
76. Chaudron LH, Jefferson JW. Mood stabilizers during breastfeeding: review. J Clin Psychiatry. 2000;61:79–90.

77. Chen G, Henter ID, Manji HK. Translational research in bipolar disorder: emerging insights from genetically based models. Mol Psychiatry. 2010;15:883–95.
78. Chiu CT, Liu G, Leeds P, Chuang DM. Combined treatment with the mood stabilizers lithium and valproate produces multiple beneficial effects in transgenic mouse models of Huntington's disease. Neuropsychopharmacology. 2011;36:2406–21.
79. Cipriani A, Barbui C, Salanti G, Rendell J, Brown R, Stockton S, et al. Comparative efficacy and acceptability of antimanic drugs in acute mania: multiple-treatments meta-analysis. Lancet. 2011;378:1306–15.
80. Cipriani A, Smith K, Burgess S, Carney S, Goodwin G, Geddes J. Lithium versus antidepressants in the long-term treatment of unipolar affective disorder. Cochrane Database Syst Rev. 2006;4:CD003492.
81. Ceron-Litvoc D, Soares BG, Geddes J, Litvoc J, de Lima MS. Comparison of carbamazepine and lithium in treatment of bipolar disorder: systematic review of randomized controlled trials. Hum Psychopharmacol. 2009;24:19–28.
82. Choe HK, Son GH, Chung S, Kim M, Sun W, Kim H, et al. Maternal stress retards fetal development in mice with transcriptome-wide impact on gene expression profiles of the limb. Stress. 2011;14:194–204.
83. Cohen LS, Friedman JM, Jefferson JW, Johnson EM, Weiner ML. Reevaluation of risk of in utero exposure to lithium. JAMA. 1994;271:146–50.
84. Cohen LS, Wang B, Nonacs R, Viguera AC, Lemon EL, Freeman MP. Treatment of mood disorders during pregnancy and postpartum. Psychiatr Clin North Am. 2010;33:273–93.
85. Colom F, Vieta E, Duban C, Pacchiarotti I, Sanchez-Moreno J. Clinical and therapeutic implications of predominant polarity in bipolar disorder. J Affect Disord. 2006;93:13–7.
86. Cox DR, Oakes D. Analysis of Survival Data. London: Chapman & Hall; 2001.
87. Crossley NA, Bauer M. Acceleration and augmentation of antidepressants with lithium for depressive disorders: two meta-analyses of randomized, placebo-controlled trials. J Clin Psychiatry. 2007;68:935–40.
88. Cruz N, Sanchez-Moreno J, Torres F, Goikolea JM, Valenti M, Vieta E. Antipsychotic drugs in acute bipolar depression. Int J Neuropharmacol. 2010;13:5–14.
89. Cuffí ML, Artells R, Navarro A, Ciruela F, Carbonell L. Regulation of α-adrenoceptor gene expression by chronic lithium treatment in rat brain. Methods Find Exp Clin Pharmacol. 2010;2010:721–5.
90. Curia G, Aracri P, Sancini G, Mantegazza M, Avanzini G, Franceschetti S. Protein-kinase C-dependent phosphorylation inhibits the effect of the antiepileptic drug topiramate on the persistent fraction of sodium currents. Neuroscience. 2004;127:63–8.
91. Dave M, Langbart MM. Nadolol for lithium tremor in the presence of liver damage. Ann Clin Psychiatry. 1994;6:51–2.
92. Dayer A, Aubry JM, Roth L, Ducrey S, Bertschy G. A theoretical reappraisal of mixed states: dysphoria as a third dimension. Bipolar Disord. 2000;2:316–24.
93. Deb AA, Wilson SS, Rove KO, Kumar B, Koul S, Lim DD, et al. Potentiation of mitomycin C tumoricidal activity for transitional cell carcinoma by histone deacetylase inhibitors in vitro. J Urol. 2011;186:2426–33.
94. De Fruyt J, Deschepper E, Audenaert K, Constant E, Floris M, Pitchot W, et al. Second-generation antipsychotics in the treatment of bipolar depression: systematic review and meta-analysis. J Psychopharmacol. 2011 [Epub ahead of print, 22 Sept].
95. Denicoff KD, Meglathery SB, Post RM, Tandeciarz SI. Efficacy of carbamazepine compared with other agents: a clinical practice survey. J Clin Psychiatry. 1994;55:70–6.
96. Denicoff KD, Smith-Jackson EE, Disney ER, Ali SO, Leverich GS, Post RM. Comparative prophylactic efficacy of lithium, carbamazepine, and the combination in bipolar disorder. J Clin Psychiatry. 1997;58:470–8.
97. Dennison U, Clarkson M, O'Mullane J, Cassidy EM. Incidence and clinical correlates of lithium toxicity: retrospective review. Ir J Med Sci. 2011;180:661–75.
98. Devaki R, Shankar Rao S, Nadgir SM. Effect of lithium on the adrenoceptor-mediated second messenger system in rat brain. J Psychiatry Neurosci. 2006;31:246–52.

99. Diazgranados N, Ibrahim L, Brutsche NE, Newberg A, Kronstein P, Khalife S, et al. Randomized add-on trial of an N-methyl-D-aspartate antagonist in treatment-resistant bipolar depression. Arch Gen Psychiatry. 2010;67:793–802.

100. Dou G, Aleshin O, Datta A, Yuan C. Automated enzymatic assay for measurement of lithium ions in human serum. Clin Chem. 2005;51:1989–91.

101. Dubé S, Tollefson GD, Thase ME, Briggs SD, Van Campen LE, Case M, et al. Onset of antidepressant effect of olanzapine and olanzapine/fluoxetine combination in bipolar depression. Bipolar Disord. 2007;9:618–27.

102. Dumlu K, Orhon Z, Özerdem A, Tural Ü, Ulas H, Tunca Z. Treatment-induced manic switch in the course of unipolar depression can predict bipolarity: cluster-analysis based evidence. J Affect Disord. 2011;134:91–101.

103. Duncan RE, Bazinet RP. Brain arachidonic acid uptake and turnover: implications for signaling and bipolar disorder. Curr Opin Clin Nutr Metab Care. 2010;13:130–8.

104. Einarson A. Safety of psychotropic drug use during pregnancy: review. MedGenMed. 2005;7:3–12. www.medscape.com/medgenmed.

105. El-Mallakh RS, Salem MR, Chopra A, Mickus GJ, Penagaluri P, Movva R. Blinded, randomized comparison of immediate-release and extended-release carbamazepine capsules in manic and depressed bipolar subjects. Ann Clin Psychiatry. 2010;22:3–8.

106. Emrich HM, von Zerssen D, Kissling W, Möller HJ, Windorfer A. Effect of sodium valproate on mania. The GABA-hypothesis of affective disorders. Arch Psychiatr Nervenkr. 1980;229:1–16.

107. Faedda GL, Baldessarini RJ, Suppes T, Tondo L, Becker I, Lipschitz D. Pediatric-onset bipolar disorder: a neglected clinical and public health problem. Harv Rev Psychiatry. 1995;3:171–95.

108. Faedda GL, Baldessarini RJ, Tohen M, Strakowski SM, Waternaux C. Episode sequence in bipolar disorder and response to lithium treatment. Am J Psychiatry. 1991;148:1237–9.

109. Faedda GL, Tondo L, Baldessarini RJ. Lithium discontinuation: uncovering latent bipolar disorder? Am J Psychiatry. 2001;158:1337–9.

110. Faedda GL, Tondo L, Baldessarini RJ, Suppes T, Tohen M. Outcome after rapid vs. gradual discontinuation of lithium treatment in bipolar disorders. Arch Gen Psychiatry. 1993;50:448–55.

111. Fassi G, Igoa A, Liste OA. [Valproate-induced hyperammonemic encephalopathy: review of cases in the psychiatric setting (Spanish)]. Vertex. 2008;19:371–7.

112. Fava M, Alpert J, Nierenberg A, Lagomasino I, Sonawalla S, Tedlow J, et al. Double-blind study of high-dose fluoxetine versus lithium or desipramine augmentation of fluoxetine in partial responders and nonresponders to fluoxetine. J Clin Psychopharmacol. 2002;22:379–87.

113. Fieve RR, Kumbaraci T, Dunner DL. Lithium prophylaxis of depression in bipolar I, bipolar II, and unipolar patients. Am J Psychiatry. 1976;133:925–9.

114. Finley PR, Warner MD, Peabody CA. Clinical relevance of drug interactions with lithium. Clin Pharmacokinet. 1995;29:172–91.

115. Fleisher AS, Truran D, Mai JT, Langbaum JB, Aisen PS, Cummings JL, et al. Chronic divalproex sodium use and brain atrophy in Alzheimer disease. Neurology. 2011;77:1263–71.

116. Focosi D, Azzarà A, Kast RE, Carulli G, Petrini M. Lithium and hematology: established and proposed uses. J Leukoc Biol. 2009;85:20–8.

117. Forlenza OV, Diniz BS, Radanovic M, Santos FS, Talib LL, Gattaz WF. Disease-modifying properties of long-term lithium treatment for amnestic mild cognitive impairment: randomized controlled trial. Br J Psychiatry. 2011;198:351–6.

118. Frassetto F, Tourneur Martel F, Barjhoux CE, Villier C, Bot BL, Vincent F. Goiter in a newborn exposed to lithium in utero. Ann Pharmacother. 2002;36:1745–8.

119. Frye MA, Ketter TA, Leverich GS, Huggins T, Lantz C, Denicoff KD, et al. Increasing use of polypharmacotherapy for refractory mood disorders: 22 years of study. J Clin Psychiatry. 2000;61:9–15.

120. Fumarola A, Di Fiore A, Dainelli M, Grani G, Calvanese A. Medical treatment of hyperthyroidism: state of the art. Exp Clin Endocrinol Diabetes. 2010;118:678–84.

121. Galbally M, Roberts M, Buist A. Mood-stabilizers in pregnancy: systematic review. Aust N Z J Psychiatry. 2010;44:967–77.
122. Garbutt JC, Gillette GM. Apparent complex partial seizures in a bipolar patient after withdrawal of carbamazepine. J Clin Psychiatry. 1988;49:410–1.
123. Garrod AB. Treatise on the Nature and Treatment of Gout and Rheumatic Gout. London: Walton & Maberly; 1859.
124. Gaynes BN, Lux LJ, Lloyd SW, Hansen RA, Gartlehner G, Keener P, et al., editors. Nonpharmacologic Interventions for Treatment-Resistant Depression in Adults. Rockville, MD: US Agency for Healthcare Research and Quality, AHRQ Comparative Effectiveness Reviews, Report No. 11-EHC056-EF; 2011.
125. Geddes JR, Burgess S, Hawton K, Jamison K, Goodwin GM. Long-term lithium therapy for bipolar disorder: systematic review and meta-analysis of randomized controlled trials. Am J Psychiatry. 2004;161:217–22.
126. Geddes JR, Goodwin GM, Rendell J, Azorin JM, Cipriani A, Ostacher MJ, et al. Lithium plus valproate combination therapy versus monotherapy for relapse prevention in bipolar I disorder (BALANCE): a randomized open-label trial. Lancet. 2010;375:385–95.
127. Gelenberg AJ, Kane JM, Keller MB, Lavori P, Rosenbaum JF, Cole K, et al. Comparison of standard and low serum levels of lithium for maintenance treatment of bipolar disorder. N Engl J Med. 1989;321:1489–93.
128. George MS, Ward Jr HE, Ninan PT, Pollack M, Nahas Z, Anderson B, et al. Pilot study of vagus nerve stimulation (VNS) for treatment-resistant anxiety disorders. Brain Stimul. 2008;1:112–21.
129. Gershon S, Chengappa KNR, Malhi GS. Lithium specificity in bipolar illness: a classic agent for the classic disorder. Bipolar Disord. 2009;11 (Suppl 2):34–44.
130. Ghaemi SN, Ostacher MM, El-Mallakh RS, Borrelli D, Baldassano CF, Kelley ME, et al. Antidepressant discontinuation in bipolar depression: a Systematic Treatment Enhancement Program for Bipolar Disorder (STEP-BD) randomized clinical trial of long-term effectiveness and safety. J Clin Psychiatry. 2010;71:372–80.
131. Ghaemi SN, Wingo AP, Filkowski MA, Goodwin FK, Baldessarini RJ. Effectiveness of long-term antidepressant treatment in bipolar disorder: meta-analysis. Acta Psychiatr Scand. 2008;118:347–56.
132. Ghannoum M, Lavergne V, Yue CS, Ayoub P, Perreault MM, Roy L. Successful treatment of lithium toxicity with sodium polystyrene sulfonate: retrospective cohort study. Clin Toxicol. 2010;48:34–41.
133. Gibbons RD, Hur K, Brown CH, Mann JJ. Relationship between antiepileptic drugs and suicide attempts in patients with bipolar disorder. Arch Gen Psychiatry. 2009;66:1354–60.
134. Gijsman HJ, Geddes JR, Rendell JM, Nolen WA, Goodwin GM. Antidepressants for bipolar depression: systematic review of randomized, controlled trials. Am J Psychiatry. 2004;161:1537–47.
135. Gitlin MJ, Cochran SD, Jamison KR. Maintenance lithium treatment: side effects and compliance. J Clin Psychiatry. 1989;50:127–31.
136. Glazer WM, Sonnenberg JG, Reinstein MJ, Akers RF. Novel, point-of-care test for lithium levels: description and reliability. J Clin Psychiatry. 2004;65:652–5.
137. González-Pinto A, Reed C, Novick D, Bertsch J, Haro JM. Assessment of medication adherence in a cohort of patients with bipolar disorder. Pharmacopsychiatry. 2010;43: 263–70.
138. Goodwin FK, Fireman B, Simon GE, Hunkeler EM, Lee J, Revicki D. Suicide risk in bipolar disorder during treatment with lithium and divalproex. JAMA. 2003;290:1467–73.
139. Goodwin FK, Jamison KR, editors. Manic Depressive Illness. 2nd ed. New York: Oxford University Press; 2007.
140. Goodwin FK, Murphy DL, Dunner DL, Bunney Jr WE. Lithium response in unipolar versus bipolar depression. Am J Psychiatry. 1972;129:44–7.
141. Goodwin FJ, Whitham EA, Ghaemi SN. Maintenance treatment study designs in bipolar disorder: do they demonstrate that atypical neuroleptics (antipsychotics) are mood stabilizers? CNS Drugs. 2011;25:819–27.

142. Gottberg E, Grondin L, Reader TA. Acute effects of lithium on catecholamines, serotonin, and their major metabolites in discrete brain regions. J Neurosci Res. 1989;22:338–45.

143. Gould TD, Quiroz JA, Singh J, Zarate CA, Manji HK. Emerging experimental therapeutics for bipolar disorder: insights from the molecular and cellular actions of current mood stabilizers. Mol Psychiatry. 2004;9:734–55.

144. Grandjean EM, Aubry J-M. Lithium: updated human knowledge using an evidence-based approach (parts I, II, III). CNS Drugs. 2009;23:225–40; 331–49; 397–418.

145. Greenhill SD, Jones RS. Diverse antiepileptic drugs increase the ratio of background synaptic inhibition to excitation and decrease neuronal excitability in neurones of the rat entorhinal cortex in vitro. Neuroscience. 2010;167:456–74.

146. Grof P, Muller-Oerlinghausen B. Critical appraisal of lithium's efficacy and effectiveness in the last 60 years. Bipolar Disord. 2009;11 (Suppl 2):10–9.

147. Grünfeld JP, Rossier BC. Lithium nephrotoxicity revisited. Nat Rev Nephrol. 2009;5:270–6.

148. Grunze HC. Anticonvulsants in bipolar disorder. J Ment Health. 2010;19:127–41.

149. Gu Y, Huang LY. Gabapentin actions on N-methyl-D-aspartate receptor channels are protein kinase C-dependent. Pain. 2001;93:85–92.

150. Guzzetta F, Tondo L, Centorrino F, Baldessarini RJ. Lithium treatment reduces suicide risk in recurrent major depressive disorder. J Clin Psychiatry. 2007;68:380–3.

151. Haag H, Heidorn A, Haag M, Greil W. Sequence of affective polarity and lithium response: preliminary report on Munich sample. Prog Neuropsychopharmacol Biol Psychiatry. 1987;11:205–8.

152. Hammond WA. A Treatise on Diseases of the Nervous System. New York: D. Appleton & Co.; 1871. p. 381.

153. Hanlon LW, Romaine M. Lithium chloride as a substitute for sodium chloride in the diet; observations on its toxicity. J Am Med Assoc. 1949;139:688–92.

154. Hantouche EG, Akiskal HS, Lancrenon S, Chatenêt-Duchêne L. Mood stabilizer augmentation in apparently "unipolar" MDD: predictors of response in the naturalistic French national EPIDEP study. J Affect Disord. 2005;84:243–9.

155. Harden CL, Pennell PB, Koppel BS, Hovinga CA, Gidal B, Meador KJ, et al. Management issues for women with epilepsy: focus on pregnancy, an evidence-based review. Neurology. 2009;73:142–9.

156. Harden CL, Pennell PB, Koppel BS, Hovinga CA, Gidal B, Meador KJ, et al. Management issues for women with epilepsy: focus on pregnancy, an evidence-based review. Epilepsia. 2009;50:1247–55.

157. Huxley N, Parikh SU, Baldessarini RJ. Effectiveness of psychosocial treatments in bipolar disorder: State of the evidence. Harv Rev Psychiartry. 2000;8:126–40.

158. Harvey PD, Wingo AP, Burdick KE, Baldessarini RJ. Cognition and disability in bipolar disorder: lessons from schizophrenia research. Bipolar Disord. 2010;12:364–75.

159. Heit S, Nemeroff CB. Lithium augmentation of antidepressants in treatment-refractory depression. J Clin Psychiatry. 1998;59 (Suppl 6):28–34.

160. Hennen J, Baldessarini RJ. Reduced suicidal risk during treatment with clozapine: meta-analysis. Schizophr Res. 2005;73:139–45.

161. Henry TR. History of valproate in clinical neuroscience. Psychopharmacol Bull. 2003;37 (Suppl 2):5–16.

162. Herceg M, Muzinić L, Jukić V. Can we prevent blood dyscrasia (leucopenia, thrombocytopenia) and epileptic seizures induced by clozapine? Psychiatr Danub. 2010;22:85–9.

163. Hetmar O, Bolwig TG, Brun C, Ladefoged J, Larsen S, Rafaelsen OJ. Lithium: long-term effects on the kidney. Acta Psychiatr Scand. 1986;73:574–81.

164. Hetmar O, Clemmesen L, Ladefoged J, Rafaelsen OJ. Lithium: long-term effects on the kidney. Acta Psychiatr Scand. 1987;75:251–8.

165. Hilas O, Charnetski L. Lamotrigine-induced Stevens–Johnson syndrome. Am J Health Syst Pharm. 2007;64:273–5.

166. Holmes LB, Baldwin EJ, Smith CR, Habecker E, Glassman L, Wong SL, et al. Increased frequency of isolated cleft palate in infants exposed to lamotrigine during pregnancy. Neurology. 2008;70:2152–8.

167. Holtzheimer PE, Kelley ME, Gross RE, Filkowski MM, Garlow SJ, Barrocas A, et al. Subcallosal cingulate deep brain stimulation for treatment-resistant unipolar and bipolar depression. Arch Gen Psychiatry. 2012;69:150–8.
168. Howland RH, Shutt LS, Berman SR, Spotts CR, Denko T. Emerging use of technology for the treatment of depression and other neuropsychiatric disorders. Ann Clin Psychiatry. 2011;23:48–62.
169. Hu X, Wang J, Dong W, Fang Q, Hu L, Liu C. Meta-analysis of polycystic ovary syndrome in women taking valproate for epilepsy. Epilepsy Res. 2011;97:73–82.
170. Huot RL, Brennan PA, Stowe ZN, Plotsky PM, Walker EF. Negative affect in offspring of depressed mothers is predicted by infant cortisol levels at 6 months and maternal depression during pregnancy, but not postpartum. Ann N Y Acad Sci. 2004;1032:234–6.
171. Inoue T, Abekawa T, Nakagawa S, Suzuki K, Tanaka T, Kitaichi Y, et al. Long-term naturalistic follow-up of lithium augmentation: relevance to bipolarity. J Affect Disord. 2011;129:64–7.
172. Irvine RF. Inositol lipids in cell signaling. Curr Opin Cell Biol. 1992;4:212–9.
173. Ivković M, Damjanović A, Jovanović A, Cvetić T, Jasović-Gašić M. Lamotrigine versus lithium augmentation of antidepressant therapy in treatment-resistant depression: efficacy and tolerability. Psychiatr Danub. 2009;21:187–93.
174. Jefferson JW, Greist JH. Lithium and Bipolar Disorder: A Guide. Madison, WI: Madison Institute of Medicine; 2004.
175. Jentink J, Bakker MK, Nijenhuis CM, Wilffert B, De Jong-van den Berg LT. Does folic acid use decrease the risk for spina bifida after in utero exposure to valproic acid? Pharmacoepidemiol Drug Saf. 2010;19:803–7.
176. Johannessen SI, Landmark CJ. Antiepileptic drug interactions: principles and clinical implications. Curr Neuropharmacol. 2010;8:254–67.
177. Joffe H. Reproductive biology and psychotropic treatments in premenopausal women with bipolar disorder. J Clin Psychiatry. 2007;68 (Suppl 9):10–5.
178. Joffe RT, MacQueen GM, Marriott M, Trevor Young L. Prospective, longitudinal study of percentage of time spent ill in patients with bipolar I or bipolar II disorders. Bipolar Disord. 2004;6:62–6.
179. Jonathan-Ryves W, Dalton EC, Harwood AJ, Williams RS. GSK-3 activity in neocortical cells is inhibited by lithium but not carbamazepine or valproic acid. Bipolar Disord. 2005;7:260–5.
180. Jones RM, Arlidge J, Gillham R, Reagu S, van den Bree M, Taylor PJ. Efficacy of mood stabilizers in the treatment of impulsive or repetitive aggression: systematic review and meta-analysis. Br J Psychiatry. 2011;198:93–8.
181. Judd LL, Akiskal HS, Schettler PJ, Endicott J, Maser J, Solomon DA, et al. Long-term natural history of the weekly symptomatic status of bipolar I disorder. Arch Gen Psychiatry. 2002;59:530–7.
182. Kakkar AK, Rehan HS, Unni KE, Gupta NK, Chopra D, Kataria D. Comparative efficacy and safety of oxcarbazepine vs. divalproex sodium in the treatment of acute mania: pilot study. Eur Psychiatry. 2009;24:178–82.
183. Kammerer M, Brawek B, Freiman TM, Jackisch R, Feuerstein TJ. Effects of antiepileptic drugs on glutamate release from rat and human neocortical synaptosomes. Naunyn Schmiedebergs Arch Pharmacol. 2011;383:531–42.
184. Kammerer M, Rassner MP, Freiman TM, Feuerstein TJ. Effects of antiepileptic drugs on GABA release from rat and human neocortical synaptosomes. Naunyn Schmiedebergs Arch Pharmacol. 2011;384:47–57.
185. Kaplan EL, Meier P. Nonparametric estimation from incomplete observations. J Am Stat Assoc. 1958;53:457–81.
186. Katz RI, Chase TN, Kopin IJ. Evoked release of norepinephrine and serotonin from brain slides: inhibition by lithium. Science. 1968;162:466–7.
187. Keck Jr PE, Manji HK. Current and emerging treatment for acute mania and long-term prophylaxis for bipolar disorder, Chap. 77. In: Davis KL, Charney D, Coyle JT, Nemeroff C,

editors. Neuropsychopharmacology–Fifth Generation of Progress. New York: Lippincott Williams & Wilkins; 2008. p. 1009–118.

188. Keck Jr PR, McElroy SL. Clinical pharmacodynamics and pharmacokinetics of antimanic and mood-stabilizing medications. J Clin Psychiatry. 2002;63 (Suppl 4):3–11.

189. Keck Jr PE, McElroy SL, Bennett JA. Pharmacologic loading in the treatment of acute mania. Bipolar Disord. 2000;2:42–6.

190. Kemp DE, Ganocy SJ, Brecher M, Carlson BX, Edwards S, Eudicone JM, et al. Clinical value of early partial symptomatic improvement in the prediction of response and remission during short-term treatment trials in 3369 subjects with bipolar I or II depression. J Affect Disord. 2011;130:171–9.

191. Kennedy SH, Milev R, Giacobbe P, Ramasubbu R, Lam RW, Parikh SV, et al. Canadian Network for Mood and Anxiety Treatments (CANMAT) clinical guidelines for the management of major depressive disorder in adults: neurostimulation therapies. J Affect Disord. 2009;117 (Suppl 1):S44–53.

192. Ketter RA. Handbook of Diagnosis and Treatment of Bipolar Disorder. Arlington, VA: American Psychiatric Publishing; 2010.

193. Ketter TA, Citrome L, Wang PW, Culver JL, Srivastava S. Treatments for bipolar disorder: can number needed to treat/harm help inform clinical decisions? Acta Psychiatr Scand. 2011;123:175–89.

194. Ketter TA, Wang PW, Chandler RA, Alarcon AM, Becker OV, Nowakowska C, et al. Dermatology precautions and slower titration yield low incidence of lamotrigine treatment-emergent rash. J Clin Psychiatry. 2005;66:642–5.

195. Khalsa HMK, Salvatore P, Hennen J, Tohen M, Baldessarini RJ. Suicidal events and accidents in 216 first-episode bipolar-I disorder patients: predictive factors. J Affect Disord. 2008;106:179–84.

196. Kidd SK, Schneider JS. Protective effects of valproic acid on the nigrostriatal dopamine system in a 1-methyl-4-phenyl-1,2,3,6-tetrahydropyridine mouse model of Parkinson's disease. Neuroscience. 2011;194:189–94.

197. Koren G. Pharmacokinetics in pregnancy; clinical significance. J Popul Ther Clin Pharmacol. 2011;18:e523–7.

198. Kortenoeven ML, Li Y, Shaw S, Gaeggeler HP, Rossier BC, Wetzels JF, et al. Amiloride blocks lithium entry through the sodium channel thereby attenuating the resultant nephrogenic diabetes insipidus. Kidney Int. 2009;76:44–53.

199. Koukopoulos A, Sani G, Koukopoulos AE, Manfredi G, Pacchiarotti I, Girardi P. Melancholia agitata and mixed depression. Acta Psychiatr Scand. 2007;117 (Suppl 433):50–7.

200. Koukopoulos A, Serra G, Koukopoulos AE, Reginaldi D, Serra G. Sustained mood-stabilizing effect of memantine in the management of treatment resistant bipolar disorders: findings from a 12 month naturalistic trial. J Affect Disord. 2011;136:163–6.

201. Kovacsics CE, Gottesman II, Gould TD. Lithium's antisuicidal efficacy: elucidation of neurobiological targets using endophenotype strategies. Annu Rev Pharmacol Toxicol. 2009;49:175–98.

202. Kozma C. Neonatal toxicity and transient neurodevelopmental deficits following prenatal exposure to lithium: another clinical report and a review of the literature. Am J Med Genet A. 2005;132:441–4.

203. Krauthammer C, Klerman GL. Secondary mania: manic syndromes associated with antecedent physical illness or drugs. Arch Gen Psychiatry. 1978;35:1333–9.

204. Kukopulos A, Reginaldi D, Girardi P, Tondo L. Course of manic-depressive recurrences under lithium. Compr Psychiatry. 1975;16:517–24.

205. Kukuljan M, Vergara L, Stojilkovic SS. Modulation of the kinetics of inositol 1,4,5-trisphosphate-Induced $[Ca^{2+}]i$ oscillations by calcium entry in pituitary gonadotrophs. Biophys J. 1997;72:698–707.

206. Kushner SF, Khan A, Lane R, Olson WH. Topiramate monotherapy in the management of acute mania: results of four double-blind, placebo-controlled trials. Bipolar Disord. 2006;8:15–27.

207. Lam RW, Chan P, Wilkins-Ho M, Yatham LN. Repetitive transcranial magnetic stimulation for treatment-resistant depression: systematic review and meta-analysis. Can J Psychiatry. 2008;53:621–31.
208. Lambert PA, Carraz G, Borselli S, Carbel S. [Neuropsychotropic action of a new anti-epileptic agent: depamide (French)]. Ann Med Psychol (Paris). 1966;124:707–10.
209. Lasser RA, Baldessarini RJ. Thyroid hormones in depressive disorders: reappraisal of clinical utility. Harv Rev Psychiatry. 1997;4:291–305.
210. Lauterbach EC, Mendez MF. Psychopharmacological neuroprotection in neurodegenerative diseases. J Neuropsychiatry Clin Neurosci. 2011;23:242–60.
211. Lauterbach EC, Victoroff J, Coburn KL, Shillcutt SD, Doonan SM, Mendez MF. Psychopharmacological neuroprotection in neurodegenerative disease: assessing the preclinical data. J Neuropsychiatry Clin Neurosci. 2010;22:8–18.
212. Lazarus JH. Lithium and thyroid. Best Pract Res Clin Endocrinol Metab. 2009;23:723–33.
213. Lepkifker E, Iancu I, Horesh N, Strous RD, Kotler M. Lithium therapy for unipolar and bipolar depression among the middle-aged and older adult patient subpopulation. Depress Anxiety. 2007;24:571–6.
214. Levin GM, Grum C, Eisele G. Effect of over-the-counter dosages of naproxen sodium and acetaminophen on plasma lithium concentrations in normal volunteers. J Clin Psychopharmacol. 1998;18:237–40.
215. Levy RH, Mattson RH, Meldrum BS, Perucca E, editors. Antiepileptic Drugs. 5th ed. Philadelphia: Lippincott Williams & Wilkins; 2002.
216. Livingstone C, Rampes H. Lithium: review of its metabolic adverse effects. J Psychopharmacol. 2006;20:347–55.
217. Lorenzo LS, Vázquez GH, Zaratiegui RM, Baldessarini RJ. Antidepressants for bipolar disorder. J Affect Disord. Hum Psychopharmacol 2012; in press.
218. Machado-Vieira R, Manji HK, Zarate Jr CA. Role of lithium in the treatment of bipolar disorder: convergent evidence for neurotrophic effects as a unifying hypothesis. Bipolar Disord. 2009;11 (Suppl 2):92–109.
219. Machado-Vieira R, Salvadore G, Diaz-Granados N, Ibrahim L, Latov D, Wheeler-Castillo C, et al. New therapeutic targets for mood disorders. ScientificWorldJournal. 2010;13:713–26.
220. Mack A. Examination of the evidence for off-label use of gabapentin. J Manag Care Pharm. 2003;9:559–68.
221. MacNamara JO. Pharmacotherapy of the epilepsies, Chap. 21. In: Brunton LL, Chabner BA, Knollmann BC, editors. Goodman and Gilman's the Pharmacological Basis of Therapeutics. New York: McGraw Hill; 2011. p. 582–607.
222. Maj M, Starace F, Nolfe G, Kemali D. Minimum plasma lithium levels required for effective prophylaxis in DSM III bipolar disorder: a prospective study. Pharmacopsychiatry. 1986;19:420–3.
223. Manji HK, Moore GJ, Chen G. Bipolar disorder: leads from the molecular and cellular mechanisms of action of mood stabilizers. Br J Psychiatry. 2001;41(Suppl):s107–19.
224. Marangell LB, Dennehy EB, Wisniewski SR, Bauer MS, Miyahara S, Allen MH, et al. Case-control analyses of the impact of pharmacotherapy on prospectively observed suicide attempts and completed suicides in bipolar disorder. J Clin Psychiatry. 2008;69:916–22.
225. Marangell LB, Martinez M, Jurdi RA, Zboyan H. Neurostimulation therapies in depression: review of new modalities. Acta Psychiatr Scand. 2007;116:174–81.
226. Markham JA, Koenig JI. Prenatal stress: role in psychotic and depressive diseases. Psychopharmacology. 2011;214:89–106.
227. Mathew SJ, Manji HK, Charney DS. Novel drugs and therapeutic targets for severe mood disorders. Neuropsychopharmacology. 2008;33:2080–92.
228. Mazza M, Di Nicola M, Martinotti G, Taranto C, Pozzi G, Conte G, et al. Oxcarbazepine in bipolar disorder: critical literature review. Expert Opin Pharmacother. 2007;8:649–56.
229. McIntyre RS, Yoon J, Jerrell JM, Liauw SS. Aripiprazole for the maintenance treatment of bipolar disorder: review of available evidence. Neuropsychiatr Dis Treat. 2011;7:319–23.

230. McKnight RF, Adida M, Budge K, Stockton S, Boodwin GM, Geddes JR. Lithium toxicity profile: systematic review and meta-analysis. Lancet. 2012;379:721–8.
231. McNamara J. Pharmacotherapy of the epilepsies, Chap. 21. In: Brunton L, Chabner B, Knollman B, editors. Goodman and Gilman's the Pharmacological Basis of Therapeutics. 12th ed. New York: McGraw-Hill; 2011. p. 782–807.
232. Mellor H, Parker PJ. The extended protein kinase C superfamily. Biochem J. 1998;332:281–92.
233. Meltzer HY, Alphs L, Green AI, Altamura AC, Anand R, Bertoldi A, et al. Clozapine treatment for suicidality in schizophrenia. Arch Gen Psychiatry. 2003;60:82–91.
234. Mendels J. Role of lithium as an antidepressant. Mod Probl Pharmacopsychiatry. 1982;18:138–44.
235. Meyer JM. Pharmacotherapy of psychosis and mania, Chap. 16. In: Brunton L, Chabner B, Knollman B, editors. Goodman and Gilman's the Pharmacological Basis of Therapeutics. 12th ed. New York: McGraw-Hill; 2011. p. 417–55.
236. Meyer JM, Dollarhide A, Tuan IL. Lithium toxicity after switch from fosinopril to lisinopril. Int Clin Psychopharmacol. 2005;20:115–8.
237. Meunier H, Carraz G, Meunier Y, Eymard P, Aimard M. Propriétés pharmacodynamiques de l'acidee n-dipropylacetique. Therapie. 1963;18:435–8.
238. Miklowitz DJ. Fuctional impairment, stress, and psychosial intervention in bipolar disorder. Curr Psychiatry Rep. 2011;13:504–12.
239. Mirin SM, Schatzberg AF, Creasey DE. Hypomania and mania after withdrawal of tricyclic antidepressants. Am J Psychiatry. 1981;138:87–9.
240. Mischoulon D, Nierenberg AA, Kizilbash L, Rosenbaum JF, Fava M. Strategies for managing depression refractory to selective serotonin reuptake inhibitor treatment: a survey of clinicians. Can J Psychiatry. 2000;45:476–81.
241. Modell JG, Lenox RH, Weiner S. Inpatient clinical trial of lorazepam for the management of manic agitation. J Clin Psychopharmacol. 1985;5:109–13.
242. Mohr P, Rodriguez M, Slavíčková A, Hanka J. Application of vagus nerve stimulation and deep brain stimulation in depression. Neuropsychobiology. 2011;64:170–81.
243. Monk C, Fitelson EM, Werner E. Mood disorders and their pharmacological treatment during pregnancy: is the future child affected? Pediatr Res. 2011;69:3R–10R.
244. Moore CM, Demopulos CM, Henry ME, Steingard RJ, Zamvil L, Katic A, et al. Brain-to-serum lithium ratio and age: in vivo magnetic resonance spectroscopy study. Am J Psychiatry. 2002;159:1240–2.
245. Müller-Oerlinghausen B, Felber W, Berghöfer A, Lauterbach E, Ahrens B. Impact of lithium long-term medication on suicidal behavior and mortality of bipolar patients. Arch Suicide Res. 2005;9:307–19.
246. Müller-Oerlinghausen B, Lewitzka U. Lithium reduces pathological aggression and suicidality: a mini-review. Neuropsychobiology. 2010;62:43–9.
247. Mojtabai R, Olfson M. National trends in psychotropic medication polypharmacy in office-based psychiatry. Arch Gen Psychiatry. 2010;67:26–36.
248. Muzina DJ. Pharmacologic treatment of rapid cycling and mixed states in bipolar disorder: an argument for the use of lithium. Bipolar Disord. 2009;11 (Suppl 2):84–91.
249. Muzina DJ, Gao K, Kemp DE, Khalife S, Ganocy SJ, Chan PK, et al. Acute efficacy of divalproex sodium versus placebo in mood stabilizer-naive bipolar I or II depression: double-blind, randomized, placebo-controlled trial. J Clin Psychiatry. 2011;72:813–9.
250. Narayan V, Haddad PM. Antidepressant discontinuation manic states: critical review of the literature and suggested diagnostic criteria. J Psychopharmacol. 2011;25:306–13.
251. Newport DJ, Stowe ZN, Viguera AC, Calamaras MR, Juric S, Knight B, et al. Lamotrigine in bipolar disorder: efficacy during pregnancy. Bipolar Disord. 2008;10:432–6.
252. Nguyen HT, Sharma V, McIntyre RS. Teratogenesis associated with antibipolar agents. Adv Ther. 2009;26:281–94.
253. Novick DM, Swartz HA, Frank E. Suicide attempts in bipolar I and bipolar II disorder: review and meta-analysis of the evidence. Bipolar Disord. 2010;12:1–9.

254. Oedegaard KJ, Dilsaver SC, Hundal O, Riise T, Lund A, Akiskal HS, et al. Are migraine and bipolar disorders comorbid phenomena?: findings from a pharmacoepidemiological study using the norwegian prescription database. J Clin Psychopharmacol. 2011;31:734–9.
255. Okada M, Yoshida S, Zhu G, Hirose S, Kaneko S. Biphasic actions of topiramate on monoamine exocytosis associated with both soluble N-ethylmaleimide-sensitive factor attachment protein receptors and Ca^{2+}-induced Ca^{2+}-releasing systems. Neuroscience. 2005;134:233–46.
256. Okuma T. [History of investigation on the mood-stabilizing effect of carbamazepine in Japan (Japanese)]. Seishin Shinkeigaku Zasshi. 2002;104:647–55.
257. Okuma T, Kishimoto A, Inoue K, Matsumoto H, Ogura A. Anti-manic and prophylactic effects of carbamazepine (Tegretol) on manic depressive psychosis: preliminary report. Folia Psychiatr Neurol Jpn. 1973;27:283–97.
258. Oquendo MA, Galfalvy HC, Currier D, Grunebaum MF, Sher L, Sullivan GM, et al. Treatment of suicide attempters with bipolar disorder: randomized clinical trial comparing lithium and valproate in the prevention of suicidal behavior. Am J Psychiatry. 2011;168:1050–6.
259. Ösby U, Brandt L, Correia N, Ekbom A, Sparén P. Excess mortality in bipolar and unipolar disorder in Sweden. Arch Gen Psychiatry. 2001;58:844–50.
260. Ouyang DY, Ji YH, Saltis M, Xu LH, Zhang YT, Zha QB, et al. Valproic acid synergistically enhances the cytotoxicity of gossypol in DU145 prostate cancer cells: an iTRTAQ-based quantitative proteomic analysis. J Proteomics. 2011;74:2180–93.
261. Ozerdem A, Schmidt ME, Manji HK, Risinger RC, Potter WZ. Chronic lithium administration enhances noradrenergic responses to intravenous administration of the alpha-2 antagonist idazoxan in healthy volunteers. J Clin Psychopharmacol. 2004;24:150–4.
262. Pacchiarotti I, Valentí M, Bonnin CM, Rosa AR, Murru A, Kotzalidis GD, et al. Factors associated with initial treatment response with antidepressants in bipolar disorder. Eur Neuropsychopharmacol. 2011;21:362–9.
263. Pacchiarotti I, Valentí M, Colom F, Rosa AR, Nivoli AM, Murru A, et al. Differential outcome of bipolar patients receiving antidepressant monotherapy versus combination with an antimanic drug. J Affect Disord. 2011;129:321–6.
264. Pacchiarotti I, Mazzarini L, Kotzalidis GD, Valentí M, Nivoli AM, Sani G, et al. Mania and depression: mixed, not stirred. J Affect Disord. 2011;133:105–13.
265. Pasquali L, Busceti CL, Fulceri F, Paparelli A, Fornai F. Intracellular pathways underlying the effects of lithium. Behav Pharmacol. 2010;21:473–92.
266. Pasquini M, Picardi A, Speca A, Orlandi V, Tarsitani L, Morosini P, et al. Combining an SSRI with an anticonvulsant in depressed patients with dysphoric mood: open study. Clin Pract Epidemiol Ment Health. 2007;3:3–15.
267. Patorno E, Bohn RL, Wahl PM, Avorn J, Patrick AR, Liu J, Scheeweiss S. Anticonvulsant medication and the risk of suicide, attempted suicide, or acident death. JAMA. 2010;303:1401–9.
268. Paykel ES, Abbott R, Morriss R, Hayhurst H, Scott J. Sub-syndromal and syndromal symptoms in the longitudinal course of bipolar disorder. Br J Psychiatry. 2006;189:118–23.
269. Peng L, Li B, Du T, Wang F, Hertz L. Does conventional anti-bipolar and antidepressant drug therapy reduce NMDA-mediated neuronal excitation by downregulating astrocytic GluK2 function? Pharmacol Biochem Behav. 2012;100:712–25.
270. Pennell PB, Peng L, Newport DJ, Ritchie JC, Koganti A, Holley DK, et al. Lamotrigine in pregnancy: clearance, therapeutic drug monitoring, and seizure frequency. Neurology. 2008;70:2130–6.
271. Pérez J, Baldessarini RJ, Cruz N, Salvatore P, Vieta E. Andrés Piquer-Arrufat (1711–1772): contributions of an eighteenth-century Spanish physician to the concept of manic-depressive illness. Harv Rev Psychiatry. 2011;19:68–77.
272. Perlis RH, Sachs GS, Lafer B, Otto MW, Faraone SV, Kane JM, et al. Effect of abrupt change from standard to low serum levels of lithium: reanalysis of double-blind lithium maintenance data. Am J Psychiatry. 2002;159:1155–9.
273. Perris C. Distinction between unipolar and bipolar mood disorders: 25-years perspective. Encéphale. 1992;18 (Suppl 1):9–13.

274. Perrott J, Murphy NG, Zed PJ. L-carnitine for acute valproic acid overdose: systematic review of published cases. Ann Pharmacother. 2010;44:1287–93.

275. Pichot P. Tracing the origins of bipolar disorder. J Affect Disord. 2006;96:145–8.

276. Pompili M, Baldessarini RJ. Risk of suicidal behavior with antiepileptic drugs. Nat Rev Neurol. 2010;6:651–3.

277. Popovic D, Reinares M, Amann B, Salmero M, Vieta E. Number needed to treat analyses of drugs used for maintenance treatment of bipolar disorder. Psychopharmacology. 2011;213:657–67.

278. Popovic D, Reinares M, Goikolea HM, Bonnin CM, Gonzalez-Pinto A, Vieta E. Polarity index of pharmacological agents used for maintenance treatment of bipolar disorder. Eur Neuropsychopharmacol. 2012;22:339–46.

279. Post RM, Altshuler LL, Frye MA, Suppes T, Keck Jr PE, McElroy SL, et al. Complexity of pharmacologic treatment required for sustained improvement in outpatients with bipolar disorder. J Clin Psychiatry. 2010;71:1176–86.

280. Post RM, Denicoff KD, Leverich GS, Altshuler LL, Frye MA, Suppes TM, et al. Morbidity in 258 bipolar outpatients followed for 1 year with daily prospective ratings on the NIMH life chart method. J Clin Psychiatry. 2003;64:680–90.

281. Post RM, Frye MA, Denicoff KD, Leverich GS, Dunn RT, Osuch EA, et al. Emerging trends in the treatment of rapid cycling bipolar disorder: selected review. Bipolar Disord. 2000;2:305–15.

282. Post RM, Ketter TA, Uhde T, Ballenger JC. Thirty years of clinical experience with carbamazepine in the treatment of bipolar illness: principles and practice. CNS Drugs. 2007;21:47–71.

283. Post RM, Leverich GS, Kupka RW, Keck Jr PE, McElroy SL, Altshuler LL, et al. Early-onset bipolar disorder and treatment delay are risk factors for poor outcome in adulthood. J Clin Psychiatry. 2010;71:864–72.

284. Quiroz JA, Gould TD, Manji HK. Molecular effects of lithium. Mol Interv. 2004;4:259–72.

285. Rao JS, Rapoport SI. Mood-stabilizers target the brain arachidonic acid cascade. Curr Mol Pharmacol. 2009;2:207–14.

286. Rapiejko PJ, Northup JK, Evans T, Brown JE, Malbon CC. G-proteins of fat-cells: role in hormonal regulation of intracellular inositol-1,4,5-trisphosphate. Biochem J. 1986;240:35–40.

287. Rapoport SI. Arachidonic acid and the brain. J Nutr. 2008;138:251S–20.

288. Rapoport SI, Basselin M, Kim HW, Rao JS. Bipolar disorder and mechanisms of action of mood stabilizers. Brain Res Rev. 2009;61:185–209.

289. Rosa AR, Fountoulakis K, Siamouli M, Gonda X, Vieta E. Is anticonvulsant treatment of mania a class effect? Data from randomized clinical trials. CNS Neurosci Ther. 2011;17:167–77.

290. Rybakowski JK. Bipolarity and inadequate response to antidepressant drugs: clinical and psychopharmacological perspective. J Affect Disord. 2012;136:e13–9.

291. Sachs GS, Nierenberg AA, Calabrese JR, Marangell LB, Wisniewski SR, Gyulai L, et al. Effectiveness of adjunctive antidepressant treatment for bipolar depression. N Engl J Med. 2007;356:1711–22.

292. Salvatore P, Baldessarini RJ, Centorrino F, Egli S, Albert M, Gerhard A, et al. Weygandt's, The Manic-Depressive Mixed States: translation and commentary on its significance in the evolution of the concept of bipolar manic-depressive disorder. Harv Rev Psychiatry. 2002;10:255–75.

293. Salvatore P, Baldessarini RJ, Tohen M, Khalsa HMK, Perez Sanchez-Toledo J, Zarate Jr CA, et al. Two-year stability of DSM-IV diagnoses in 500 first-episode psychotic disorder patients. J Clin Psychiatry. 2009;70:458–66.

294. Salvatore P, Baldessarini RJ, Tohen M, Khalsa HM, Sanchez-Toledo JP, Zarate Jr CA, et al. Two-year stability of ICD-10 diagnoses in 500 first-episode psychotic disorder patients. J Clin Psychiatry. 2011;72:183–93.

295. Sanches M, Newberg AR, Soares JC. Emerging drugs for bipolar disorder. Expert Opin Emerg Drugs. 2010;15:453–66.

296. Sanchez-Moreno J, Martinez-Aran A, Tabarés-Seisedos R, Torrent C, Vieta E, Ayuso-Mateos JL. Functioning and disability in bipolar disorder: an extensive review. Psychotherapy Psychosom. 2009;28:285–97.

297. Sarris J, Mischoulon D, Schweitzer I. Omega-3 for bipolar disorder: meta-analyses of use in mania and bipolar depression. J Clin Psychiatry. 2012;73:81–6.

298. Sastre E, Nicolay A, Bruguerolle B, Portugal H. Effect of lithium on norepinephrine metabolic pathways. Life Sci. 2005;77:758–67.

299. Schioldann J. "On periodical depressions and their pathogenesis" by Carl Lange (1886). Hist Psychiatry. 2011;22:108–30.

300. Schöttle D, Huber CG, Bock T, Meyer TD. Psychotherapy for bipolar disorder: review of the most recent studies. Curr Opin Psychiatry. 2011;24:549–55.

301. Sedler MJ. Falret's discovery: the origin of the concept of bipolar affective illness. Am J Psychiatry. 1983;140:1127–33.

302. Segredon I, Xenitides K, Panagiotopoulos M, Bochtsou V, Antoniadou O, Livaditis M. Group psychosocial interventions for adults with schizophrenia and bipolar illness: the evidence base in light of publications between 1985 and 2006. Int J Soc Psychitry. 2012;58:229–38.

303. Seo HJ, Chiesa A, Lee SJ, Patkar AA, Han C, Masand PS, et al. Safety and tolerability of lamotrigine: results from 12 placebo-controlled clinical trials and clinical implications. Clin Neuropharmacol. 2011;34:3947.

304. Serretti A, Drago A, De Ronchi D. Lithium pharmacodynamics and pharmacogenetics: focus on inositol mono phosphatase (IMPase), inositol polyphosphatase (IPPase) and glycogen synthase kinase-3-beta (GSK-3-beta). Curr Med Chem. 2009;16:1917–48.

305. Schou M. Forty years of lithium treatment. Arch Gen Psychiatry. 1997;54:9–15.

306. Shaldubina A, Agam G, Belmaker RH. Mechanism of lithium action: state of the art, ten years later. Prog Neuropsychopharmacol Biol Psychiatry. 2001;25:855–66.

307. Shaltiel G, Mark S, Kofman O, Belmaker RH, Agam G. Effect of valproate derivatives on human brain myoinositol-1-phosphate (MIP) synthase activity and amphetamine-induced rearing. Pharmacol Rep. 2007;59:402–7.

308. Shelton RC. Mood-stabilizing drugs in depression. J Clin Psychiatry. 1999;60 (Suppl 5): 37–42.

309. Shorter E. History of lithium therapy. Bipolar Disord. 2009;11 (Suppl 2):4–9.

310. Sidor MM, MacQueen GM. Antidepressants for the acute treatment of bipolar depression: systematic review and meta-analysis. J Clin Psychiatry. 2011;72:156–67.

311. Siegel AJ, Baldessarini RJ, Klepser MB, McDonald JC. Primary and drug-induced disorders of water homeostasis in psychiatric patients: principles of diagnosis and management. Harv Rev Psychiatry. 1998;6:190–200.

312. Singh R, Mukhopadhyay K. Survival analysis in clinical trials: basics and must-know areas. Perspect Clin Res. 2011;2:145–8.

313. Sirivella S, Gielchinsky I. Surgery of the Ebstein's anomaly: early and late outcomes. J Card Surg. 2011;26:227–33.

314. Smith LA, Cornelius VR, Azorin JM, Perugi G, Vieta E, Young AH, et al. Valproate for the treatment of acute bipolar depression: systematic review and meta-analysis. J Affect Disord. 2010;122:1–9.

315. Smith FE, Cousins DA, Thelwall PE, Ferrier IN, Blamire AM. Quantitative lithium magnetic resonance spectroscopy in the normal human brain on a 3-T clinical scanner. Magn Reson Med. 2011;66:945–9.

316. Stoll AL, Severus WE, Freeman MP, Rueter S, Zboyan HA, Diamond E, et al. Omega-3 fatty acids in bipolar disorder: preliminary double-blind, placebo-controlled trial. Arch Gen Psychiatry. 1999;56:407–12.

317. Stoner SC, Nelson LA, Lea JW, Marken PA, Sommi RW, Dahmen MM. Historical review of carbamazepine for the treatment of bipolar disorder. Pharmacotherapy. 2007;27:68–88.

318. Soufia M, Aoun J, Gorsane MA, Krebs MO. [SSRIs and pregnancy: a review of the literature (French)]. Encéphale. 2010;36:513–6.

319. Suppes T, Baldessarini RJ, Faedda GL, Tohen M. Risk of recurrence following discontinuation of lithium treatment in bipolar disorder. Arch Gen Psychiatry. 1991;48:1082–8.

320. Svendsen KB, Bech JN, Sørensen TB, Pedersen EB. A comparison of the effects of etodolac and ibuprofen on renal haemodynamics, tubular function, renin, vasopressin and urinary excretion of albumin and alpha-glutathione-S-transferase in healthy subjects: placebo-controlled cross-over study. Eur J Clin Pharmacol. 2000;56:383–8.
321. Swann AC. The strong relationship between bipolar and substance-use disorder. Ann N Y Acad Sci. 2010;1187:276–93.
322. Swartz HA, Thase ME. Pharmacotherapy for the treatment of acute bipolar II depression: current evidence. J Clin Psychiatry. 2011;72:356–66.
323. Tamura M, Sajo M, Kakita A, Matsuki N, Koyama R. Prenatal stress inhibits neuronal maturation through downregulation of mineralocorticoid receptors. J Neurosci. 2011;31:11505–14.
324. Tariot PN, Schneider LS, Cummings J, Thomas RG, Raman R, Jakimovich LJ, et al. Chronic divalproex sodium to attenuate agitation and clinical progression of Alzheimer disease. Arch Gen Psychiatry. 2011;68:853–61.
325. Tarr GP, Glue P, Herbison P. Comparative efficacy an acceptability of mood-stabilizer and second-generation antipsychotic monotherapy for acute mania: systematic review and meta-analysis. J Affect Disord. 2011;134:14–9.
326. Thase ME, Corya SA, Osuntokun O, Case M, Henley DB, Sanger TM, et al. Randomized, double-blind comparison of olanzapine/fluoxetine combination, olanzapine, and fluoxetine in treatment-resistant major depressive disorder. J Clin Psychiatry. 2007;68:224–36.
327. Thomas SP, Nandhra HS, Jayaraman A. Systematic review of lamotrigine augmentation of treatment resistant unipolar depression (TRD). J Ment Health. 2010;19:168–75.
328. Tomson T, Battino D. Pharmacokinetics and therapeutic drug monitoring of newer antiepileptic drugs during pregnancy and the puerperium. Clin Pharmacokinet. 2007;46:209–19.
329. Tomson T, Battino D. Teratogenic effects of antiepileptic medications. Neurol Clin. 2009;27:993–1002.
330. Trede K, Viguera AC, Bottéro A, Baldessarini RJ. Treatise on insanity in pregnant, postpartum, and lactating women by Louis-Victor Marcé (1858): commentary. Harv Rev Psychiatry. 2009;17:157–65.
331. Tohen M, Zarate Jr CA, Hennen J, Khalsa HM, Strakowski SM, Gebre-Medhin P, et al. The McLean-Harvard First-Episode Mania Study: prediction of recovery and first recurrence. Am J Psychiatry. 2003;160:2099–107.
332. Tondo L, Baldessarini RJ, Floris G, Tondo L, Baldessarini RJ. Long-term clinical effectiveness of lithium maintenance treatment in types I and II bipolar disorders. Br J Psychiatry. 2001;178 (Suppl 40):184–90.
333. Tondo L, Baldessarini RJ, Vázquez G, Lepri B, Visioli C. Clinical responses to treatment in acutely depressed patients with bipolar versus unipolar major affective disorders. Acta Psychiatr Scand. 2012; in press.
334. Tondo L, Hennen J, Baldessarini RJ. Reduced suicide risk with long-term lithium treatment in major affective illness: meta-analysis. Acta Psychiatr Scand. 2001;104:163–72.
335. Tondo L, Hennen J, Baldessarini RJ. Meta-analysis of treatment responses of rapid-cycling and non-rapid-cycling bipolar disorder patients. Acta Psychiatr Scand. 2003;104:4–14.
336. Tondo L, Isacsson G, Baldessarini RJ. Suicide in bipolar disorder: risk and prevention. CNS Drugs. 2003;17:491–511.
337. Tondo L, Lepri B, Baldessarini RJ. Risks of suicidal ideation, attempts and suicides among 2826 men and women with types I and II bipolar, and recurrent major depressive disorders. Acta Psychiatr Scand. 2007;116:419–28.
338. Tondo L, Lepri B, Cruz N, Baldessarini RJ. Age at onset in 3014 Sardinian bipolar and major depressive disorder patients. Acta Psychiatr Scand. 2010;121:446–52.
339. Tondo L, Vázquez G, Baldessarini RJ. Mania associated with antidepressant-treatment: comprehensive meta-analytic review. Acta Psychiatr Scand. 2010;121:404–14.
340. Trede K, Salvatore P, Baethge C, Gerhard A, Maggini C, Baldessarini RJ. Manic-depressive illness: evolution in Kraepelin's textbook, 1883–1926. Harv Rev Psychiatry. 2005;13:155–78.
341. Tripathi BM, Majumder P. Lactating mother and psychotropic drugs. Mens Sana Monogr. 2010;8:83–95.

342. Tsai AC, Rosenlicht NZ, Jureidini JN, Parry PI, Spielman GI, Healy D. Systematic search of the literature reveals limited evidence to support use of aripiprazole, a second-generation antipsychotic medication, in maintenance therapy of bipolar disorder, despite widespread use. PLoS Med. 2011;8:e1000434.

343. Undurraga J, Baldessarini RJ, Valenti M, Pacchiarotti I, Tondo L, Vázquez G, Vieta E. Bipolar depression: clinical correlates of receiving antidepressants. J Affect Disord. 2012; 139:89–93.

344. Underwood TW, Frye CB. Drug-induced pancreatitis. Clin Pharm. 1993;12:440–8.

345. Vajda FJ, Graham J, Roten A, Lander CM, O'Brien TJ, Eadie M. Teratogenicity of the newer antiepileptic drugs: the Australian experience. J Clin Neurosci. 2012;19:57–9.

346. Valenti M, Benabarre A, Garcia-Amador M, Molina O, Bernardo M, Vieta E. Electroconvulsive therapy in the treatment of mixed states in bipolar disorder. Eur Psychiatry. 2008;23:53–6.

347. Van Amelsvoort T, Bakshi R, Devaux CB, Schwabe S. Hyponatremia associated with carbamazepine and oxcarbazepine therapy: review. Epilepsia. 1994;35:181–8.

348. van der Loos ML, Mulder P, Hartong EG, Blom MB, Vergouwen AC, van Noorden MS, et al. Long-term outcome of bipolar depressed patients receiving lamotrigine as add-on to lithium with the possibility of the addition of paroxetine in nonresponders: randomized, placebo-controlled trial with a novel design. Bipolar Disord. 2011;13:111–7.

349. van Melick EJ, Wilting I, Souverein PC, Egberts TC. Differences in lithium use patterns in the Netherlands: comparing middle-aged and older patients in a database study. Am J Geriatr Pharmacother. 2012 [Epub ahead of print, March 12].

350. Vasudev K, Macritchie K, Geddes J, Watson S, Young A. Topiramate for acute affective episodes in bipolar disorder. Cochrane Database Syst Rev. 2008;1:CD003384.

351. Vasudev A, Macritchie K, Watson S, Geddes JR, Young AH. Oxcarbazepine in the maintenance treatment of bipolar disorder. Cochrane Database Syst Rev. 2008;1:CD005171.

352. Vázquez G, Baldessarini RJ, Yildiz A, Tamayo J, Tondo L, Salvatore P. Multi-site international collaborative clinical trials in mania: commentary. Int J Neuropsychopharmacol. 2011;14:1013–6.

353. Vázquez G, Tondo L, Baldessarini RJ. Comparison of antidepressant responses in patients with bipolar vs. unipolar depression: meta-analytic review. Pharmacopsychiatry. 2011;44: 21–6.

354. Ventriglio A, Vincenti A, Centorrino F, Talamo A, Fitzmaurice G, Baldessarini RJ. Use of mood-stabilizers for hospitalized adult psychotic and bipolar disorder patients. Int Clin Psychopharmacol. 2011;25:88–95.

355. Vestergaard P. Skeletal effects of central nervous system active drugs: anxiolytics, sedatives, antidepressants, lithium and neuroleptics. Curr Drug Saf. 2008;3:185–9.

356. Vieta E. Bipolar mixed states and their treatment. Expert Rev Neurother. 2005;5:63–8.

357. Vieta E, Cruz N, García-Campayo J, de Arce R, Manuel Crespo J, Vallès V, et al. Double-blind, randomized, placebo-controlled prophylaxis trial of oxcarbazepine as adjunctive treatment to lithium in the long-term treatment of bipolar I and II disorder. Int J Neuropsychopharmacol. 2008;11:445–52.

358. Vieta E, Günther O, Locklear J, Ekman M, Miltenburger C, Chatterton ML, et al. Effectiveness of psychotropic medications in the maintenance phase of bipolar disorder: meta-analysis of randomized controlled trials. Int J Neuropsychopharmacol. 2011;14:1029–49.

359. Vieta E, Locklear J, Günther O, Ekman M, Miltenburger C, Chatterton ML, et al. Treatment options for bipolar depression: systematic review of randomized, controlled trials. J Clin Psychopharmacol. 2010;30:579–90.

360. Vieta E, Owen R, Baudelet C, McQuade RD, Sanchez R, Marcus RN. Assessment of safety, tolerability and effectiveness of adjunctive aripiprazole to lithium orvalproate in bipolar mania: 46-week, open-label extension following a 6-week double-blind study. Curr Med Res Opin. 2010;26:1485–96.

361. Vigo DV, Baldessarini RJ. Anticonvulsants in the treatment of major depressive disorder: an overview. Harv Rev Psychiatry. 2009;17:231–41.

362. Viguera AC, Baldessarini RJ, Hegarty JM, Van Kammen D, Tohen M. Risk of discontinuing maintenance medication in schizophrenia. Arch Gen Psychiatry. 1997;54:49–55.
363. Viguera AC, Cohen LS, Baldessarini RJ, Nonacs R. Managing bipolar disorder during pregnancy: weighing the risks and benefits. Can J Psychiatry. 2002;47:426–36.
364. Viguera AC, Cohen LS, Whitfield T, Reminick AM, Bromfield E, Baldessarini RJ. Perinatal use of anticonvulsants: differences in attitudes and recommendations among neurologists and psychiatrists. Arch Womens Ment Health. 2010;13:175–8.
365. Viguera AC, Newport DJ, Ritchie J, Stowe Z, Whitfield T, Mogielnicki J, et al. Lithium in breast milk and nursing infants: clinical implications. Am J Psychiatry. 2007;164:342–5.
366. Viguera AC, Nonacs R, Cohen LS, Tondo L, Murray A, Baldessarini RJ. Risk of discontinuing lithium maintenance in pregnant vs. nonpregnant women with Bipolar Disorders. Am J Psychiatry. 2000;157:179–84.
367. Viguera AC, Whitfield T, Baldessarini RJ, Newport DJ, Stowe Z, Reminick A, et al. Risk of recurrence in women with bipolar disorder during pregnancy: prospective study of mood stabilizer discontinuation. Am J Psychiatry. 2007;164:1817–24.
368. Weller M, Gorlia T, Cairncross JG, van den Bent MJ, Mason W, Belanger K, et al. Prolonged survival with valproic acid use in the EORTC/NCIC temozolomide trial for glioblastoma. Neurology. 2011;77:1156–64.
369. Yatham LN, Maj M, editors. Bipolar Disorder. Hoboken, NJ: Wiley-Blackwell; 2010.
370. Yeragani VK, Gershon S. Hammond and lithium: historical update. Biol Psychiatry. 1986;21:1101–2.
371. Yerby MS. Teratogenicity and antiepileptic drugs: potential mechanisms. Int Rev Neurobiol. 2008;83:181–204.
372. Yildiz A, Guleryuz S, Ankerst DP, Ongür D, Renshaw PF. Protein kinase C inhibition in the treatment of mania: a double-blind, placebo-controlled trial of tamoxifen. Arch Gen Psychiatry. 2008;65:255–63.
373. Yildiz A, Vieta E, Leucht S, Baldessarini RJ. Efficacy of antimanic treatments: meta-analysis of randomized, controlled trials. Neuropsychopharmacology. 2011;36:375–89.
374. Yildiz A, Vieta E, Tohen M, Baldessarini RJ. Factors modifying drug and placebo responses in randomized trials for bipolar mania. Int J Neuropsychopharmacol. 2011;14:863–75.
375. Yonkers KA, Vigod S, Ross LE. Diagnosis, pathophysiology, and management of mood disorders in pregnant and postpartum women. Obstet Gynecol. 2011;117:961–77.
376. Wadzinski J, Franks R, Roane D, Bayard M. Valproate-associated hyperammonemic encephalopathy. J Am Board Fam Med. 2007;20:499–502.
377. Weissel M. [Administration of thyroid hormones in therapy of psychiatric illnesses (German)]. Acta Med Austriaca. 1999;26:129–31.
378. Wijeratne C, Draper B. Reformulation of current recommendations for target serum lithium concentration according to clinical indication, age and physical comorbidity. Aust N Z J Psychiatry. 2011;45:1026–32.
379. Williams RS, Cheng L, Mudge AW, Harwood AJ. Common mechanism of action for three mood-stabilizing drugs. Nature. 2002;417:292–5.
380. Wingo AP, Harvey PD, Baldessarini RJ. Effects of lithium on cognitive performance: meta-analysis. J Clin Psychiatry. 2009;70:1588–97.
381. Wilting I, Egberts AC, Movig KL, Laarhoven JH, Heerdink ER, Nolen WA. Association between concomitant use of serotonergic antidepressants and lithium-induced polyuria: multicenter medical chart review study. Pharmacopsychiatry. 2008;41:129–33.
382. Wong MM. Management of bipolar II disorder. Indian J Psychol Med. 2011;33:18–28.
383. World Health Organization (WHO). International Classification of Diseases and Related Health Problems, tenth revision (ICD-10). Geneva: World Health Organization; 1992.
384. Wyffels L, Muccioli GG, De Bruyne S, Moerman L, Sambre J, Lambert DM, et al. Synthesis, in vitro and in vivo evaluation, and radiolabeling of aryl anandamide analogues as candidate radioligands for in vivo imaging of fatty acid amide hydrolase in the brain. J Med Chem. 2009;52:4613–22.

385. Xu X, Müller-Taubenberger A, Adley KE, Pawolleck N, Lee VW, Wiedemann C, et al. Attenuation of phospholipid signaling provides a novel mechanism for the action of valproic acid. Eukaryot Cell. 2007;6:899–906.
386. Zarate Jr C, Machado-Vieira R, Henter I, Ibrahim L, Diazgranados N, Salvadore G. Glutamatergic modulators: the future of treating mood disorders? Harv Rev Psychiatry. 2010;18:293–303.
387. Zarate CA, Manji HK. Protein kinase C inhibitors: rationale for use and potential in the treatment of bipolar disorder. CNS Drugs. 2009;23:569–82.
388. Zarate Jr CA, Singh JB, Carlson PJ, Quiroz J, Jolkovsky L, Luckenbaugh DA, et al. Efficacy of a protein kinase C inhibitor (tamoxifen) in the treatment of acute mania: a pilot study. Bipolar Disord. 2007;9:561–70.
389. Zarin DA, Pass TM. Lithium and the single episode. When to begin long-term prophylaxis for bipolar disorder. Med Care. 1987;25(Suppl 12):S76–84.

Chapter 4
Antidepressant-Antianxiety Agents

Melancholy, indeed, should be diverted by every means but drinking.

Samuel Johnson

Introduction

Background

The complex title of this chapter reflects several basic points about this important class of medicines. They are now the most widely used psychotropic drugs, and even more in general medicine than in psychiatry [171, 384, 576]. Their origins in the 1950s were aimed specifically at a relatively complex group of depressive disorders prior to consolidation and broad acceptance of the concept "major depressive disorder." Over the past half-century, indications for the use of these nominally "antidepressant" drugs have broadened extraordinarily. Some agents in this category, old and new, could well have been developed primarily as antianxiety agents. Indeed some that are used internationally as antidepressants are approved by the US FDA only for anxiety disorders (e.g., fluvoxamine and milnacipran). The broad use of such drugs includes various types of clinical depression, anxiety-associated disorders, attention disorders, some pain syndromes, and even urinary incontinence.

The wide use of antidepressants has arisen along with modern broadening of the concept of clinical depression to replace disorders previously considered to be limited to typically severe, melancholic, or sometimes psychotic forms of depression. Modern "major depression" includes disorders formerly labeled "psychoneuroses" and not always readily differentiated from anxiety disorders and reactions to stressors. These dissimilar types of disorders with depressive symptoms became combined into an expanded concept of "major depressive disorder," as developed in the late 1970s and sustained since then, as reflected in the standard diagnostic systems of the American Psychiatric Association [APA] [12] and the World Health

R.J. Baldessarini, *Chemotherapy in Psychiatry: Pharmacologic Basis of Treatments for Major Mental Illness*, DOI 10.1007/978-1-4614-3710-9_4,
© Springer Science+Business Media New York 2013

Organization [WHO] [605]. The concept can include episodes of major depression encountered in bipolar disorders, but distinctions between bipolar and unipolar depression nosologically and therapeutically remain controversial and unsettled, as is discussed in Chapter 3. The development of a major depression concept defined a huge potential market for drugs with efficacy against depressions ranging from mild to severe, as well as against anxiety symptoms that commonly accompany depressive symptoms and are ubiquitous components of many other psychiatric disorders. Development of relatively safe and easier-to-use antidepressants following the introduction of the first serotonin-reuptake inhibitors (SRIs) zimelidine (withdrawn for toxicity) and then fluoxetine in the late 1980s led to an explosive expansion of the use of modern antidepressants by a wide range of nonpsychiatric as well as psychiatric clinicians in the 1990s and 2000s [33, 35, 238].

A problem associated with, and possibly related to, the expansive application of modern antidepressant drugs for a broad range of illnesses is that their efficacy has not been easy to demonstrate, particularly in depressive illnesses of mild or moderate intensity. Indeed, drug-placebo differences (effect-size) for this and other types of psychotropic agents have been declining for several decades [295, 570]. Some recent reviews suggest that the efficacy of antidepressants in acute major depression may be quite limited and barely distinguishable from placebo treatment, especially illnesses of mild initial severity. Such cases are likely to be heavily represented in modern, placebo-controlled trials among ambulatory patient-subjects as increasing difficulties of subject-recruitment are dealt with [180, 262, 300, 379]. It is also striking that success in demonstrating statistical differences between drug and placebo often appears to depend more on responses to placebo than to active antidepressant drugs [300, 449, 451, 570]. Efforts to pool data from controlled trials in major depression, and more generally trials of other classes of drugs, are frustrated by lack of direct access to raw data of commercial trials submitted for regulatory review, by selective publication of findings that favor pharmaceutical products, and by inconsistent and incomplete access to some unpublished trials, especially those in which antidepressants have failed to outperform placebo and those conducted without industrial support and so not under as strict regulatory control [262, 300, 568]. Efforts to evaluate long-term, potentially prophylactic, effects of continuous antidepressant treatment also are limited by probable selection biases that limit generalization of observed effects. Notably, it is common to conduct initial, short-term trials in acute illnesses before moving on to a longer-term continuation trials. Such designs are attractive and seem to be an efficient use of patient-subjects, study personnel, and costs, but they also "enrich" sampling of patients who are responsive to a particular treatment prior to longer-term treatment that is usually associated with discontinuation of an effective treatment [582]. That is, such "enriched" trial designs involve effects of discontinuing a treatment already shown to be effective, and the findings may or may not be a fair representation of long-term prophylactic effects. Notably, long-term trials of antidepressants for anxiety disorders are far less common than for depression, leaving considerable gaps in the research basis of long-term treatment. Overall, evidence of limited or variable short-term efficacy in major depressive disorder, and very limited proof of long-term prophylactic effects, as well as limited consideration of the entire range of potential indications may seem surprising for these extraordinarily widely employed psychotropic agents.

Discovery of Antidepressants

Dating from work of Boerhaave and Auenbrugger in the seventeenth and eighteenth centuries, camphor (1,7,7-trimethylbicyclo[2.2.1]heptan-2-one) was used as a stimulant and mood-elevating agent as well as to induce seizures that were used to treat mania and psychosis; camphor also was used experimentally as recently as the early twentieth century to treat depression [412]. In the 1930s, a chemically unrelated stimulant, pentylenetetrazol, (6,7,8,9-tetrahydro-5H-tetrazolo[1,5-a]azepine), was introduced by Meduna to induce seizures and was employed briefly to treat severe depression [173]. However, the uncontrollable nature of chemically induced seizures led to abandonment of the method in favor of electroconvulsive therapy (ECT) introduced in 1938 by Cerletti and Bini in Rome [104, 280, 500]. Various opiates also were commonly used to treat depression until the 1950s, and interest in them as experimental antidepressants continues, particularly for depression that remains resistant to standard treatments [77, 392]. The stimulant amphetamines evolved from ephedrine (R,S-2-[methylamino]-1-phenylpropan-1-ol), an active principle of the Chinese folk medicine *má huáng*, prepared from the *Ephedra sinica* plant. Ephedrine was isolated from the plant in Japan by Nagai and synthesized in Berlin by Edeleanu in the same year, 1887 [157, 482]. Amphetamines were used from the 1930s to reduce fatigue and during World War II to enhance combat performance. During the same era, amphetamines were employed to treat clinical depression, sometimes combined with sedative barbiturates, and this practice continued until it was displaced by modern antidepressants in the 1960s [147, 270].

The first modern pharmaceutical agent to gain regulatory approval for the treatment of depression (in 1958) was *iproniazid* (N'-isopropylisonicotinohydrazide; Marsalid®), the isopropyl analog of the older antituberculosis drug isoniazid, which dates back to 1912 [238]. Iproniazid was synthesized in 1951 at the Hoffman-La Roche Corporation in Switzerland and New Jersey [178, 486]. It was soon found to inhibit oxidase enzymes that metabolized monoamines [620]. Within the next two years, there was clinical interest in its behavioral effects in Paris among clinical investigators who developed chlorpromazine [136], as well as in the United States. In America, iproniazid was first observed clinically to be associated with excessive arousal and activation among patients in tuberculosis sanatoria, but also suspected of being of potential value for the treatment of withdrawn or depressed psychiatric patients, possibly based on inhibition of the oxidative breakdown of monoamines [122, 465, 489, 509]. Shortly thereafter, Kline and his colleagues at Rockland State Hospital in New York reported on more formal therapeutic trials, initially in withdrawn psychotic-disorder patients and then in depressed patients, and the agent was termed a "psychic energizer" [303]. The hepatic-toxic effects of iproniazid, a risk of many hydrazines, led to its removal from clinical use by the FDA in 1958, but many other molecules were developed with similar pharmacological and clinical properties, and some continue in clinical use to the present time [466].

A second important development was the discovery of antidepressant properties of imipramine, the first tricyclic antidepressant. Imipramine was synthesized in a series of antihistamines derived from iminodibenzyl, first synthesized in 1898 [551], and reviewed for possible pharmaceutical development in 1954 [479]. One of these, the Geigy product G-22150 (imipramine) was placed in preliminary clinical trials and found initially to lack any obvious clinical effects in psychiatric patients. One of the initial investigators, Roland Kuhn, a psychiatrist at the Münsterlingen Hospital near Konstanz in Switzerland, and colleagues at Giegy Corporation in Basle, were stimulated by the growing success of chlorpromazine, an apparent structural analog, and carried out further clinical trials, initially in psychotic disorder patients, again finding little evidence of an antipsychotic effect. Clinical observations arising from that experience led to the impression that mood-elevating and energizing effects might be present, leading to trials in 1957–1958 in depressed patients, for whom it was considered to be beneficial, leading to coining of the term "thymoleptic" for the effect [238, 315–317]. Credit for the initial concept of a potential antidepressant effect of imipramine remains uncertain; candidates include assistants working with Kuhn and colleagues at Geigy Corporation [238, 317]. Of note, the previous experience with ECT and with iproniazid appears to have had limited impact on conceiving that a medicinal treatment for severe depressive illness might be feasible. Only the evolution of international clinical experience and a series of controlled trials led to gradual acceptance of the concept by the early 1960s and introduction of a large series of tricyclic-type antidepressant-anxiolytic agents [33, 115, 254, 302].

A mystery arising from early work on the pharmacodynamic actions of both MAO inhibitors and tricyclic antidepressants was whether effects on both norepinephrine and serotonin were essential to antidepressant effects, as experimental studies of actions of antidepressant agents left some ambiguity. Neurotransmission by both of these monoamines was facilitated by the pharmacodynamic actions of most early (largely tertiary-amine) tricyclics (which blocked inactivation by neuronal re-uptake) and the MAO inhibitors (which inhibited the predominant form of the oxidase in nerve endings to potentiate endogenous monoamines). As understanding emerged of the relatively selective noradrenergic effects of secondary amine tricyclic metabolites, such as desipramine from imipramine, and their efficacy as antidepressant drugs in their own right, it appeared initially as if noradrenergic effects were sufficient. Unclear, however, was whether potentiation of central serotonergic neurotransmission might also produce antidepressant or antianxiety effects. In late 1960s and 1970s, several investigators encouraged more direct testing of the clinical potential of selective serotonergic actions of novel drugs [95, 96, 321]. In influential study in New York found that an inhibitor (p-chlorophenylalanine) of tryptophan hydroxylase, the rate-limiting enzyme in the pathway from L-tryptophan to 5-hydroxytrypamine (serotonin), could reverse the mood-elevating effects of the MAO inhibitor tranylcypromine and of imipramine in depressed patients, providing strong direct support for an important contribution of serotonin [497].

This work led directly to the development of the first selective serotonin-reuptake inhibitor (SRI) as a rationally predicted, novel type of antidepressant based on pharmacological theory. It was an indole (serotonin-like) derivative *indalpine* (3-[2-piperidin-4-ylethyl]-1H-indole), developed by Uzan and Le Fur in a Parisian pharmaceutical company, Pharmuka [571]. Introduced in 1977, indalpine became a rapidly and internationally accepted treatment for depression until its risks of agranulocytosis and other toxic effects led to its abrupt removal from the market in 1987, and to abandonment of other serotonergic agents in development at Pharmuka [246]. A successor was developed from antihistamine compounds with weak effects on monoamine uptake, which were modified chemically until a selective serotonin reuptake inhibiting compound was obtained. The first such agent was *zimelidine* (3-[4-bromophenyl]-*N,N*-dimethyl-3-[pyridin-3-yl]prop-2-en-1-amine), introduced in 1983 by Astra Pharmaceuticals in Sweden with the help of Carlsson and his colleagues at the University of Göteborg, based on chemical modification of the antihistamine compound chlorpheniramine. However, it, too, was withdrawn from clinical development due to toxic febrile reactions with several cases of ascending paralysis of the Guillain–Barré type [246]. However, soon thereafter appeared the first successful SRI, *fluvoxamine* ([*E*]-5-methoxy-1-[4-(trifluoromethyl)phenyl]pentan-1-one-*O*-2-aminoethyl oxime; Luvox®), developed by the Duphar division of Solvay Corporation in Belgium in 1986. It continues in clinical use, largely as an antianxiety and anti-obsessional agent, although it was withdrawn for concerns about a possible relationship to aggressive behavior, and later reintroduced into the American market through Jazz Pharmaceuticals in California. The first highly successful SRI was developed at Eli Lilly Laboratories in Indiana. It was fluoxetine (*N*-methyl-3-phenyl-3-[4-(trifluoromethyl)-phenoxy]propan-1-amine; Prozac®), developed in 1974 from chemical modifications of antihistaminic precursors (particularly diphenhydramine) along with a series of agents selective for serotonin or norepinephrine reuptake or both. Fluoxetine was released commercially in 1987 to become the first billion dollar a year psychopharmaceutical product, based on an extraordinarily rapid expansion of the antidepressant market into general medicine as well as efforts in psychiatry to develop safer and easier to use antidepressants [238, 603].

Types of Antidepressants

A wide variety of types of molecules have been identified and developed as antidepressant drugs (Table 4.1), and their historical evolution is summarized in Fig. 4.1. Representative agents and their structures include several major classes (Figs. 4.2, 4.3, 4.4, 4.5, and 4.6). Older tricyclic-type drugs include some compounds with "tetracyclic" structures (e.g., amoxapine, maprotiline), but very similar pharmacology to traditional tricyclics such as imipramine. This tricyclic-like group is usefully subdivided by the level of alkyl substitution of terminal amino atoms on their side chains (or the distal nitrogen in the piperazine ring of amoxapine), into compounds with a tertiary (thrice substituted) or secondary (doubly substituted) terminal amino nitrogen atom (Fig. 4.1). The tertiary amine tricyclic antidepressants affect the

Table 4.1 Antidepressants: generic and brand names

Agent	Brand names®
Tricyclic-type agents (TCAs)	
Amitriptyline	Elavil, etc.
Amoxapine	Ascendin
Clomipramine	Anafranil
Desipramine	Norpramin, pertofrane, etc.
Doxepin	Sinequan, etc.
Imipramine	Tofranil, etc.
Nortriptyline	Aventyl, pamelor, etc.
Maprotiline	Ludiomil
Protriptyline	Vivactil
Trimipramine	Surmontil
Serotonin-reuptake inhibitors (SRIs)	
Citalopram	Celexa
S-Citalopram	Lexapro
Fluoxetine	Prozac, Serafem, etc.
Fluvoxamine	Luvox
Paroxetine	Paxil
Sertraline	Zoloft
Vilazodone	Viibryd
Serotonin- and norepinephrine-reuptake inhibitors (SNRIs)	
Desvenlafaxine	Pristiq
Duloxetine	Cymbalta
Milnacipran	Savella
Venlafaxine	Effexor
Norepinephrine-reuptake Inhibitor (NRI)	
Atomoxetine	Strattera
Atypical agents	
Bupropion	Wellbutrin, Zyban, etc.
Mirtazapine	Remeron
Nefazodone	Serzone
Trazodone	Desyrel, etc.

These 26 agents are used in the United States, where fluvoxamine and mirtazapine are FDA approved for anxiety disorders, and atomoxetine for attention disorders

inactivation by neuronal reuptake of both norepinephrine and serotonin, and so might be considered "SNRIs," whereas secondary amine agents are virtually norepinephrine-selective reuptake inhibitors or NRIs. These subgroups differ pharmacologically in several ways. Notably, the tertiary amine tricyclics have mixed interactions with central noradrenergic and serotonergic systems, whereas the secondary amine congeners, which are often produced by hepatic metabolism from tertiary parent compounds (e.g., desipramine from imipramine and nortriptyline from amitriptyline) are more selectively noradrenergic, and somewhat less anticholinergic. Serotonin-reuptake inhibitors (SRIs) are also often termed "selective serotonin reuptake inhibitors" (SSRIs) and include several of the most widely

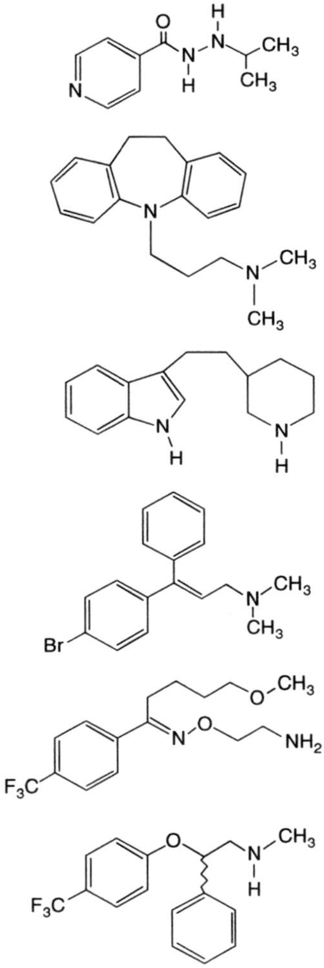

Fig. 4.1 Evolution of
antidepressants. Chemical
structures are shown of drugs
that include the first
monoamine oxidase (MAO)
inhibitor (iproniazid), the first
tricyclic antidepressant
(imipramine), and two early
selective serotonin-reuptake
inhibitors (SRIs; indalpine
and zimelidine) that were
withdrawn from clinical use
due to toxicity, as was
iproniazid, and the two first
successful SRIs (fluvoxamine
and fluoxetine) that remain in
clinical use following the end
of their patent protection, as
is imipramine. Years of their
clinical introduction are
shown

Iproniazid (1954)

Imipramine (1957)

Indalpine (1982)

Zimelidine (1982)

Fluvoxamine (1984)

Fluoxetine (1987)

employed modern antidepressants (Fig. 4.2). Several agents have dual actions
(also shared with tertiary amine tricyclics) of inhibiting neuronal reuptake of
both norepinephrine and serotonin (SNRIs; Fig. 4.3). A group of atypical agents
whose pharmacology remains incompletely developed but appears less to involve
inhibiting neurotransmitter inactivation by neuronal reuptake of norepinephrine
or serotonin than other antidepressants (Fig. 4.4). Several monoamine oxidase
(MAO) inhibitors (the original antidepressants discovered in the 1950s) and stimulants continue to be employed clinically (Fig. 4.5). MAO inhibitors are no longer
commonly used clinically to treat depression, but are sometimes tried after other,
simpler and safer standard antidepressants have failed [79]. Stimulants, including
the amphetamines, methylphenidate, and others, had been employed to treat depression from the 1930s into the 1960s, but were almost entirely displaced by more

TRICYCLIC-TYPE ANTIDEPRESSANTS

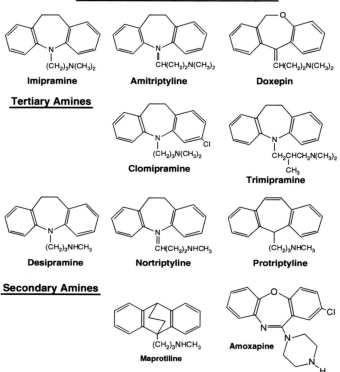

Fig. 4.2 Tricyclic-type antidepressants. Agents are separated into tertiary and secondary amines, based on the level of alkyl substitution of terminal side chain amino moieties. The tertiary amine group tends to be less selective in having effects on both noradrenergic and serotonergic systems as well as potent anticholinergic actions, whereas the secondary amine drugs are much more selective for noradrenergic neurons and are less anticholinergic

effective first-generation antidepressants (MAO inhibitors and tricyclics), although tranylcypromine has both MAO inhibiting and amphetamine-like properties and bupropion is also a mild stimulant. The main current uses of stimulants is for attention disorders, narcolepsy, and occasionally and empirically in geriatric conditions with elements of depression or demoralization with physical illnesses, and early dementia, which uses are best supported by short-term studies.

In addition, benzodiazepines, especially those of high potency, are sometimes used for anxiety, agitation, and insomnia in depressive illness, but are not considered true antidepressants [10, 423, 470]. In primary-care practice, the use of benzodiazepines is very prevalent, sometimes in lieu of an antidepressant, especially in mildly depressed general medical patients [140]. Another novel agent in development is agomelatine, which is a melatonin receptor agonist that benefits sleep and circadian rhythms, as well as acting as a serotonin 5HT-2C receptor antagonist that increases release of central catecholamines by blocking inhibitory heteroceptors on dopaminergic

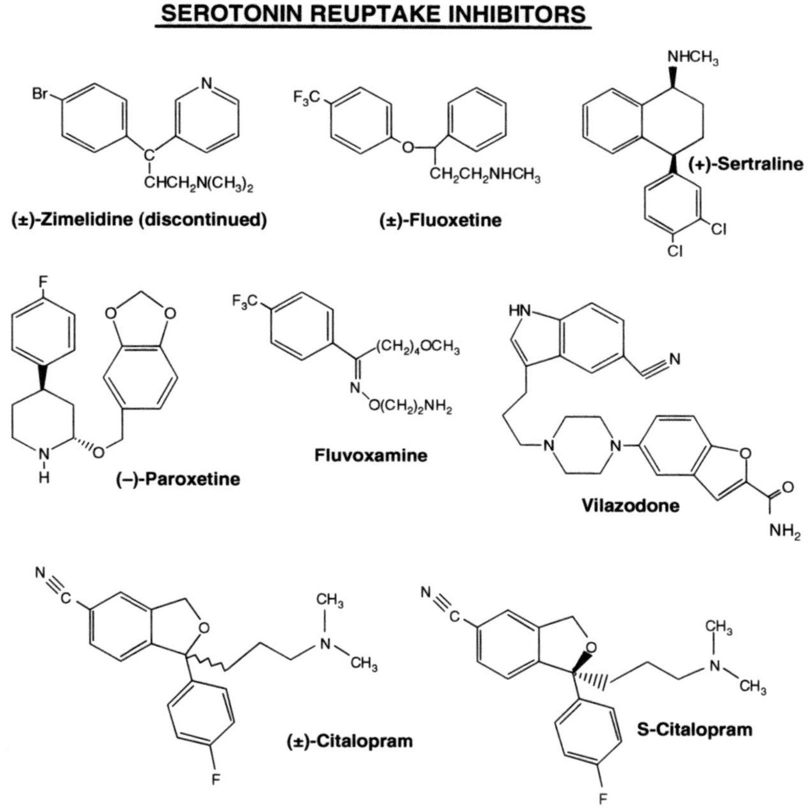

SEROTONIN REUPTAKE INHIBITORS

Fig. 4.3 Serotonin-reuptake inhibitors (SRIs). There are eight compounds with FDA approval. All are used to treat both major depressive and some anxiety disorders, except that fluvoxamine lacks an approved indication for depression in the US

and noradrenergic neurons [139, 229, 468]. Direct dopamine agonists (such as pramipexole) developed for the treatment of hyperprolactinemia and Parkinson's disease may also have useful mood-elevating effects requiring further study [7, 112].

Several natural products also have been tested as potential antidepressants, at least for illnesses of milder severity, including in small numbers of promising placebo-controlled, randomized trials. They include the endogenous metabolite S-adenosyl-L-methionine (SAMe) as well as long-chain unsaturated fatty acids such as docosahexaenoic acid (DHA), and plant products including extracts of *Hypericum perforatum* ("St. John's wort"), *Crocus sativus* (saffron stamen or flower), *Echium vulgare* ("Vipers' bugloss"), and perhaps *Lamiaceae* (lavender) [30, 97, 155, 358, 404, 405, 538]. Efforts have also been made to develop antagonists of hypothalamic corticotrophin releasing hormone (CRF), which controls release of adrenocorticotrophic hormone (ACTH), and so alters release of adrenocorticoid stress steroids. Such compounds have not been proved effective for the treatment of major depression [403]. Additional techniques employed for the treatment of severe or otherwise treatment-resistant

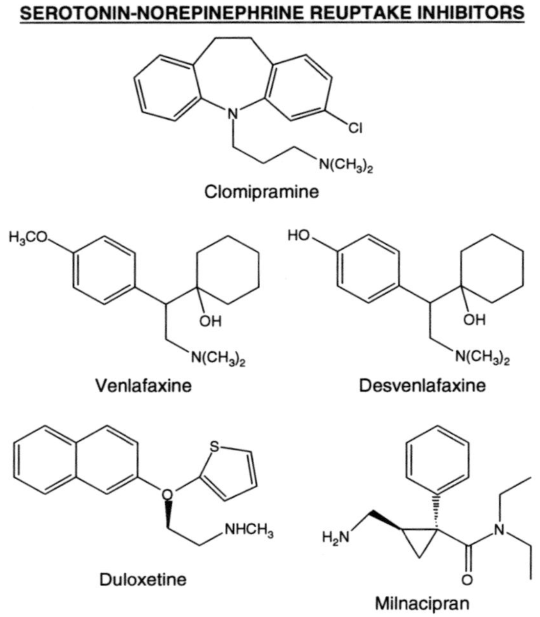

SEROTONIN-NOREPINEPHRINE REUPTAKE INHIBITORS

Clomipramine

Venlafaxine

Desvenlafaxine

Duloxetine

Milnacipran

Fig. 4.4 Serotonin–norepinephrine-reuptake inhibitors (SNRIs). All five agents have FDA approval for use in treating major depressive and some anxiety disorders, except that milnacipran lacks an approved indication for depression in the US

depression include ECT, recently FDA-approved vagal nerve stimulation (VNS), repeated transcranial magnetic stimulation (rTMS), and experimental deep-brain electrical stimulation techniques [62, 253, 257, 290, 320, 345, 363, 422, 437, 508].

Finally, various experimentally supported, formalized psychotherapies designed specifically to treat mood disorders (including cognitive-behavioral and interpersonal methods), as well as supportive interventions, are routinely combined with antidepressants, and have been found to be approximately as effective themselves as the antidepressant drugs, at least in mild-to-moderate depression lacking melancholic features or a family history, and if treatment is relatively prolonged [90, 94, 248, 251, 351, 492].

Actions of Antidepressants

Focus on Monoamine Neurotransmission

Theories of pharmacodynamic action mechanisms of antidepressants have focused on their ability to increase the physiological activity or availability of monoamine neurotransmitters in the central nervous system (CNS), at least acutely. This focus of research attention is understandable since virtually all clinically employed

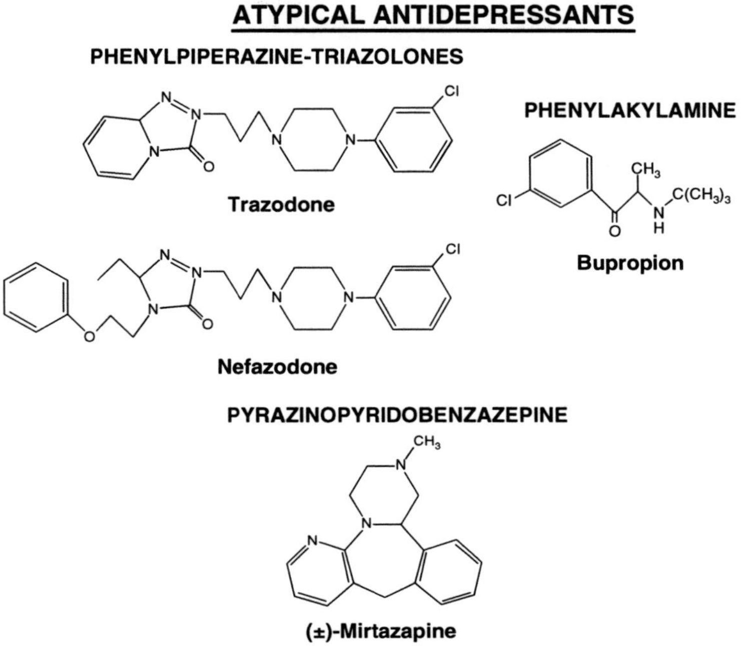

ATYPICAL ANTIDEPRESSANTS

PHENYLPIPERAZINE-TRIAZOLONES

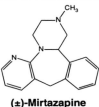

Trazodone

PHENYLAKYLAMINE

Bupropion

Nefazodone

PYRAZINOPYRIDOBENZAZEPINE

(±)-Mirtazapine

Fig. 4.5 Atypical antidepressants. These agents with FDA approval for treatment of major depression, with only partially characterized pharmacodynamic actions. The two phenylpiperazine-triazolones include trazodone, which is mainly used as a sedative-hypnotic, and nefazodone, which is rarely used due to risk of severe hepatic toxicity. Bupropion has some dopaminergic as well as noradrenergic effects, with mild stimulant-like properties. Mirtazapine is a highly sedating antidepressant with complex effects on autoreceptors in both central noradrenergic and serotonergic systems

antidepressants have actions that appear to enhance the cerebral activity of either norepinephrine or serotonin, usually with lesser or variable effects on dopamine and other synaptic transmitter molecules. Soon after discovery of antidepressant effects of iproniazid and imipramine, there were indications that MAO inhibitors inhibit an enzyme crucial for the metabolism of monoamines including norepinephrine and serotonin [620], and that tricyclic antidepressants block the neuronal transport of radiolabeled norepinephrine, initially in cardiac tissue containing sympathetic nerve terminals and later in brain tissue [27, 589]. As noted, these initial findings were followed by efforts to distinguish contributions of each monoamine, and particularly, of serotonin, to the actions of antidepressant drugs [95]. These initial discoveries arose along with firm establishment of norepinephrine, dopamine, and serotonin (5-hydroxytryptamine) as neurotransmitters in discrete neuronal systems within the CNS, relying heavily on modeling based on knowledge of the peripheral sympathetic nervous system [191, 211, 255].

However, there may be a degree of circularity in this focus on potentiation of norepinephrine, serotonin, or both since the development of virtually all antidepressants following the serendipitous discovery of iproniazid and imipramine in the

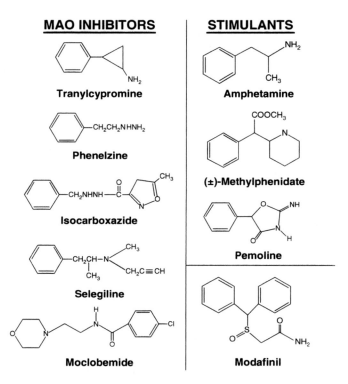

Fig. 4.6 Monoamineoxidase (MAO) inhibitors and stimulants. MAO inhibitors include older, long-acting agents (tranylcypromine, phenelzine, and isocarboxazide) that inhibit both major molecular forms of the enzyme (A and B). Isocarboxazide is no longer marketed. Selegiline is partially selective for type B MAO at low doses and is available as a skin patch. Mocobemide is a short-acting and selective antagonist of MAO-A that is not FDA approved. Stimulants include amphetamines, methylphenidate and pemoline, currently used mainly to treat attention disorders. Modafinil (and its active *R*-enantiomer, armodafinil [not shown]) are mild stimulant-like agents used primarily to treat narcolepsy

1950s has been strongly influenced by theories of the actions of these lead agents. It is increasingly appreciated that these initial actions are far from the entire picture of complex, secondary, and other downstream adaptations to altered monoaminergic interneuronal synaptic conduction. Moreover, the circular process of identifying potential antidepressants by the actions of lead agents, though an effective practical strategy for drug development, may well have limited consideration of alternative mechanisms and innovative treatments, particularly without the guidance of a tissue pathology or even a compelling pathophysiology for mood disorders [32]. A similar phenomenon was discussed above in Chapter 2 regarding antipsychotic drugs as dopamine receptor antagonists.

In addition, as in a hyperdopaminergic hypothesis concerning mania or psychosis, a strong effort was made to establish a monoamine deficiency theory of depression. Despite decades of experimental effort, evidence that there is a cerebral deficiency of either catecholamines or serotonin in depressed persons has proved to be highly elusive [32, 194, 478]. Lack of a coherent hypothesis concerning the pathophysiology or etiology of depressive illnesses probably should not be surprising, particularly since the emergence of the extraordinarily broad and heterogeneous concept of "major depression" in the 1980s as a seemingly unsatisfactory phenotype [12, 347]. Lack of a biomedical theory of depression did not gainsay empirical development of more antidepressants of the kinds already available by imitation of the chemical structure or the pharmacodynamics of established agents. However, proof that the monoamine-potentiation strategy is either necessary or sufficient to account for the clinical effects of antidepressant-anxiolytic agents remains elusive, as is further speculation that deficiencies of monoaminergic transmission may be an essential component of the pathophysiology of depression.

It is has been very difficult and largely unsuccessful to try to develop agents as potential antidepressants that do not interact with noradrenergic or serotonergic neurotransmission. Alternative approaches have considered agents that interact with cerebral peptide systems, pituitary hormones including corticotrophin, glutamate transmission, and other physiological systems [86, 252, 348, 475]. And yet, most clinically employed antidepressants, particularly those developed since the 1980s, are remarkably similar pharmacologically and virtually all of them interact with cerebral neurotransmission mediated by serotonin and variably with norepinephrine. Since most available drugs of this class are limited in efficacy, hope continues that improved antidepressants based on different pharmacodynamic theories may yet be forthcoming.

Cerebral Monoamine Systems

The strong focus on central monoaminergic systems to account for pharmacodynamic actions of antidepressant drugs requires consideration of the organization and functions of these systems, particularly those involving the catecholamine norepinephrine (NE) and the indoleamine 5-hydroxytryptamine ([5-HT], serotonin). Both systems arise from groups of neuronal cell bodies located in the pons or brainstem (Fig. 4.7). Noradrenergic neurons are located principally in the loci cœrulei, bilaterally symmetrical, pigmented nuclei containing melanin, located in the lateral floor of the fourth cerebral ventricle at the level of the pons. The locus cœruleus (or ceruleus) has sometimes been described as the head ganglion of the sympathetic nervous system. Its neurons produce bifurcated axons, one limb of which descends to synapse in the intermediolateral column of neuronal groups in the thoracolumbar spinal cord. These cells are muscarinic acetylcholinergic neurons of the preganglionic component of the peripheral sympathetic nervous system. These neurons project to paravertebral, and more peripheral or collateral, sympathetic ganglia where the postganglionic, noradrenergic neurons of the peripheral sympathetic nervous system arise. In turn, the postganglionic neurons project broadly to influence the physiological activity of many visceral targets, including cardiac and vascular tissue, eye, smooth muscle, and glandular tissues. Their pattern of innervation is relatively diffuse, as increasingly fine, preterminal arborizations infiltrate target tissues and produce synapses in passing (en passant) rather than at precise terminal targets.

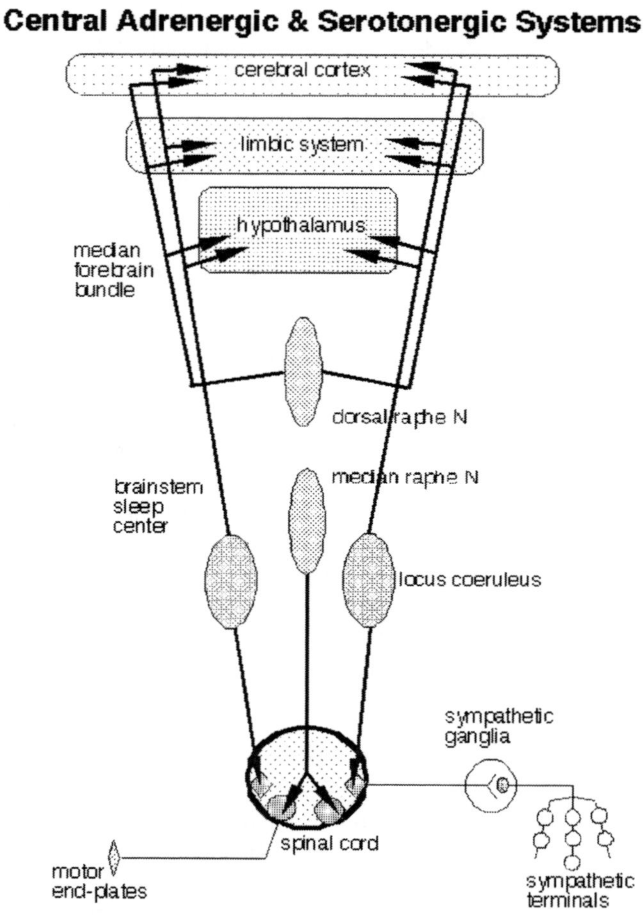

Central Adrenergic & Serotonergic Systems

Fig. 4.7 Central adrenergic and serotonergic systems implicated in the actions of antidepressant drugs. Both types of monoamergic systems arise from neuronal cell bodies in brainstem. Norepinephrine-producing neurons arise mainly in the bilaterally located, pigmented locus coeruleus, with bifurcated axons that include descending projections to the lateral intermediate nuclei to regulate the preganglionic, acetylcholinergic neurons of the peripheral adrenergic sympathetic nervous system, which, in turn, synapse in the peripheral sympathetic ganglia with postganglionic noradrenergic projections to cardiovascular, smooth muscle, glandular, and other peripheral tissues. They also project anteriorly to innervate the hypothalamus, limbic system, and cerebral cortex, where their terminals are organized much like that of the peripheral postganglionic sympathetic neurons. The serotonin-producing neurons arise in two major sets of centrally located dorsal and median raphe nuclei, which, respectively project mainly to hypothalamus and forebrain, or descend to the ventral horn cells of the peripheral acetylcholinergic motor neuron system which innervate the motor end plates of skeletal muscle

These neurons tend to exert physiologically opposite effects to the muscarinic cholinergic parasympathetic division of the peripheral autonomic nervous system. An equivalent neuroanatomical scheme is found in the anterior projections of the bifurcated axons of the locus cœruleus, as well, as they course through the median forebrain bundle to innervate the hypothalamus, limbic system, and cerebral cortex, again in a diffuse way. This organizational scheme is well suited for setting the background physiological tonality of other neuronal systems, and so is quite analogous to a sympathetic nervous system, but with the hypothalamus and forebrain as target tissues rather than peripheral muscle and glandular tissues.

The serotonergic system arises from seven principal neuronal cell groups located along the midline of the midbrain and pons. Two of these groups are particularly important for the present discussion (Fig. 4.7). The median raphe nucleus projects its axons to the ventral horn cells of the spinal cord to exert regulatory influences on the nicotinic-cholinergic lower motor neurons that innervate skeletal muscle by projecting to the motor endplates of receptive muscle cells. Anterior projections of serotonergic neurons arise principally in the dorsal raphe nucleus, again coursing through the median forebrain bundle to innervate hypothalamus, limbic system, and cerebral cortex widely and diffusely.

The anterior projections of both the noradrenergic and serotonergic systems are well designed to influence a range of autonomic functions that are closely involved with the physiology of emotions, including appetitive, ingestive, reproductive, and other autonomic functions associated with the viscera and cardiovascular tissues. They as well as hypothalamic centers including the suprachiasmatic nucleus are also involved in regulating circadian (24-h) and ultradian (typically 90-min) rest-activity, sleep–wake cycles, and other phasic bodily functions. Many of these functions can be disturbed in severe, especially "endogenous" (apparently spontaneous) or melancholic forms of major clinical depression.

A major clue to the importance of the monoaminergic fibers passing through the bilaterally symmetrical medial forebrain bundles is the phenomenon of self-stimulation [113, 396]. Electrodes placed in this structure can elicit strong bar-pressing behavior in laboratory animals for electrical self-stimulation. Such stimuli have powerful reinforcing properties that have been linked theoretically to brain mechanisms underlying addictive as well as emotional states that may include depression [113, 237, 310]. The medial forebrain bundle has been considered a promising target site for developing deep brain stimulation as a potential treatment of depression [477].

An influential series of clinical experiments has involved alterations of the synthesis of norepinephrine and serotonin by reducing the dietary availability of their essential precursor aromatic amino acids, L-tyrosine and L-tryptophan, or administration of clinically tolerated inhibitors of the rate-limiting enzymes required for the synthesis of norepinephrine or serotonin, notably α-methyl-p-tyrosine (AMPT) to inhibit tyrosine hydroxylase, or p-chlorophenylalanine (PCPA) to inhibit trypto-phan hydroxylase [109, 137, 356, 464, 498]. Pretreatment with AMPT has reduced or reversed the clinical effects of tricyclic antidepressants, and PCPA has exerted similar effects, selectively, on serotonin-reuptake inhibitors. Conversely, there is also evidence that co-treatment with a serotonin autoreceptor antagonist (such as pindolol) can facilitate initial responses to serotonin reuptake inhibitor antidepressants, and that a noradrenergic α_2 autoreceptor antagonist (such as phentolamine) can potentiate the effects of selective norepinephrine reuptake inhibitor antidepressants [71, 266, 415, 491]. Such effects have provided strong support for the theory that many antidepressants require functioning norepinephrine or serotonin systems to produce antidepressant effects. Note, however, that such experiments do not indicate whether deficiencies of the neurotransmitter amines may be involved in the pathophysiology of clinical depression [83, 459].

Organization of Noradrenergic and Serotonergic Nerve Terminals

The monoaminergic neural systems of particular interest regarding actions of antidepressant drugs—noradrenergic and serotonergic—are highly specialized but are organized in very similar ways, based on the presence of specific proteins unique to each cell type. These include the cytoplasmic, rate-limiting oxidative (hydroxyl-group adding) enzymes required for the first step in their synthesis from essential aromatic amino acids, respectively, L-tyrosine and L-tryptophan (Fig. 4.8). The next metabolic step in both types of neurons requires a widely available, cytoplasmic, L-aromatic amino acid decarboxylase that was also considered in Chapter 2 regarding synthesis of dopamine. In noradrenergic neurons, in contrast to dopaminergic cells, an additional oxidase (dopamine-β-hydroxylase) located within presynaptic storage vesicles is required to convert dopamine to the final product, norepinephrine. Both norepinephrine and serotonin, like dopamine, are stored in presynaptic, membrane-enclosed vesicles that contain chromogranin proteins, ATP, and Ca^{++}. With depolarizing electrical discharges of the monoaminergic neurons, the presynaptic vesicle membranes and the neuronal cell membranes fuse and allow the stored neurotransmitters to escape by the process of exocytosis. The amines so-released into the synaptic cleft exert excitatory actions at postsynaptic receptors on nearby receptive cells. In the noradrenergic system, these include α_1- and β-adrenergic receptors, and for serotonin there are at least seven types of receptor proteins. As in the dopaminergic system, these receptors are linked functionally with effector enzymes regulated by guanosine triphosphate (GTP) associated, or G-proteins, to produce second messengers that influence the functioning of receptive cells.

Physiological inactivation of the released amines is mediated mainly by active reuptake into the presynaptic terminals, for re-storage and recycling, with excesses in the presynaptic cytoplasm controlled by monoamineoxidase (MAO), acting in concert with extraneuronal catechol-O-methyltransferase (COMT) in the metabolism of norepinephrine. The reuptake process in each neuron type is mediated by highly specific membrane protein gene products, the norepinephrine transporter (NET) and serotonin transporter (SERT), respectively. Amine re-uptake is an energy-consuming process that involves simultaneous extrusion of Na^+ ions, using energy provided by ATP produced in nearby mitochondria [493]. The transporters

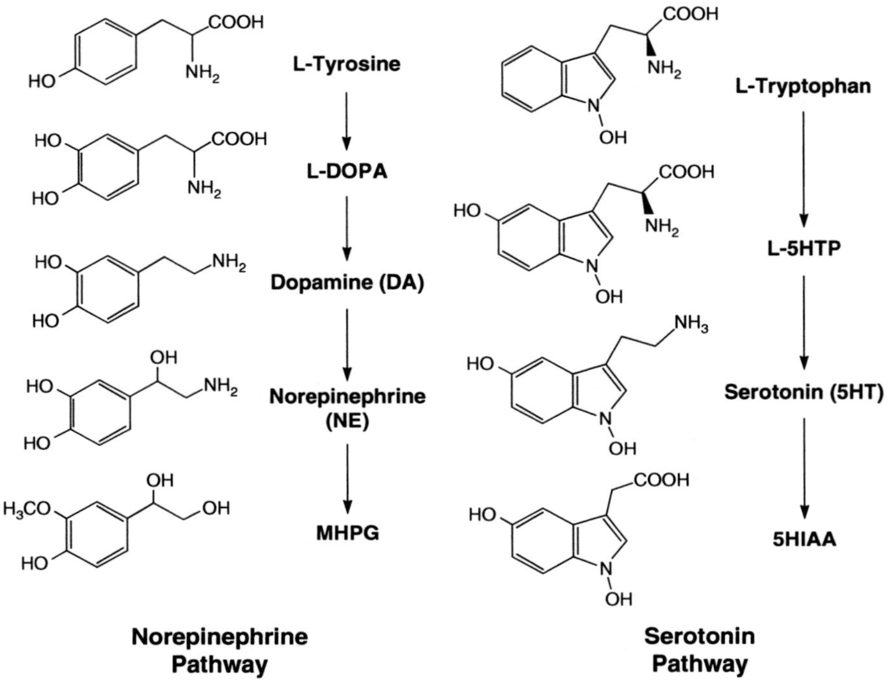

Fig. 4.8 Metabolic pathways of norepinephrine and serotonin. Norepinephrine is synthesized in noradrenergic nerve terminals from the essential amino acid (dietary and cannot be produced in the body) L-tyrosine, which is converted by tyrosine hydroxylase to dihydroxyphenylalanine (DOPA), in turn decarboxylated by L-aromatic amino acid decarboxylase to dopamine (DA). DA is converted to norepinephrine (NE) in presynaptic storage vesicles by dopamine-β-hydroxylase. Mitochondrial type-A MAO oxidatively deaminates NE to produce inactive 3,4-dihydroxyphenylacetic (DOPacetic) acid, which can be secondarily O-methylated by extraneuronal catecholamine-O-methyltransferase (COMT) and reduced to yield 3-methoxy-4-hydroxyphenethylene glycol (MHPG; the phenethyleneglycol 1-[3-methoxy-4-hydroxyphenyl]ethane-1,2-diol) as a final metabolite. Serotonin follows parallel biochemical production steps from the essential amino acid L-tryptophan, which is oxidized in serotonin neurons to L-5-hydroxytryptophan (5HTP), in turn, decarboxylated to serotonin (5-hydroxytrypamine, 5HT). Mitochondrial MAO-A deaminates 5HT to produce the inactive metabolite 5-hydroxyindolacetic acid (5HIAA)

are sites of initial actions of antidepressant and stimulant drugs, although most stimulants, but few antidepressants, also exert inhibitory effect on dopamine reuptake. An exception is that at the NET-selective agent (atomoxetine), and possibly also desipramine, can prevent reuptake of both and norepinephrine and dopamine in portions of cerebral cortex [92, 408]; another is that mirtazapine, by acting on $5HT_{1A}$ heteroceptors on dopaminergic neurons, can increase release of dopamine [223, 371]. Monoaminergic nerve terminals have type-A MAO located in presynaptic mitochondria, whereas most other tissues have type B (e.g., human platelets) or both (e.g., liver). It is believed that inhibition of type-A MAO, a specific gene product typical of monoaminergic nerve terminal mitochondria, is more likely to produce mood-elevating effects [598].

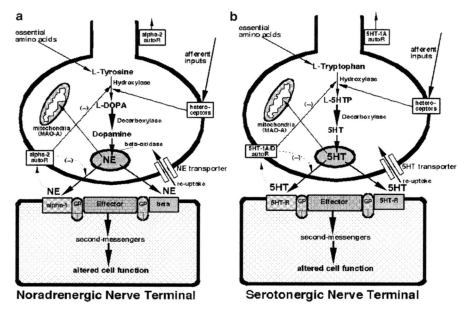

Fig. 4.9 Noradrenergic (**a**) and serotonergic (**b**) nerve terminals. Schematic organization of noradrenergic and serotonergic nerve terminals is provided, although terminals are actually swellings or varicosities in the complex, perterminal arborizations of neuronal axons to form synaptic relations with nearby cells in passing. In noradrenergic neurons, norepinephrine (NE) is synthesized from L-tryrosine by the oxidative action of rate limiting and highly regulated tyrosine hydroxylase to form L-dihydroxyphenylalanine (DOPA). In serotonergic neurons, serotonin (5-hydroxytryptamine [5HT]) is produced from L-tryptophan by another rate-limiting oxidase, tryptophan hydroxylase, unique to serotoninergic neurons to form 5-hydroxytryptophan (5HTP). Both DOPA and 5HTP are readily decarboxylated by unibiquitous L-aromatic amino acid decarboxylase to yield the amines dopamine or 5-hydroxytryptamine, respectively. In noradrenergic neurons, dopamine is further side-chain oxidized by dopamine-β-hydroxyase, an enzyme located within presynaptic storage vesicles to yield norepinephrine, complexed with Ca^{++}, ATP (adenosine triphosphate), and chromogranin proteins [84]; storage is prevented by reserpine and tetrabenazines. In both terminals, the monoamines are released from membrane-protected presynaptic storage vesicles into the synaptic cleft by a process of membrane fusion (exocytosis), to exert actions on various postsynaptic receptors (α_1 and β adrenergic, and seven or more 5HT receptors) associated with various molecular effector mechanisms, to influence the functional state of the receptive cells. Inactivation of released monoamines is principally effected by neuron-specific, membrane transporter proteins to carry the amines back into the presynaptic terminals for restorage and reuse, by a process that exchanges amines for Na^+ ions and requires ATP for energy. Reuptake is blocked by stimulant (including dopamine transport) and antidepressant drugs. Excesses of presynaptic monoamines in the cytoplasm are metabolized by type-A monoamineoxidase located in presynaptic mitochondria. Important regulatory effects (negative feedback) on both the rate-limiting oxidative enzymes and the amine release process are exerted by presynaptic autoreceptors (auto-R: α_2 in the noradrenergic system at terminals as well as cell bieds (perikarya) and their dendrites, and $5HT_{1A}$ at serotonin terminals and probably more dominant $5HT_{1D}$ receptors on serotonin perikarya and dendrites; additional regulatory effects are mediated by heteroreceptors for other neurotransmitters, including α_2 NE receptors on 5HT neurons, and 5HT receptors (probably including $5HT_{1D}$) on NE neurons. In addition, there is some evidence that NE terminals in the cerebral cortex can take up DA, which can be potentiated by antidepressants that block the NET process [305]. The schemes are adapted from a review by Baldessarini [33] of decades of experimental contributions

Both noradrenergic and serotonergic neurons are under the control of various regulatory mechanisms. Prominent among them are the actions of presynaptic autoreceptors, respectively, α_2, and serotonin types 1A and 1D and others [621]. Autoreceptors regulate the neurophysiological activity of monoaminergic neurons, control the activity of their rate-limiting synthesizing enzymes largely through modulating their activation by phosphorylation, and exert control over the transmitter release process of exocytosis [174, 266]. Schematic organization of noradrenergic and serotonergic nerve terminals is shown in Fig. 4.9.

Pharmacodynamics of Antidepressants

A basic neuropharmacological characteristic of antidepressants is their affinity or potency to interact with membrane protein monoamine transporters that mediate the re-uptake, physiological inactivation process and are characteristic gene products of specific types of neurons. Affinities (as inhibitory constants, K_i, in units of nanomolar or 10^{-9} moles/L) of antidepressants selective for norepinephrine (NET) or serotonin (SERT) are summarized in Tables 4.2 and 4.3. The most potent NET-selective antidepressant is desipramine ($K_i = 0.83$ nM), whereas the most selective at NET vs. SERT transporters is the experimental agent oxyprotiline (800-fold; Table 4.2). The most

Table 4.2 Transporter potency (K_i, nM): norepinephrine-selective antidepressants

Drug	NET	SERT	DAT	SERT vs. NET selectivity
Desipramine	0.83	17.5	3,200	21.1
Protriptyline	1.40	19.6	2,130	14.0
Atomoxetine	3.50	43.0	1,270	12.3
Norclomipramine	2.50	41.0	–	16.4
Nortriptyline	4.35	18.5	1,140	4.25
Oxaprotiline	5.00	4,000	4,350	800
Lofepramine	5.30	71.4	18,500	13.5
Reboxetine	7.14	58.8	11,500	8.24
Maprotiline	11.1	5,900	1,000	532
Nomifensine	15.6	1,000	55.6	64.1
Amoxapine	16.1	58.5	4,360	3.63
Doxepin	29.4	66.7	12,200	2.27
Mianserin	71.4	4,000	9,100	56.0
Viloxazine	156	17,000	100,000	109
Mirtazapine	4,760	100,000	100,000	21.0

Shown are for affinity (inhibition constant, K_i in nanomolar [nM] concentrations) at norepinephrine (NET), serotonin (SERT), and dopamine (DAT) membrane transporter proteins, with NET vs. SERT selectivity, and agents ranked by NET affinity. Data are adapted from Owens et al. [400], Tatsumi et al. [541], and Richelson [448], in experiments involving cultured cell lines transfected with human transporter genes. Note that desipramine is the most potent at NET, and that the experimental agent oxaprotiline is most selective for NET over SERT. Most antidepressants have low affinity for dopamine transporters, although bupropion, nefazodone, paroxetine, and sertraline have moderate affinity (25–500 nM). Data in Tables 4.2, 4.3, 4.4, and 4.5 are adapted from sources including reviews by Richelson [447, 448] and Baldessarini [33]

Table 4.3 Transporter affinity (K_i, nM): serotonin-selective antidepressants

Drug	NET	SERT	DAT	SERT vs. NET selectivity
Paroxetine	40.0	0.12	500	333
Vilazodone	–	0.20	–	–
Clomipramine	3.50	0.28	2,200	132
Sertraline	37.0	0.29	25.0	1,440
Fluoxetine	417	0.81	3,600	515
S-Citalopram	7,840	1.10	>10,000	7,130
R,S-Citalopram	5,100	1.38	28,000	3700
Imipramine	37.0	1.41	8,300	26.2
Duloxetine	11.2	1.60	–	7.00
Fluvoxamine	1,300	2.22	9,100	586
Amitriptyline	34.5	4.33	3,200	7.97
Norcitalpram	780	7.40	–	105
Dothiepin	45.5	8.33	5,300	5.46
Milnacipran	83.3	9.10	71,400	9.15
Nor$_2$-citalopram	1,500	24.0	–	62.5
Norfluoxetine	410	25.0	1,100	16.4
Ziprasidone	>10,000	30.0	80.0	>333
Desvenlafaxine	558	38.0	25,000	14.7
Norsertraline	420	76.0	440	5.53
Venlafaxine	1,060	83.0	9,100	12.8
Zimelidine	5,180	152	12,000	34.1
Trazodone	8,300	160	7,140	51.9
Nefazodone	360	200	360	1.80
Trimipramine	2,400	1,500	10,000	1.60
Bupropion	52,600	9,100	526	5.78

Methods, measures, and references are as for Table 4.1; selectivity is for SERT vs. NET, with compounds in descending order of SERT potency. Includes some TCAs, SNRIs, and metabolites. *Nor* and *des* indicate removal of an akyl radical from terminal amino nitrogen atoms of more fully substituted parent compounds (citalopram, fluoxetine, sertraline, venlafaxine); *nor$_2$* indicates removal of two substituents (nor$_2$-citalopram). Note that paroxetine is the most potent agent at SERT and that *S*-citalopram is most SERT selective

potent antidepressant at SERT transporters is paroxetine (K_i=0.12 nM), and the most selective for SERT vs. NET transporters is citalopram, especially as its active *S*-enantiomer (optical isomer; 7,130-fold; Table 4.3). Among "atypical" antidepressants that seem not to act primarily by inhibiting monoamine transporters, affinities for both NET and SERT transporters are relatively weak (such as the last five agents in Table 4.3). Whereas most tricyclic and SRI type antidepressants have negligible affinity for dopamine transporters (DAT; Tables 4.2 and 4.3), at least moderate DAT affinity, and slight SERT-over-NET selectivity, are found with nefazodone, sertraline, its active metabolite norsertraline, and bupropion, as well as the second-generation antipsychotic drug ziprasidone, which has some mood-elevating activity (Table 4.3).

An additional pharmacological activity of antidepressants that probably contributes to their complex, varied, and still incompletely defined action mechanisms is

Table 4.4 Affinity of antidepressants at monoamine receptors

High	Moderate	Low
α_1 Noradrenergic		
Doxepin	Imipramine	Citalopram
Nefazodone	Desipramine	Paroxetine
Amitriptyline	Sertraline	Bupropion
Clomipramine	Mirtazapine	Fluoxetine
Amoxapine		Fluvoxamine
		Venlafaxine
$5HT_{2A}$ serotonergic		
Amoxapine	Fluoxetine	Citalopram
Nefazodone	Desipramine	Fluvoxamine
Mirtazapine		Sertraline
Doxepin		Paroxetine
Clomipramine		Bupropion
Amitriptyline		Venlafaxine
Imipramine		
ACh_m acetylcholinergic		
Amitriptyline	Paroxetine	Fluoxetine
Clomipramine	Desipramine	Citalopram
Doxepin	Sertraline	Nefazodone
Imipramine	Mirtazapine	Fluvoxamine
	Amoxapine	Bupropion
		Venlafaxine
H_1 histaminic		
Mirtazapine	Desipramine	Fluoxetine
Doxepin		Bupropion
Amitriptyline		Sertraline
Nefazodone	Citalopram	Fluvoxamine
Amoxapine		Venlafaxine
Clomipramine		

Ki (nM) values in cultured cells transfected with genes for human receptor proteins: *High*: ≤10 nM, *moderate*: 100–1,000 nM, *low*: ≥1,000 nM. Most antidepressants have low affinity for α_2 and β noradrenergic and $5HT_{1A}$ and $5HT_{1D}$ receptors. Most tricyclic antidepressants have very low affinity for α_2 autoreceptors. Note also that most SRI antidepressants have low $5HT_{2A}$ affinity. Relatively sedating agents have high H_1 affinity. Affinities correspond closely with potency as inhibitors of the respective receptor types

their tendency to interact with various monoamine receptors. A partial summary of this incompletely defined set of properties is provided in Table 4.4. Tricyclic antidepressants and mirtazapine have relatively high or at least moderate affinity at α_1 noradrenergic receptors, whereas most SRI antidepressants have very little (Table 4.4). Most tricyclic antidepressants have very low affinity at α_2 noradrenergic autoreceptors, though mirtazapine has substantial affinity at α_2 heteroceptors on serotonin neurons, with uncertain effects on serotonergic transmission [221, 355, 608]. It is likely that having at least moderate affinity at α_1 sites provides a degree of protection of them from desensitization and down-regulation (decreased genetic

Table 4.5 Antidepressant actions at noradrenergic and serotonergic neuronal systems

- *Immediate response*: neuronal reuptake of norepinephrine (NE) or serotonin (5HT; or both) or intraneuronal mitochondrial MAO-A is rapidly inhibited by antidepressants; this inhibition continues indefinitely, possibly with eventual decrease in the abundance of membrane transporter proteins, and with loss of MAO protein until resynthesized in the absence of MAO inhibitors
- *Initial adaptation*: Presynaptic autoreceptor mechanisms sensitive to released neurotransmitters in the synaptic cleft reduce presynaptic neurophysiological activity as well as production and release of NE or 5HT
- *Adaptation of autoreceptors over 1–4 weeks*: Sensitivity and abundance of autoreceptors decrease at overstimulated autoreceptors (α_2 and $5HT_{1A}$ and probably $5HT_{1D}$), at which antidepressants have relatively low affinity, leaving them exposed to increased synaptic abundance of transmitter amines. This adaptation is associated with return of presynaptic neurophysiological activity as well as production and release of NE or 5HT
- *Functions and adaptations of heteroceptors*: NE and 5HT neuronal systems cross-regulate each other, through α_2 heteroceptors on 5HT neurons, and $5HT_{1A}$ or other serotonergic receptors on NE neurons. Some drugs act directly to inhibit heteroceptors (e.g., mirtazapine at α_2 and possibly 5HT autoreceptors); their tonic inhibitory mechanisms probably become less sensitive to result in increased production and release of NE or 5HT
- *Adaptation of postsynaptic receptors over 1–4 weeks*: Receptors to which antidepressants have substantial affinity are likely to remain at baseline abundance and function or even increase in sensitivity (possibly α_1 noradrenergic receptors, and probably some postsynaptic 5HT receptors), whereas those at which antidepressants have little affinity are exposed to transmitter amines and become less sensitive and abundant (notably, β noradrenergic and 2A serotonin receptors)
- *Outcomes*: As clinical antidepressant effects emerge over 1–4 weeks, decreased inactivation of NE or 5HT by reuptake continues and presynaptic regulatory mechanisms have adapted to allow production and release of the neurotransmitters to return to or somewhat above baseline status, with transmission possible through various postsynaptic NE and 5HT-receptor mediated mechanisms that are not fully identified

expression of their synthesis) when synaptic concentrations of norepinephrine are increased by the actions of tricyclic antidepressants. Conversely, most SRI antidepressants have little affinity for $5HT_{2A}$ receptors, which can desensitize and down-regulate with long-term SRI treatment [124, 213, 417]. Antidepressants vary greatly in their affinity for central muscarinic acetylcholine receptors: amitriptyline, clomipramine, doxepin, and imipramine being about the most potent, and most SRIs having little interaction, with the notable exception of paroxetine and perhaps sertraline (Table 4.4). Interactions with histamine H_1 receptors to produce central antihistaminic effects with notable sedation are most likely with mirtazapine, doxepine, amitriptyline, nefazodone, amoxapine, and clomipramine (Table 4.4).

Scheme of Antidepressant Actions

Table 4.5 presents a summary scheme of an interpretation of the actions of antidepressant drugs over time, based on decades of studies of the tricyclic antidepressants and more recent studies of the serotonin-reuptake inhibitors [33, 271]. Immediate effects of tricyclic and serotonin- or serotonin–norepinephrine-reuptake inhibitors include their ability to block the reuptake of norepinephrine, serotonin, or both, and antidepressant MAO inhibitors inhibit intraneuronal MAO-A. These actions are not known to be lost with prolonged treatment, and indeed may be reinforced by later antidepressant-specific decreases in availability of NET or SERT transporter proteins through complex and varied mechanisms, whereas MAO-A remains inhibited by most MAO inhibitors until the protein can be re-synthesized after removal of the inhibitor [58, 65, 301, 622, 625].

Soon after starting treatment with tricyclic, serotonin- or serotonin–norepinephrine-reuptake inhibitor antidepressants, there are striking compensatory changes in the neuronal systems perturbed such that regulatory mechanisms appear to respond to increased concentrations of transmitter amines (norepinephrine or serotonin), in the synaptic cleft and lead to rapid reduction in physiological activity of the presynaptic, aminergic neurons. In addition, there is evidence that synthesis and release of the neurotransmitters are reduced, as if to compensate for overstimulation of perisynaptic mechanisms. The basis of these early regulatory changes is not entirely clear, and may involve long-loop, possibly multineuronal, circuits that feed back upon neuronal cell bodies in the locus cœruleus or raphe nuclei, as well as effects mediated by autoreceptors at nerve terminals or cell bodies and dendrites as key regulators of neuronal activity [438, 562, 584]. More is known about the actions of inhibitory α_2 noradrenergic receptors, which have been localized to both the preterminal axons of noradrenergic neurons as well as to their dendrites and perikarya in the locus cœruleus [559]. These autoreceptors are associated with inhibition of neuronal activity and they may have effects to reduce exocytosis or release of norepinephrine at nerve terminals. Another important element of their actions is to reduce the activity of tyrosine hydroxylase, the rate-limiting step in the synthesis of norepinephrine, possibly by inhibiting adenylate cyclase and the production of cyclic-AMP, which is required for phosphorylation activation of tyrosine hydroxylase. Autoregulation of the serotonergic neurons also occurs, but appears to be more complex, and to involve various serotonin type-1 receptor subtypes, including 1A, 1B, and 1D, which have varied distributions at serotonergic terminals, cell bodies, and dendrites [23, 250].

It may seem that the value of inhibiting transport by NET or SERT is defeated by these efficient, compensatory, feedback mechanisms. However, over several weeks, remarkably paralleling the time for clinically appreciable improvements in depressive symptoms to emerge, additional adaptive changes have been reported. Possibly as a reflection of the lack of major interactions of tricyclic, serotonin, or serotonin-norepinephrine potentiating antidepressants with autoreceptors, including α_2 and $5HT_{1A}$ types, these receptors remain open to increased stimulation by the monoamine neurotransmitters. A typical response to such overstimulation is for the sensitivity and abundance, and eventually sometimes the synthesis of autoreceptors and other receptors to decline as if to restore homeostasis. However, the loss of autoreceptor feedback leads to a gradual return of neuronal activity, production, and release of norepinephrine or serotonin. Other receptors to which there is little interaction by antidepressants, notably the postsynaptic β noradrenergic receptors, also gradually become less sensitive and abundant. Some pharmacologists have proposed that this change may be essential for the actions of many (but not all) antidepressant treatments [534]. However, if this were the case, one would expect centrally active β-adrenergic antagonists such as propranolol or atenolol to have antidepressant effects. Instead, they lack such effects and may even worsen depression in some susceptible persons [333, 597]. In the serotonin system, $5HT_{2A}$ receptors are little affected directly by antidepressant drugs, and they, too, often become less sensitive or abundant over time [108, 123, 346]. Changes in the complex array of more than a half-dozen other serotonin receptors are not fully documented.

Tricyclic antidepressants have some antagonistic interactions with $α_1$ noradrenergic receptors, and such effects may contribute to postural hypotension associated with their clinical use [276]. Despite the possible protective effect of such interactions, recent findings indicate inconsistently that $α_1$ receptors can become more or less abundant after prolonged treatment with various types of antidepressants, including fluoxetine as well as tricyclics [149, 436]. Given the decreased availability of postsynaptic β noradrenergic receptors, and uncertain status of $α_1$ receptors following long-term treatment with antidepressants, an adequate explanation of the final means of responding to increased synaptic availability of norepinephrine remains elusive, as does an adequate explanation of the outcome of changes in serotonergic neurons induced by antidepressants.

The actions of so-called atypical antidepressants are not well defined. Bupropion, derived from a family of mild-stimulant-like appetite suppressing agents, has some effects that may increase the abundance of dopamine as well as norepinephrine, as well as altering the functioning of serotonergic neurons [205]. One of the few dopaminergic antidepressants to have been used clinically in the past was nomifensine (R,S-2-methyl-4-phenyl-1,2,3,4-tetrahydroisoquinolin-8-amine; Merital®), which blocks reuptake of both dopamine and norepinephrine and was effective in attention deficit disorders as well as depression, but withdrawn owing to associated risk of hemolytic anemia [188]. The pharmacodynamics of mirtazapine also are incompletely defined, but it may act on mechanisms that involve cross-relationships between noradrenergic and serotonergic neurons, both of which express heteroreceptors for the opposite transmitter, to provide for regulatory interactions. Mirtazapine has well-established antagonistic effects at $α_2$ auto- and heteroreceptors that probably allow for increased release of norepinephrine, and less certainly of serotonin [608]. Additional, analogous, interactions with other serotonin receptors may also contribute to increased release of norepinephrine or serotonin. As noted above, the potent H_1 antihistaminic actions of mirtazapine probably contribute to its notable clinical sedative effects. This complex series of effects and adaptations of noradrenergic and serotonergic neuronal systems to the actions of antidepressants are summarized in Table 4.5.

The preceding picture of actions of antidepressants is oversimplified and incomplete. Additional complex interactions between noradrenergic, serotonergic, and even dopaminergic neuronal systems follow antidepressant treatments [110, 536]. Repeated treatment with antidepressants can affect other neuronal systems, including glutamatergic and GABAergic neurons, as well as altering the genetic expression of various neuronal growth factors and other regulatory proteins, and modifying the formation of dendrites and synapses [13, 15, 18, 21, 81, 174, 298, 341, 383, 480, 483, 487, 624]. A number of novel mechanisms, most related to the metabolism or functioning of central monominergic or amino acid neurotransmission, have been proposed or preliminarily explored, but none has yet reached to point of acceptance as clinically useful treatments for depression [107, 249, 342, 394, 435, 599, 619].

Disposition of Antidepressants

Many antidepressant drugs are metabolized initially to additional, pharmacologically active derivatives. A common metabolic route is by oxidative dealkylation of terminal, substituted, amino groups by a family of hepatic microsomal cytochrome-P450-associated oxidase enzymes. These CYP450 isozymes decompose and inactivate a wide range of foreign (xenobiotic), and some endogenous organic molecules. The evident strategy of metabolic management of most foreign substances that are sufficiently lipophilic to pass the blood–brain diffusion barrier is to inactivate them and to increase their water solubility for easier elimination. The structures of the P450 drug-metabolizing enzymes include large, complex heme molecules with a porphyrin ring containing an atom of iron. A characteristic physicochemical property is to absorb light in spectrophotometry at 450 nM wavelength, leading to the term CYP450.

Most tertiary amine tricyclic antidepressants are converted by these metabolic systems to secondary amine derivatives, some of which have been developed as antidepressants in their own right (e.g., desipramine from imipramine, and nortriptyline from amitriptyline). Initially, there was some consideration that the secondary amine derivatives were crucial, pharmacologically active species and might even be faster acting or more effectively antidepressant than the tertiary amine parent compounds, although this speculation was not sustained. Generally, the initial dealkylation step (to a *nor* or *desmethyl* derivative) is relatively rapid. Subsequent metabolism of the secondary amine or other metabolites, such as by ring hydroxylation, is slower than the initial dealkylation steps. Therefore, the overall elimination half-life of most tertiary amine tricyclic and some other antidepressants (notably, amoxapine, fluoxetine, fluvoxamine, venlafaxine) is based on at least two metabolic rates. In such circumstances, two approximate half-life values are provided (Table 4.6). The final metabolism of many antidepressants involves glucuronide conjugation mediated by uridine-5'-diphosphoglucuronosyltransferases (UGT) to form a water-soluble derivative prior to renal elimination [623]. Table 4.6 provides the preferred CYP450 metabolic pathways for most antidepressant drugs, along with approximate plasma elimination half-lives, and typical plasma drug concentrations encountered at clinically effective and tolerated doses.

A particularly striking example of the difference in clearance of a parent antidepressant compound and an *N*-dealkylated (*desmethyl* or *nor*) metabolite is

Table 4.6 Pharmacokinetics of antidepressants

Agent	Preferred CYP450	Elimination half-life (hours)	Plasma concentration (ng/mL)
Tertiatry amine tricyclic type			
Amitriptyline	2C9, 2C19. 2D6	16+30	100–200
Clomipramine	2C19, 2D6	32+70	150–500
Doxepin	2D6	17+50	150–250
Imipramine	1A2, 2C19, 2D6	12+30	175–300
Trimipramine	2C19, 2D6	14+30	100–300
Secondary amine tricyclic type			
Amoxapine	2D6	8+30	200–500
Desipramine	2D6	30	120–300
Maprotiline	2D6	48	200–400
Nortriptyline	2D6	30	60–150
Protriptyline	2D6	85	100–250
SRIs or SNRIs			
Citalopram	2C19	35	75–150
S-Citalopram	2C19	30	35–75
Desvenlafaxine	3A4	11	100–200
Duloxetine	1A2, 2D6	12	80–200
Fluoxetine	2C9, 2D6	50+240	100–500
Fluvoxamine	1A2, 2D6	20	100–250
Paroxetine	2D6	22	30–100
Sertraline	2B6, 3A4, 3C19	25+66	25–50
Venlafaxine	2D6	5+11	100–400
Vilazodone	3A4	22	100–200
NRI			
Atomoxetine	2C19, 2D6	5	—
Atypical			
Atomoxetine	2C19, 2D6	5	—
Bupropion	2B6	14	75–100
Mirtazapine	2D6, 3A4	30	—
Nefazodone	3A4	3	
Trazodone	3A4	6	800–1,600

TCAs have similar structure or pharmacology to tricyclics. *NRI* norepinephrine reupake inhibitor; *SRI* serotonin-reuptake inhibitor; *SNRI* serotonin and norepinephrine uptake inhibitor. Atomoxetine is indicated for hyperactivity. Preferred routes of metabolism by cytochrome-P450 hepatic microsomal oxidases (CYP450) are shown (may differ from pathways with which these agents can interfere). Inhibitory interactions are especially high with fluoxetine vs. CYP-1A2 and CYP-2C, paroxetine and fluvoxamine vs. CYP-2D6, and of nefazodone vs. CYP-3A/4. Agents with two half-life values have active metabolites with slower elimination than the parent compounds. Data are based on manufacturer's product information reports, de Leon et al. [138], Harvard Health Newsletter [233], and Porcelli et al. [431]

fluoxetine, which is oxidized to the long-acting, pharmacologically active metabolite norfluoxetine, with an extraordinarily long elimination half-life, estimated to be about 10 days. This property confers very prolonged action and very slow elimination of norfluoxetine, markedly limiting its risk of inducing the physiological withdrawal syndrome associated with other short-acting serotonin-reuptake inhibitor antidepressants but increasing risks of potentially toxic interactions with other drugs, such as tricyclic antidepressants and MAO inhibitors, days after its discontinuation.

The clearance of most antidepressants is very rapid in children and slows with age, presumably with declining functionality of hepatic enzymatic mechanisms [100, 142, 458, 595]. Optimal plasma concentrations of antidepressants were formerly commonly considered to guide clinical practice, but only a few, mainly for tricyclic antidepressants, have reasonably well-established guidelines that account for more than 50% of the variance in clinical responses. These include: amitriptyline (index concentrations: 90–140 ng/mL), nortriptyline (60–150 ng/mL), imipramine (175–250 ng/mL), and desipramine (120–350 ng/mL). Of these typical circulating concentrations, those for nortriptyline are among the best established, and suggest an optimal range below or above which less favorable responses are likely [418]. Plasma concentrations of tricyclic antidepressants, at least, are highly consistent and predictable within individuals, e.g., following a small, safe, test dose of drug [117]. This individual predictability strongly suggests genetic control of drug disposition and has again become of interest in the search for genetic factors that can predict clinical responses, tolerability, and optimal doses of antidepressants in an effort to improve their clinical value [287, 330, 430, 431]. To date, however, this very plausible, individualized, approach to improving the efficacy of antidepressants has not yielded practical, widely clinically accepted methods [503].

Clinical Treatment with Antidepressants

Introduction

Antidepressants are among the most commonly prescribed and used drugs of all types, and were first among persons aged 18–64 in the mid-2000s, being prescribed at a rate of nearly 7% of all physician visits, and amounting to 10–15 billion US dollars in sales annually worldwide (HCNS 2007; [236]). Older antidepressants, including tricyclics and MAO inhibitors are rarely prescribed as first-line choices as serotonin- and serotonin–norepinephrine-reuptake inhibitors and other second-generation agents have come to dominate clinical practice worldwide since the 1980s. Several of these agents (e.g., bupropion, citalopram, fluoxetine, fluvoxamine, mirtazapine, sertraline, venlafaxine) no longer have patent protection and are available as less expensive generic products, making them more available to patients of limited financial means and attractive to insurers and managed care programs.

The pattern of clinical application of antidepressants changed markedly in the 1990s, following introduction of the serotonin-reuptake inhibitors and other modern antidepressants. Although evidence is lacking that second-generation antidepressants are appreciably more effective than older agents including tricyclics and MAO inhibitors, they are considerably better tolerated, easier to use, and much less likely to produce severe or potentially fatal outcomes if taken in overdoses, as had been common in suicide attempts with first-generation antidepressants. The relative simplicity of using serotonin-reuptake inhibitors led to an explosive expansion of their use, not only by mental health professionals, but by primary-care physicians and clinicians in a range of specialties, sometimes relatively indiscriminately for unclear indications [35, 43, 364]. Despite this increased use of modern antidepressants, the rates of recognition, diagnosis, and adequate treatment of depressive illnesses have remained inadequate. Even in the present era, these rates often fall below 10% of persons with diagnosable major depression [10, 150, 398, 424, 588, 600].

Evidence of the clinical efficacy of most antidepressants is less compelling than for most other classes of psychotropic medicines [300, 428, 550]. Virtually all types of antidepressants have outperformed a placebo in approximately two-thirds of comparisons in published findings from controlled trials. However, this imperfect record may be even less strong than it appears since there is evidence that many trials with "negative" (lack of a statistical difference between test drug and a placebo control) or "failed" (neither test agent nor a standard older drug outperform a placebo) outcomes are not published [568]. Sometimes relatively favored results are reported in published reports, even though they may not be the predefined end-points specified in the initial design of a trial [428]. Moreover, many such trials are not readily accessed owing to several characteristics of unpublished findings. These include: (a) information considered proprietary being unavailable from manufacturers or regulators, (b) potentially inconsistent recording of drug trials and posting them on corporate websites, or (c) lack of interest in reporting negative findings by unregulated academic investigators working without pharmaceutical support. However, inclusion in reviews or meta-analyses of such unpublished, largely negative, findings as are at least partially available leads to substantially weaker drug-placebo differences than are found in published findings [300]. The difference is accounted for almost entirely by higher placebo-associated responses in unpublished antidepressant trials, leading to drug-placebo differences (effect-sizes) that are as much as three times greater in published than in unpublished trials [300].

Dosing and Time Course of Clinical Response

For commonly clinically employed antidepressants, their generic and US trade names, as well as usual daily doses, extreme doses, and relative potencies (based on ratios of median, typical doses to that of imipramine as a standard comparator) are provided (Table 4.7).

In general, dose–response relationships for antidepressants are not securely established and few trials involve prospective, random assignment to specific doses [80, 215, 432]. Adequate evaluation of dose–effect relationships usually requires assessments at metabolic steady state, at which daily drug intake and elimination

Table 4.7 Antidepressant doses

Agent	Brand names®	Usual dose	Extreme doses	Equivalent doses	Potency ratio
Tricyclic-type agents (TCAs)					
Amitriptyline	Elavil, etc.	100–200	25–400	150	1.00
Amoxapine	Ascendin	200–300	50–600	250	0.60
Clomipramine	Anafranil	100–150	25–250	125	1.20
Desipramine	Norpramin, Pertofrane, etc.	100–200	25–300	150	1.00
Doxepin	Sinequan, etc.	100–200	25–400	150	1.00
Imipramine	Tofranil, etc.	100–200	25–400	150	1.00
Nortriptyline	Aventyl, Pamelor, etc.	75–150	25–200	112	1.34
Maprotiline	Ludiomil	100–150	25–225	125	1.20
Protriptyline	Vivactil	15–40	10–60	28	5.36
Trimipramine	Surmontil	50–250	25–300	128	1.17
Serotonin-reuptake inhibitors (SRIs)					
Citalopram	Celexa	20–40	10–60	30	5.00
S-Citalopram	Lexapro	10–25	5–30	20	7.50
Fluoxetine	Prozac, Serafem, etc.	20–40	10–80	30	5.00
Fluvoxamine	Luvox	100–200	25–300	150	1.00
Paroxetine	Paxil	20–40	10–50	30	5.00
Sertraline	Zoloft	100–175	25–200	140	1.07
Vilazodone	Viibryd	20–40	10–40	60	2.50
Serotonin- and norepinephrine-reuptake inhibitors (SNRIs)					
Desvenlafaxine	Pristiq	50	25–400	50	3.00
Duloxetine	Cymbalta	60–100	20–120	80	1.88
Milnacipran	Savella	25–50	25–100	38	3.95
Venlafaxine	Effexor	75–250	25–350	160	0.94
Norepinephrine-reuptake inhibitor (NRI)					
Atomoxetine	Strattera	40–80	5–100	60	2.50
Atypical agents					
Bupropion	Wellbutrin, zyban, etc.	200–300	100–450	250	0.60
Mirtazapine	Remeron	15–40	15–45	30	5.00
Nefazodone	Serzone	200–400	100–600	300	0.50
Trazodone	Desyrel, etc.	150–200	50–600	175	0.86
Monoamineoxidase inhibitors (MAOIs)					
Phenelzine	Nardil	45–60	15–90	52	2.88
R(–)-Selegiline	Eldepryl, Emsam	6–10	6–20	8	18.8
Tranylcypromine	Parnate	20–30	10–60	25	6.00

Doses are in mg/day. Dosing equivalents (mg/day) are based on comparison of median typical daily doses to 150 mg/day for imipramine as a standard comparator. Data are based on manufacturers' product information bulletins

are in equilibrium (typically in 5 or 6 elimination half-lives, or less than a week for agents with a half-life of 20–30 h). For some agents with unusually prolonged elimination times (e.g., fluoxetine, protriptyline [Table 4.6]), the duration of observation may not be sufficiently prolonged as to provide steady-state conditions for in dose–effect assessments. Most recommended doses are relatively conservative estimates, usually negotiated by manufacturers and regulatory agencies to provide a balance between being high enough to demonstrate efficacy but low enough to limit risks of intolerable adverse effects.

An important early study of the effects of imipramine treatment in panic disorder found a clear relationship between greater clinical benefits and higher doses, across a spectrum of randomized dosing from placebo to nearly 200 mg/day of imipramine. However, there was also a dose-dependent risk of discontinuing, presumably owing largely to poor tolerability, as doses increased above 50 mg/day [344]. These findings reinforce the general pharmacological point that all dose–response relationships are continuous and probabilistic, and that dose–benefit and dose–adverse effect relationships are well described by S-shaped functions relating effects to the logarithm of daily dose. In such relationships, dose–benefit curves lie to the left and are separated by the "therapeutic index," margin of safety, or ratio of median doses for beneficial vs. adverse effects. For first-generation antidepressants, this index was approximately 10, so that up to a 10-day supply could be dispensed with reasonable assurance that an acute overdose would not be fatal if the drug had not been boarded; the margin of safety of modern antidepressants is much larger. Adverse effects are commonly referred to, somewhat euphemistically, as "side" effects.

A striking example of the trade-off between too little and too much in antidepressant dosing is the case of bupropion, which was nearly held off the market due to an association with epileptic seizures at the upper end of the dosing range employed in its initial clinical trials [31]. As a compromise, it was licensed based on lack of a statistically significant increased risk of seizures at doses below 450 mg/day, even though considerably higher doses had been employed in trials and even though dose–risk relationships are continuous and probabilistic. The relatively low recommended doses may contribute to the clinical reputation of bupropion as having a relatively low risk of inducing mania in bipolar depressed patients [235, 556]. Although larger doses of antidepressants often are associated with improved responses, there also are reports of complex relationships between drug dose and antidepressant effects, with superior benefits as well as better tolerability at more moderate doses [91, 358, 404, 406].

Effective practice calls for routine use of daily doses of antidepressants at the equivalent of about 150 mg/day or imipramine or 20 mg/day of fluoxetine (daily imipramine- or fluoxetine-equivalent doses). Higher doses, up to 5 mg of impramine per kg of body weight, or its equivalent in children, and up to 200–300 mg/day imipramine-equivalent in healthy adults, can be tried when lesser doses appear to be inadequate and higher doses are tolerated. It is often possible, following an initial period of gradual exposure, to administer an entire day's dose of an antidepressant at one time (commonly at bedtime), but this is most safely done with young adults at daily doses below the equivalent of 200 mg/day of imipramine or 30 mg/day of fluoxetine. Moreover, some agents with short half-life (e.g., bupropion, trazodone) or with adverse effects on sleep (e.g., fluoxetine, bupropion) are best avoided at bedtime.

The timing of clinically observable and substantial beneficial effects of most antidepressants, in contrast to sedative-anxiolytics and stimulants, is typically delayed for several weeks, perhaps reflecting the complex secondary pharmacodynamic and physiological adaptations to antidepressant treatment described above [214]. Remission in acute major depression is typically nonlinear any many weeks of treatment may be required to reach sustained recovery. Based on scores obtained with standard symptom rating scales, most improvement typically occurs by 4–6 weeks, at early rates of perhaps 10% per week, with very little additional response with placebo treatment thereafter and only modest additional gains (often 1–2% per week) with active drugs up to 12 weeks, though with greater drug-placebo differences after 4 weeks (e.g., [91, 522]). Explanations for this pattern of relatively slow clinical response to antidepressants have remained elusive, but secondary physiological adjustments in molecular mechanisms initiated by antidepressant drug treatment in the CNS as described above may be involved. Such delays can be frustrating to patients and clinicians and present risks, including of suicidal behavior. Such risks call for particularly close clinical supervision early in treatment. Some clinicians may offer a sedative or hypnotic agent to obtain some subjective benefits early in treatment.

There is good evidence that some degree of improvement of depressive symptoms observed within the first 1 or 2 weeks of antidepressant treatment may be a good predictor of eventual response or clinical remission, or that its absence augurs a poor outcome [53, 202, 243, 289, 318, 537]. Lack of any indication of improvement within even such a short time may encourage adjusting or changing treatments, particularly in circumstances of relatively high risk of adverse responses, such as in elderly, medically unwell, depressed patients [263, 292]. It follows that continuing antidepressant treatment in the upper range of tolerated doses for at least 2 or 3 months may yield responses that would otherwise be missed, whereas lack of any improvement within 2 weeks may encourage early consideration of alternative treatments or use of adjunctive agents.

The preceding considerations indicate that failure to respond to moderate doses of an antidepressant should encourage trials of higher doses and more prolonged use of tolerated doses, particularly when there are even minor early improvements, before an alternative antidepressant or more complex, multiagent, treatment regimens are considered [416]. Although typically effective doses of antidepressants for treatment of outpatients with a major depressive disorder are at the approximate equivalent of 100–150 mg/day of imipramine or 20 mg/day of fluoxetine (Table 4.7), relatively high doses may be required for some anxiety disorders, notably obsessive-compulsive disorder, which is likely to require daily doses several times greater than are typical for major depression, but with increased risk of adverse effects [75].

Antidepressant Selection; Comparative Efficacy

Clinical choices of specific antidepressants rest heavily on tolerability and acquisition cost, since it is extremely difficult to demonstrate substantial differences among specific agents or between types of antidepressants based on their apparent efficacy [16, 61, 111, 167, 168, 227, 297, 334, 337, 410, 524, 570]. Examples of findings

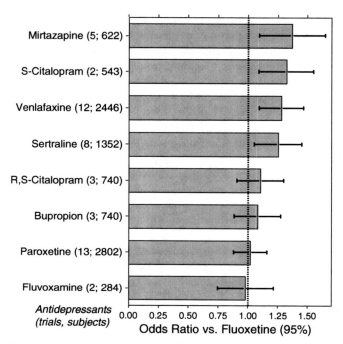

Fig. 4.10 Outcomes of randomized trials of modern antidepressants vs. fluoxetine as a standard comparator (odds ratios with 95% confidence intervals). Data are adapted from a meta-analytic review by Cipriani et al. [111], showing agents compared with fluoxetine, with numbers of trials and total numbers of depressed patient-subjects. *Vertical dotted line* is the null value of 1.0; note that most confidence intervals overlap, indicating little difference among specific agents

indicating strikingly similar, and modest, efficacy in the treatment of major depressive disorder from meta-analytic reviews of randomized trials of various antidepressants vs. placebo and against the standard agent fluoxetine are provided (Figs. 4.10 and 4.11; Tables 4.8 and 4.9).

It is important to point out that such findings based on meta-analysis, though anticipated as a route to sound, evidence-based psychiatric therapeutics to guide practice policies, medical education, and clinical treatment, can be inadequate or misleading. The data available are highly at risk for biases in their selection, analysis, interpretation, and reporting. There is growing awareness that both industrial and academic studies can be subject to biases influencing selection and presentation of findings from even well-designed and conducted therapeutic trials, notably including those involving antidepressants [300, 568].

Efforts have been made by government agencies and journal editors to increase the likelihood that findings from formal therapeutic trials will be registered and numbered at their inception, and later reported, at least in summary form. Such reporting can involve internet websites such as those managed by the US National Institutes of Health [378] or provided as a public service at corporate sites by many

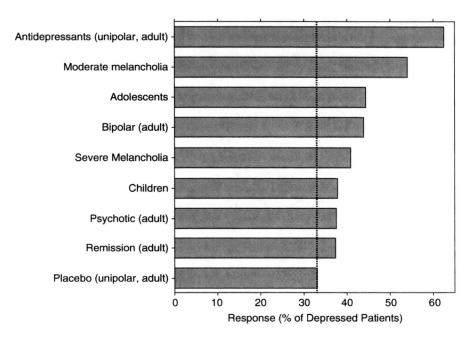

Fig. 4.11 Approximate rates of response to antidepressant or placebo treatment in major depressive disorder in various clinical groups. Data for adult responses to antidepressants and placebo are from Undurraga and Baldessarini [570]. Values for melancholia compare moderate and severe illnesses [420]. Values for psychotic depression are from Kocsis et al. [304]. Estimates for bipolar depression are highly variable and are averaging across several reports [207, 501, 557, 574, 575]. Values for juveniles are tentative, with very limited data for prepubertal children [565]

pharmaceutical manufacturers, to supplement information provided in published reports. Nevertheless, there are limits to what can be known through such services regarding details of trials whose results have not been subjected to the peer-review process and published in relatively full, archival form. Many trials with "negative" findings (lack of a statistical difference between a test agent and a control treatment) conducted by academic investigators not as components of a pharmaceutical drug-development program, risk neither being published nor even known to the general or academic public, and so not being considered in reviews. Some trials that are made known, particularly with summaries and selected presentations of data, may not be fully and optimally informative.

Efforts to gain access to databases of trial results submitted for review by regulatory bodies such as the US FDA and its international equivalents for licensing and safety monitoring often result in access only to executive summaries, and rarely to raw data, which typically are considered "proprietary" by both drug manufacturers and regulators. Moreover, some negative or failed trials that are required to be submitted for regulatory review are neither provided for independent analysis nor published. Perhaps a third of the total corpus of data submitted for regulatory review is never published following critical, expert, peer review, and most

Table 4.8 Meta-analysis of responder-rate ratios (RR) for antidepressants vs. placebo in random-ized, controlled trials reported between 1980 and 2010

Drugs	Trials (n)	Pooled RR	95%CI	p-value
Amitriptyline	11	1.74	1.50–2.01	<0.001
Mirtazapine	3	1.73	1.26–2.36	<0.001
Imipramine	17	1.58	1.37–1.83	<0.001
Citalopram	6	1.48	1.24–1.76	<0.001
Desipramine	2	1.45	1.07–1.96	<0.01
Venlafaxine	13	1.45	1.35–1.56	<0.001
Paroxetine	11	1.44	1.26–1.66	<0.001
Desvenlafaxine	8	1.41	1.16–1.72	<0.001
Escitalopram	5	1.33	1.20–1.48	<0.001
Sertraline	8	1.33	1.20–1.47	<0.001
Selegiline	3	1.33	1.07–1.65	<0.01
Fluoxetine	17	1.31	1.07–1.60	<0.01
Duloxetine	10	1.29	1.09–1.52	<0.05
Bupropion	7	1.23	1.14–1.33	<0.001

Based on random-effects meta-analyses for a total of 126 placebo-controlled trials of 19 antide-pressants, showing ratios of responder rates (usually with ≥50% improvement from baseline depression ratings for drug/placebo [RR], with 95% confidence intervals [CI]. Shown are data for 14 agents with ≥2 trials in a total of 121 trials, ranked in descending order of RR. Distance above the null value of RR = 1.0 indicates greater separation of drug from placebo (ranging from 23 to 74% superiority to placebo). The overall meta-analytically pooled RR for all 126 trials = 1.42 (CI: 1.38–1.48); ($p<0.0001$). Note that the CIs overlap between most agents, and a secular artifact of lower placebo-associated responses affects older trials with TCAs and MAOIs to increase their apparent effect-sizes. Adapted from Undurraga and Baldessarini [570]

unpublished trials that are identifiable involve negative results [300, 568]. Overall, efforts to circumvent limitations and potential biases imposed by incomplete reporting of trials outcomes, though very well intended, may lead to their own biases of access, selection, and assessment. The process of developing rational, data-based therapeutic policies and practices, therefore, continues to require a crit-ical perspective and considerable clinical judgment [156, 261, 531].

In addition to limitations of availability of data for review and meta-analysis, there are fundamental limitations to the process of meta-analysis itself. Although huge numbers of individual patient-subjects may be involved, summary analyses are similar to descriptive, clinical studies, and their number for statistical analysis is that of the trials, not of the subjects. Comparisons among specific agents or between classes of antidepressants are likely to yield minor differences for additional rea-sons. First, drug-placebo contrasts (effect-sizes), even in published reports, are noto-riously limited with most antidepressants and are highly dependent on placebo responses. In turn, placebo responses are likely to reflect characteristics of the sites and investigators involved and on the types of patients recruited [300, 570, 574]. Each trial is an exercise in averaging across potentially large variances among sites, investigators, and raters, as well as among patient-subjects. Meta-analytic pooling of data across trials involves further averaging of potentially heterogeneous trials, even with efforts to follow laudable quality standards for including studies. The results

Table 4.9 Effect-size: antidepressants vs. placebo in randomized, controlled trials

Drugs	Effect size (Hedges' g)
Venlafaxine	0.40
Paroxetine	0.37
Mirtazapine	0.35
S-Citalopram	0.30
Duloxetine	0.30
Fluoxetine	0.26
Nefazodone	0.26
Sertraline	0.26
R,S-Citalopram	0.24
Bupropion	0.17
All drugs	
FDA files	0.31
Published data	0.41

Data are for meta-analytically computed effect-size (Hedges' g = improvement with [drug–placebo]/pooled standard deviation). From review by Turner et al. [568], involving 74 trials (31% unpublished) with 12,564, patients, evaluated by FDA. Of published trials findings 72.6% were "positive" (drug significantly superior to placebo); of unpublished trials, only 4.4% were positive. Publication yielded 32% greater apparent effect-size

are very likely to represent highly regressed averaging of relatively small differences over very large heterogeneity. Such characteristics make it very challenging to categorize specific treatments as more effective or better tolerated than others.

Additional potential difficulties in comparing trials and medicines include potential nonequivalence of doses across trials involving different drugs. Such mismatching can arise from simple and prevalent lack of definitive dose–effect data. Such characterizations would involve trials that are ethical but involve strict randomization to specific doses over a sufficiently broad range as to provide substantial variance in outcomes. Such requirements are likely to impact on the challenge of retaining subjects, particularly at the high (poorly tolerated) and low (ineffective) ends of the dosing range involved. Additional pharmacological problems can arise when comparisons involve a preferred, newer, enantiomerically purified agent vs. an older racemic mixture of optical isomers (e.g., *S*- vs. *R,S*-citalopram; *R*- vs. *R,S*-fluoxetine; *d*- vs. *d,l*-methylphenidate). Enantiomerically purified drugs have become much more common in recent years as the technology for separating isomers has become more cost-effective at the industrial level, and is encouraged by hopes of limiting risks of adverse effects by *more-or-less* doubling the potency of purified products [455]. Typically such dosage comparisons assume a precise 2:1 potency ratio for enantiomer:racemate, as if half of the racemate were inert and merely being discarded. However, "inactive" isomers sometimes can affect the potency of the more active or preferred isomer to distort estimates of equivalent doses [132].

Although specific agents or classes of antidepressants are hard to distinguish by efficacy in the treatment of major depressive disorders, there have been suggestions that some types of drugs may have advantages for particular indications. For example, syndromes of chronic "hysteroid dysphoria" that may be related to borderline personality disorder, or of "atypical" depression with hypersomnia and hyperphagia may respond particularly well to MAO inhibitors [148, 242, 401, 402, 527, 547]. It has been proposed that older tricyclic antidepressants may be superior for patients with melancholic or otherwise relatively severe depression or in need of hospital-level care [410, 521]. However, such comparisons may be confounded by greater tolerability of serotonin-reuptake inhibitors compared to tricyclics and greater acceptance of less tolerable agents in severe illnesses [29, 343]. None of these suggestions regarding specific indications for particular types of antidepressants is firmly established.

Efficacy of Antidepressants: Methodological Considerations

Assessment of efficacy of antidepressant drugs has always been challenging. Appropriately, regulatory agencies have required, and reviews have relied on findings from placebo-controlled, randomized and blinded trials. However, most reported studies involving subjects with clinical depression have relied on ambulatory patients with illnesses of moderate severity, rarely complicated by psychotic features or severe comorbidity. Results also have varied markedly with outcome measures. Outcomes typically have included: (a) the proportion of depressed patients randomized to treatment with drug vs. placebo who show a "*response*" that has usually been defined at achieving at least 50% improvement in symptom severity compared to ratings at intake based on a standard symptom-rating scale (most often the Hamilton Depression Rating Scale [HDRS], which is available with 17 or 21 items, or the Montgomergy–Åsberg Depression Rating Scale [MADRS]); (b) actual *percentage-improvement* of such ratings; (c) less often, rates of achieving symptomatic *remission*, usually defined as reaching a very low score on a standard depression rating scale; or (d) least often, *clinical recovery*, usually defined as remission sustained for perhaps 2 or 3 months, although not necessarily with full return of premorbid levels of functioning. As might be expected, the rates involved decline as the rigor of the definition of response increases (from *a* to *d*), in part owing to the time-limited nature of most trials which usually range from 6 to 12 weeks duration in acute major depression. Rates of remission are typically 10–20% lower, but can be as much as 50% lower, than rates of response.

From the start of therapeutics research with antidepressant drugs in the early 1960s, rates of success in demonstrating statistical superiority of active drugs over inactive placebos, or the effects of nonpharmacological treatments, have been modest, with failure to demonstrate statistical superiority of drugs in at least a third of trials, even based on relatively liberal criteria for clinical improvement, such as category *a* ("response"). Such outcomes are considered "unsuccessful" or "negative" trials. In addition, therapeutic trials sometimes include a third randomized arm involving a standard agent of accepted clinical value; failure to distinguish a standard treatment from placebo is considered a "failed" trial. Given the substantial rates of negative or failed antidepressant trials, and modest drug-placebo differences, it would be highly risky to evaluate a proposed new antidepressant treatment only by comparison to an accepted standard treatment. That is, such trials carry a high risk of finding no apparent difference, with risk of interpretation of a new treatment being "about as good as" a standard treatment. Ironically, the less well designed and conducted the trial, the higher the risk of not finding differences between treatment conditions. Convincing evidence of similar efficacy requires particularly well-designed and -conducted trials of substantial size and statistical power. Evidence that a new treatment is actually superior to a standard treatment would be even more welcome, but examples are hard to find.

A basic characteristic of trials of antidepressants recognized since the 1960s is that differences in response between active drugs and placebos (effect-sizes) usually are quite modest. Formerly, most antidepressant studies followed a simple rule that about two-thirds of trials produced data favoring drug over placebo, and about two-thirds of depressed patients showed symptomatic benefit, and only one-third responded to an inactive placebo. Use of *active placebos*, such as agents with atropinic effects to mimic adverse effects of some antidepressants, comparisons with an anxiolytic sedative, or use of very small doses of an antidepressant, has yielded even smaller contrasts [365]. Moreover, even nominally inactive placebo treatments typically involve potentially important "nonspecific" factors such as psychological support and close clinical follow-up and sometimes use of sedative-hypnotic agents. Currently, differences in symptomatic responses between antidepressants and placebo are even less favorable than formerly, sometimes with differences on the order of 10–15% in symptomatic improvement [300, 568, 570]. It has been recognized for several decades that successful and unsuccessful therapeutic trials of antidepressants often show only minor differences in average drug-associated responses, but large differences in responses during random assignment to a placebo [451]. This phenomenon of variable responses during treatment with a placebo as a more important factor than drug-associated responses for the outcome of antidepressant trials has been noted in recent reviews as well [300, 568, 570]. In response to the realization that placebo-associated outcomes were crucial for the success of antidepressant trials, considerable attention has been directed toward identifying patients least likely to respond during treatment with a placebo vs. an active antidepressant drug. As might be expected, increased drug-placebo contrasts generally appear to parallel clinically expected differences in illness severity, such that greater severity tends to disfavor responses during placebo treatment and so favor drug-placebo contrasts [164].

An apparently important factor associated with placebo responses and trial outcomes, as noted above regarding trials in acute mania [611, 612], is that placebo-associated responses but not drug-associated responses, in antidepressant trials have been rising steadily and drug-placebo differences falling since the 1980s [570]. The trend is paralleled by increased numbers of collaborating sites and of subjects per trial, probably in an effort to overcome falling effect-size. The growth in size and complexity of trials, especially those involving recruitment in alternative cultures, may contribute to less stable and reliable diagnostic and assessment data in modern trials [574]. It is plausible to suspect, further, that increasingly "noisy" or unstable data are likely to have a greater effect on placebo-associated conditions, tending to promote "regression to the mean" effects [189, 528, 570, 612].

An unexpected and artifactual consequence of the trend toward selective increases in placebo responses from an average of approximately 30% in the 1980s to 40% or more currently is that contrasts of older drugs in older trials (e.g., tricyclic anti-depressants) yield somewhat superior apparent drug-placebo differences than are found with modern agents (e.g., serotonin-reuptake inhibitors). However, when modern placebo rates are compared to responses to older agents, differences between older and newer antidepressants disappear [570].

Efficacy of Antidepressants: Recent Findings

Given this background, it is not surprising that drug-placebo contrasts associated with antidepressants in recent reviews of their placebo-controlled trials are sufficiently limited as to have encouraged wide and highly critical public and professional discussion of their value (e.g., [20, 133, 180, 262]). Reported average differences in responses to antidepressants vs. placebos often have ranged from 11% [300] to 24% [133], but even to effect-sizes of 30–40%, in part depending on whether unpublished data are included and on the outcome measures employed ([568]; Table 4.9). Our recent review of available, peer-reviewed, published trials results dating from the 1980s indicated quite moderate overall antidepressant drug-placebo contrasts in the treatment of patients diagnosed with major depression. There was a 42% greater chance of responding to an active drug than a placebo, a modest 12.5% drug-placebo difference in improvement, and a moderate estimated number-needed-to-treat (NNT) [NNT=8] ([570]; Table 4.10). Treatments with NNT estimates below 10 patients usually are considered as at least moderately favorable, and those with NNT of 5 or less, particularly favorable [294]. The observation of somewhat larger differences between drugs and placebo with older antidepressants (tricyclics and MAO inhibitors) is, however, confounded by the rise in placebo-associated responses in recent years, as noted.

Overall, the several preceding findings indicate quite modest drug-placebo contrasts, or effect-size, and add support to the general impression that most antidepressants have similar beneficial effects in patients diagnosed with major depression, especially in moderately ill outpatients. A possible exception may be bupropion (Tables 4.8 and 4.9), perhaps owing to the decision to limit its maximum recommended doses to avoid dose-dependent risk of epileptic seizures or abnormalities of the electroencephalogram [177, 336, 523].

Optimization of Antidepressant-Placebo Contrasts

The modest differences between antidepressants and placebos raise the question of whether antidepressants are simply marginally effective agents, or if they may be particularly useful for some patients, as clinical experience appears to indicate.

Since the outcomes of controlled trials appear to depend crucially on responses associated with placebo treatment, the suggestions noted above concerning patients least likely to respond to a placebo may be instructive and clinically useful. Characteristics of patients more likely to respond clinically during treatment with an antidepressant than with a placebo include: more years from onset of depressive illness, longer current depressive episode, lack of comorbid anxiety or substance abuse, endogenous features without notable stressful immediately antecedent events, and functional impairment in employment or personal care. Most of these characteristics have been sustained in recent studies, and others added, usually in terms of characteristics associated with an unfavorable separation of antidepressants from placebos. Nevertheless, the evidence supporting such impressions is quite limited [164, 386]).

The important topic of possibly inferior responses to antidepressants in short- or long-term treatment of bipolar disorder patients was considered in Chapter 3. It continues to be a pressing, unresolved, question with conflicting research findings [151, 203, 204, 207, 235, 462, 501, 575]. Even mild indications of hypomanic or other symptoms associated with bipolar disorder may be associated with inferior response to antidepressant treatment of depression [63, 151, 462]. In addition, patients with hypersomnia, hyperphagia, or other "atypical" signs with relatively poor response to antidepressants share similarities with bipolar disorder patients [242, 401, 402, 419, 547]. If the hypothesis that bipolar features are associated with inferior response to antidepressant treatment is valid, it adds to the importance of early recognition of bipolar mood disorders, particularly of relatively subtle forms not associated with mania. The question is particularly important given the high proportion of time that bipolar disorder patients spend in depressive-dysphoric states, and the generally unsatisfactory state of treatment of these components of bipolar disorders [38, 45].

Features suggesting a bipolar diagnosis, even prior to manic or hypomanic manifestations, were discussed in Chapter 3. They include a family history of mood disorder or substance abuse, early onset age, relatively frequent recurrences, presence of prominent psychomotor retardation, "leaden paralysis," "atypical" features including hypersomnia, daytime napping, hyperphagia, or psychotic features [351, 360]. It remains uncertain whether bipolar depression is generally poorly responsive to antidepressant treatment, or if poor response, or even worsening, may be associated with the presence of such features as agitation, anger, prominent anxiety, or other suggestions of overarousal or bipolarity as critical variables in individual patients [557]. It is striking that, despite current uncertainty of their relative efficacy in bipolar depression, antidepressants remain among the most prevalent treatments given to, and often demanded by, patients diagnosed with bipolar disorder [34, 36, 45]. Currently favored in adult bipolar depression are bupropion and short-acting serotonin- or serotonin–norepinephrine-reuptake inhibitors; mirtazapine and nefazodone are also plausible options, but require further evaluation [529].

Initial illness severity also is of considerable interest in regard to effect-size in trials of antidepressants, but may have a complex relationship to outcomes. There have been many reports that higher initial symptom severity ratings on standard

Table 4.10 Efficacy of types of antidepressants in randomized, controlled trials (1980–2010)

Drug types	Trials (n)	Drug vs. placebo efficacy measures (95%CI)		
		Responders RR	%-Improvement RD	NNT
TCAs	31	1.62 (1.47–1.78)	16.2 (13.3–19.1)	6.2 (5.2–7.5)
SNRIs	30	1.40 (1.30–1.51)	9.80 (7.14–12.5)	10.2 (8.0–14.0)
MAOIs	5	1.39 (1.11–1.48)	16.0 (0.98–33.0)	6.2 (3.0–10.2)
SRIs	47	1.37 (1.27–1.48)	11.5 (8.70–14.2)	8.7 (7.0–11.5)
Atypical	11	1.25 (1.15–1.35)	12.8 (8.19–17.4)	7.8 (5.7–12.2)
All drugs	124	1.42 (1.38–1.48)	12.5 (11.0–14.1)	8.0 (7.1–9.1)

From random-effects meta-analysis of responses to antidepressants vs. placebo, based on responder (usually ≥50% improvement) rate ratios (RR [rate ratio]; null = 1.0) or differences in percentage-improvement (RD [rate difference]; null = 0%), with estimated numbers-needed-to-treat (NNT; reciprocal of RD, smaller is better), in 124 randomized, placebo-controlled trials of 19 antidepressants reported between 1980 and 2010 that met inclusion criteria. For all outcomes, differences between drug and placebo were highly statistically significant (all z-scores ≥3; all $p < 0.005$), but the RR and RD measures are not closely correlated ($r_s = +0.257$, $p = 0.565$). The 95% confidence intervals (CI) overlap between most comparisons, precluding ranking by efficacy; moreover, a major secular trend toward rising placebo-associated responses over the years led to inflation of earlier contrasts due to lower placebo rates. Agent types are tricyclic-type antidepressants (TCAs), serotonin–norepinephrine-reuptake inhibitors (SNRIs), monoamineoxidase inhibitors (MAOIs), serotonin-reuptake inhibitors (SRIs), and atypical agents (including bupropion and mirtazapine). Data are adapted from Undurraga and Baldessarini [570]

depression rating scales are associated with greater apparent effect-size, and the association appears to have been accepted as a truism [180, 300]. Much of the effect is likely to reflect the "law of initial values," such that more deviant initial status is likely to yield larger changes that are statistically more favorable [66, 596]. However, the effect of initial deviance can apply to both drug and placebo responses to complicate outcomes considered as drug-placebo differences (efficacy). Importantly, depressed patients with unusually high initial symptom ratings on a standard rating scale (e.g., above 40 on the HDRS), as well as those who have psychotic features, severe anxiety, active substance abuse, or are suicidal are very likely to be excluded from, or to avoid, controlled trials, especially those involving a placebo condition. Moreover, very severe depression ratings are also likely to be associated with more complex clinical histories, including multiple episodes, variable previous responses to treatment, hospitalization, comorbidity, and disability, as well as a longer current episode. Such features are also likely to limit responses to antidepressant treatment [169, 245, 258, 279, 370, 386]. Some studies have, indeed, found that very severe initial depression ratings were associated with poor drug-placebo differentiation [258, 372, 420]. Curiously, extreme severity may reduce drug-placebo contrasts in depressed patients with melancholic features, but increase the contrast among those without such features [420]. However, melancholia itself may not predict an inferior response to antidepressant treatment [76, 90, 409].

Psychotic features have long been recognized as an unfavorable factor in predicting response to antidepressant treatment in major depression. In a review of 12 trials involving over 1,000 depressed patients, response rates among those lacking psychotic features averaged 69 ± 6%, compared to only 28 ± 6% with such features [134].

However, the specific impact of psychosis as opposed to extreme illness severity remained unclear until a study compared depressed patients matched for initial standardized depression symptom ratings, with and without psychotic features, both compared to moderately severely ill patients [304]. As might be expected, response rates ranked: moderately depressed, not psychotic (52%) > severely depressed, not psychotic (48%) > severe and psychotic (37%), indicating that both severely symptomatic depression and psychotic features probably contributed to inferior beneficial responses to antidepressant treatment.

For severe and psychotic major depressive illnesses, electroconvulsive treatment (ECT) continues to be widely considered the treatment of choice [226], with combinations of antidepressant and antipsychotic agents as a preferred second choice. In a review of therapeutic trials of antidepressants or antipsychotics alone or in combination among patients with psychotic major depression, responses ranked: antipsychotics alone ($29.7 \pm 9.0\%$) < antidepressants alone ($42.9 \pm 10.9\%$) < combinations ($57.3 \pm 9.6\%$; [594]). Antidepressants also have been used empirically in efforts to limit the severity of depressive and negative symptoms in schizophrenia or other chronic psychotic disorders, usually with limited responses, but with very few controlled trials to test for efficacy and tolerability [54, 105, 354, 543].

Additional important predictors of inferior responses to antidepressants for major depression include depression secondary to a major general medical or neuromedical illness, cognitive impairment in the elderly, severe melancholia, multiple prior depressive episodes or chronic dysthymia, and previous failed antidepressant trials [8, 245, 263, 292, 313, 420, 515, 593].

Among psychiatric comorbidities, presence of a severe anxiety disorder appears to be an unfavorable predictor of antidepressant response in depression [258, 313, 439]. In contrast, abuse of alcohol or opiates has not been sustained as an important negative predictor in meta-analytic reviews [264, 413]. Personality disorders also are suspected of contributing to poor responses, but remain inadequately studied [313].

Age may well be an important correlate of antidepressant response: antidepressant efficacy appears generally to rise with age [279, 370]. Although information pertaining to prepubertal children is very limited, most antidepressants have not done as well in depressed juveniles of mixed ages as in adults, and even less well in depressed children considered separately [565, 583]. Some findings with serotonin-reuptake inhibitors and bupropion are encouraging for depressed adolescents [565], and tricyclics appear to be similarly effective as stimulants in pediatric attention-deficit hyperactivity disorder (ADHD; [259]). Only fluoxetine is FDA approved for juvenile depression, based on relatively large trials but modest effect-sizes. There is some evidence that older depressed adults, especially those over age 65 years may show decreased suicidal ideation and behavior when treated with an antidepressant, with lesser effects or even worsening among younger patients [36, 43, 51, 323]. Modern and older antidepressants appear to be effective in elderly as well as young–adult patients, although tricyclics and MAO inhibitors are usually avoided in the elderly for safety reasons. In general, there is a pressing need to study depression in prepubertal children and to compare antidepressant efficacy in young vs. elderly adults, especially since antidepressants have been found to reduce suicidal risk selectively in elderly depressed patients while possibly increasing risk in juveniles and young adults [51, 323]. Among demographic factors, women may be somewhat more responsive to antidepressant treatment of depression than men, perhaps particularly when treated with a serotonin-reuptake inhibitor [279, 370, 577, 609, 615].

Finally, there is a substantial but largely inconclusive research literature pertaining to the value of various biomedical testing methods to predict clinical response to treatment in patients diagnosed with major depression. These include assays of monoamine metabolites in body fluids and measures of endocrine functions, particularly in the adrenocortical and thyroid systems [24, 32, 163, 362, 368, 421, 499]. There is also evidence that initial behavioral excitation provoked by a stimulant or normalization of depression-associated shortened latency to rapid eye movement sleep early in antidepressant treatment can predict favorable responses several weeks later [22, 311]. Encephalographic (EEG) and other physiological measures to identify patients likely to respond to antidepressant treatment also are under consideration [52, 329, 373]. There is growing interest in the application of modern molecular genetic methods to identify patients most likely to respond favorably to antidepressant treatment, based on such factors as drug-metabolizing enzymes or on pharmacodynamic elements including the control of synthesis of SERT proteins [55, 138, 287, 431]. Some of these approaches represent extensions of earlier analyses of pharmacokinetic phenotypes represented by circulating concentrations of antidepressants, as was discussed above regarding antidepressant disposition. Although these various approaches have shed some light on biological aspects of depression and its treatment, very little of clinical value and utility has emerged to date.

Management of Treatment-Resistant Depression

Definitions of "treatment resistance" vary, but nonresponse and partial response are prevalent outcomes among unipolar-depressed patients [165]. Responses to initial antidepressant treatment are clinically unsatisfactory in one-third to one-half of patients diagnosed with acute major depression, and one-fifth to one-third of patients appear to be essentially treatment unresponsive [165, 514, 515]. As already noted, proposed risk factors for poor treatment responses include comorbid anxiety disorders, multiple episode recurrences with hospitalizations, very severe melancholic depression, early onset age, and prominent functional disability [446, 514, 569]. Since major depression and antidepressant treatment are among the most common conditions encountered in clinical psychiatry, it is not surprising that addressing treatment resistance in major depression patients is one of the most frequent problems addressed by contemporary consultants in clinical psychopharmacology. Given the high rates of clinically unsatisfactory outcomes of antidepressant treatment, there is intense interest in developing rational strategies for additional steps to be considered.

Table 4.11 Managing treatment-resistant depression

Probably clinically useful

   Re-evaluate diagnosis (rule out comorbidity, bipolarity) and adherence

   Increase dose (serum concentration) of a standard antidepressant, as tolerated

   Add a trial of a tested psychotherapy such as cognitive-behavioral therapy (CBT)

   Add lithium (to 0.8 mEq/L ≥6 weeks), especially with TCAs

   Add second-generation antipsychotic (aripiprazole, olanzapine, quetiapine), especially if
       agitated, psychotic or bipolar

   Switch to an MAO inhibitor and gradually increase dose, watching for toxic interactions

   ECT trial (especially if psychotic, very severe or suicidal)

   Consider vagal nerve stimulation (VNS; FDA-approved 2005)

   Consider repeated transcranial magnetic stimulation (rTMS; FDA-approved 2008)

   Discontinue antidepressant if agitated, sleepless, aggressive

Value less certain

   Add an anticonvulsant (especially lamotrigine; avoid carbamazepine)

   Switch from SRIs to an SNRI or atypical-modern antidepressant

   Switch to a tricyclic antidepressant

   Add bupropion, a stimulant or modafinil, or pramipexole, especially if sedated or dulled, or
       geriatric

   Add buprenorphine, S-adenosyl-L-methionine

   Add thyroid hormone (T3 inconsistent, TSH uncertain, T4 untested)

   Add benzodiazepine, gabapentin, especially if anxious, or buspirone, especially if apathetic

   Add pindolol to hasten onset of SRIs (evidence in treatment resistance weak)

   Consider adding acupuncture for moderate depression

   Consider transcranial direct-current stimulation (TDCS)

   Consider deep brain stimulation (DBS)

[9, 77, 99, 116, 125–127, 153, 162, 166, 196, 200, 248, 251, 253, 257, 272, 274, 307, 322, 325, 338, 363, 387, 404, 405, 427, 453, 460, 461, 467, 508, 547, 549, 563, 579, 610]

Common steps that have been tested or tried clinically are summarized in Table 4.11. They include interventions that have undergone relatively extensive testing and have both clinical and research support. A very basic initial step is to re-evaluate the diagnosis and to consider potentially relevant comorbidities. This potentially awkward initial intervention often can be managed by seeking an independent consultation with a competent colleague willing to take a fresh and open-minded approach to clinical assessment. Sometimes, unexpected endocrinological, cancer-related, or neuromedical disorders are found that require separate treatment. Elderly patients may require close differentiation of primary mood disorders from early dementia or neurological effects of concurrent illnesses or their treatments. Also common are underappreciated comorbid conditions including severe anxiety, abuse of alcohol, or other substances with mood-altering effects when abused or during withdrawal [389]. In addition, many patients are considered to have a unipolar major depressive disorder for several years before they are recognized as having a bipolar disorder, especially of type II, in which hypomanic phases may not be recognized as pathological by the patient, family members, or ascertained by clinicians; the diagnosis may affect treatment response, as discussed above. In other cases, prominent personality disorders may complicate illnesses that may appear at first to represent simple depressive disorders and limit their response to treatment. When there is confidence that unipolar major depression is the correct diagnosis or a valid component of a more complex illness, then further consideration can be given to treatment.

Increases in dose and longer treatment are simple and sometimes effective steps that should be considered before adding adjunctive or augmenting treatments or changing to other antidepressants [154, 166, 195]. Adequacy of dosing can be estimated by the appearance of physiological changes ("side effects"), and the presence of drug can be verified with serum drug concentration assays, which are available for most antidepressants [359]. Indeed, clinical experience indicates that a large proportion of "rescues" of patients considered to have treatment-resistant depression are based on determining a correct and complete diagnosis and in using a standard antidepressant systematically, with sustained and increasing doses, as tolerated. Moreover, there is little quantitative evidence than a particular next step following an initial, aggressively dosed antidepressant trial continued for at least 3 months is superior to another [195].

As additional therapeutic steps are considered, it is useful to order them so as to balance evidence supporting their value against their ease, simplicity, and costs. Before switching to dissimilar types of antidepressants, including a first-generation tricyclic or MAO inhibitor, many experienced clinicians explore use of adjunctive treatments with a standard antidepressant, given in high, tolerated, doses. An often effective intervention is to add formal psychotherapy, perhaps in collaboration with an experienced therapist [272, 274, 407]. Several methods have been formularized and subjected to experimental validation, including cognitive-behavioral therapy (CBT) and related methods, interpersonal therapy (IPT), and others [127, 248, 251, 272, 274, 427]. However, the utility and relative efficacy of psychotherapy compared to antidepressant drugs in the treatment of severe depression and the value of psychotherapy in rescuing patients who have failed antidepressant trials are not firmly established.

Among supplementations of standard antidepressant treatments, one of the most extensively tested is the addition of lithium, typically to serum concentrations as used to treat bipolar disorder, sustained for several months (Table 4.13). Clinically significant antidepressant responses have been reported in an average of nearly half of patients so treated but there is little evidence that antidepressant responses occur more rapidly with lithium added [125]. Most controlled trials that have supplemented antidepressants with lithium vs. a placebo have involved tricyclic antidepressants, but the effect of augmentation with lithium appears to extend also to modern antidepressants [125, 525]. It remains uncertain whether the effect of lithium is a direct one on depression, generally, or in a subgroup of patients, such as those with bipolar-like characteristics or current depressive episodes marked by overarousal, agitation, anger, or anxiety [241, 351, 462, 557].

Experimental evidence is emerging that supplementation of antidepressants with a modern, second-generation antipsychotic agent may enhance clinical outcomes in patients diagnosed with unipolar major depressive disorder. Indications for such treatment become more compelling when there is evidence of severe agitation,

Table 4.12 Lithium augmentation in treatment-resistant major depression

Trials	Responders/patients		Odds ratio (95% CI)
	Control	Lithium	
Heninger et al.(1983)	0/7	5.8	23.6 (1.00–556)
Kantor et al. (1986)	0/3	1/4	3.00 (0.09–102)
Zusky et al. (1988)	2/8	3/8	1.80 (0.21–15.4)
Schlöpf et al. (1989)	0/13	7/14	27.0 (1.35–542)
Browne et al. (1990)	2/10	3/7	3.00 (0.35–25.9)
Stein and Bernadt (1993)	4/18	2/16	0.50 (0.08–3.19)
Joffe et al. (1993)	3/16	9/17	4.88 (1.01–23.6)
Katona et al. (1995)	8/32	15/29	3.21 (1.09–9.48)
Baumann et al. (1996)	2/14	6/10	9.00 (1.27–63.9)
Nierenberg (2003)	3/17	2/18	0.58 (0.08–4.01)
Total (%)	24/128 (18.8%)	53/131 (40.5%)	3.11 (1.80–5.37)

Adapted from Crossley and Bauer [125]. Note that the response rate is moderate, but the effect-size large and highly significant ($z = 4.06$, $p < 0.0001$), although 2/10 trials did not find a superior effect of lithium. The effect was found with both first- and second-generation antidepressants, but there was no acceleration of onset of antidepressant effect

psychosis, or possible bipolar disorder. Agents with substantial evidence for clinical utility as adjuncts with antidepressants include aripiprazole, olanzapine, and quetiapine, although others may also be effective [126, 153, 162, 196, 307, 467, 547]. Several other adjunctive treatments have variable evidence for additional efficacy in major depression when added to a standard modern antidepressant. These include adjunctive use of various stimulants, sedative-anxiolytics, and the endogenous metabolite SAMe, as summarized in Table 4.11.

Among physical or nonpharmacological treatments, available findings indicate possible superiority of electroconvulsive treatment (ECT) to other methods including repeated transcranial magnetic stimulation (rTMS) or VNS (Table 4.11). Indeed, ECT continues to be the best established nonpharmacological treatment for severe or otherwise treatment-resistant depressive illness [226, 239, 278, 282]. ECT yielded an average response rate of 52% in seven trials involving depressed patients who had failed one or more trials of antidepressants, compared to response rates of 65–88% without having failed an antidepressant trial [134, 239]. Nevertheless, both rTMS and VNS are FDA approved as safe and effective, including for use in treatment-resistant major depression, based on substantial research evidence, and favorable findings with experimental deep-brain electrical stimulation techniques are emerging [9, 62, 199, 253, 257, 290, 345, 363, 422, 437, 508].

Finally, there is some research on the use of adjunctive treatments with antidepressants when clinical responses to antidepressants, usually with behavioral psychotherapy, have yielded unsatisfactory responses in anxiety disorders including panic disorder and obsessive-compulsive disorder. However, this area of clinical research is far less extensive than for treatment-resistant depression [48, 146, 201, 488].

Long-Term Maintenance Treatment and Discontinuation of Antidepressants

An important component of the treatment of major depression and most anxiety disorders is that they tend to recur or to follow a chronic course, making consideration of sustained, long-term treatment likely in many cases, despite generally limited and inconclusive research support for the practice [161, 187, 260, 553, 580]. Most of the experimental clinical investigation pertaining to this topic has involved depressive illnesses.

A basic assumption underlying continuous, long-term treatment is that tolerance, or loss of efficacy over time, does not occur with antidepressants. However, loss of antidepressant effects over time occurs in a rising proportion of patients with major depression treated long term with an antidepressant, and such changes may be particularly likely with long-term treatment with serotonin-reuptake inhibitors [285, 580]. However, the phenomenon is easily confounded by declining treatment adherence and rising dropouts of failed cases as well as of patients without ongoing symptoms over time, or as simple limitations of prophylactic efficacy become more apparent over time [93, 428, 625]. Nevertheless, such changes sometimes have been interpreted as evidence of drug tolerance [14, 285, 286, 454, 625], and it has been hypothesized that long-term antidepressant treatment may actually contribute to clinical worsening in some cases [130, 161]. However, it is not clear that declining antidepressant benefit is associated with declining plasma concentrations of drug or that increased dosing can restore response, as would be consistent with pharmacodynamic or pharmacokinetic tolerance.

There also is abundant evidence that discontinuing ongoing antidepressant treatment in patients with major depressive disorder, even following prolonged treatment well beyond clinical recovery from an index acute episode, is often followed by a gradually rising proportion of patients who become ill again [59, 160, 187, 228, 260, 291, 306, 309, 331, 425, 546, 580, 582]. Such observations have often been accepted as evidence that long-term or prophylactic treatment with antidepressants is necessary and effective.

Our review of long-term maintenance and discontinuation trials in major depression found major differences in outcomes following treatment discontinuation compared to continuation that was largely independent of the duration of treatment [580]. With antidepressant treatment continued vs. discontinued, median times to the start of a first recurrence of a depressive episode were 3.4 times longer (48.0 vs. 14.2 months), and the approximate rate of recurrences showed similar differences (1.85% vs. 6.24% of patients per month) within the first 12 months (Fig. 4.12c). The previous history of recurrences was strongly, and predictably, associated with risk of new recurrence after stopping antidepressant treatment, but strikingly, was unrelated to recurrence risk during continued treatment (Fig. 4.12a,b). After the first year without maintenance antidepressant treatment, there was only a minor difference in recurrence risk with and without continued treatment. Notably, even with antidepressant treatment continued, recurrences continued steadily over 5 years,

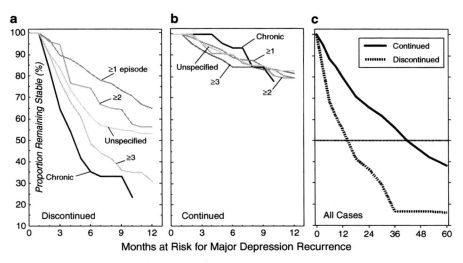

Fig. 4.12 Outcome over months of follow-up among major depression patients who continued vs. discontinued maintenance antidepressant treatment and effect of recurrence history. Data are pooled from 27 reports involving 3,037 patients diagnosed with major depressive disorder who had been treated for 5.8 months and then followed up to 60 months. (**a**) Patients who discontinued ($n=797$), or (**b**) continued ($n=2,240$) antidepressants, stratified by prior episode recurrences (≥ 1, ≥ 2, ≥ 3, chronic, or unspecified), or (**c**) risk-by-time functions for all patients continued vs. discontinued patients. There are marked and orderly increases of recurrence risk by time after discontinuation (**a**) but not with antidepressant continued (**b**), and there is a major overall difference between continued vs. discontinued treatment, involving median latencies (50% risk indicated by *horizontal dotted line*) to first recurrence of 48.0 vs. 14.2 months, with approximate recurrence rates of 1.85 vs. 6.24% per month in the first 12 months (both measures involve 3.4-fold differences, $p<0.001$), and very little change in recurrence rates thereafter. Approximately 60% of patients experienced a recurrence within 5 years, even with antidepressant treatment continued, whereas 15% had no recurrences without continued treatment (**c**). Adapted from Viguera et al. [580]

and reached 50% between 3 and 4 years, whereas only approximately 15% of major depression patients had no recurrences without antidepressant treatment between 3 and 5 years of follow-up (Fig. 4.12c). The continued risk of recurrences with continued treatment may suggest limited prophylactic benefit of antidepressant treatment, evolving pharmacological tolerance, or declining fall-off of treatment adherence, especially with longer times without illness.

In addition, data gathered in another recent review had similar findings [160]. Reviewed were 18 long-term trials reported between 1982 and 2010, involving comparisons of various antidepressants (amitriptyline, citalopram, dothiepin, fluoxetine, imipramine, mianserin, nortriptyline, paroxetine, sertraline, venlafaxine, and mixed agents) to placebo among a total of 1,785 major depression patients (aged 53 ± 14 years, $69\pm8.7\%$ women) followed-up over 24 ± 9.4 months. By random-effects meta-analysis, the pooled relative recurrence rate with antidepressant continued vs. after discontinuing to placebo was 1.99 [95% CI: 1.66–2.39; $p<0.0001$], and the pooled rate difference was 0.276 [CI: 0.208–0.345; NNT=3.6]. The apparent monthly recurrence rates averaged $2.6 \pm 1.0\%$ with placebo and $1.2 \pm 0.7\%$ with continued antidepressant. These findings indicate an approximately twofold lower recurrence risk with antidepressants continued and a favorably small estimated NNT. However, across the trials, 50–100% of recurrences occurred within the initial 6 months.

These kinds of observations have been proposed as evidence of a protective effect of long-term antidepressant treatment in recurrent unipolar major depressive disorder, analogous to that discussed in Chapter 2 regarding psychotic disorders, and Chapter 3 regarding bipolar disorders. However, interpretation of such data is complicated by the evidence, discussed earlier for antipsychotic and mood-stabilizing treatments, that many, if not most, long-term treatments with psychotropic drugs are associated with a drug discontinuation-associated tendency toward earlier or greater than expected recurrence risk. Such reactions appear not to represent simply lack of treatment, and they may represent effects of readjustments of CNS physiology on return to the unmedicated state. As noted previously, evidence for such an effect is that slow, gradual discontinuation of sedative-hypnotic agents, antipsychotic drugs, or lithium greatly reduces, and does not merely delay, the onset of a recurrence of the illnesses being treated. Similar responses are found with antidepressant drugs as well, and they appear to be largely independent of the duration of the previous long-term treatment, sometimes occurring after many months or even years of treatment, and arising within several weeks or months after treatment discontinuation ([41]; Fig. 4.13).

This phenomenon greatly complicates interpretation of long-term treatment trials that involve treatment discontinuation, particularly when medication is discontinued abruptly or over relatively short times [36, 40, 44, 46]. In most trials involving discontinuation of long-term psychotropic treatments, there is a much greater difference in recurrence rates between continued and discontinued treatments within the initial several months than at later times [160, 580]. This pattern might indicate that treatment discontinuation is a stressor, so as to confound simple interpretation of the difference in recurrence risk with vs. without antidepressant as proof of long-term prophylactic effect rather than an artifact of treatment discontinuation. Alternatively or in addition, the tendency for early excess recurrence risk may reflect individual differences in vulnerability to illness recurrence, such as may be reflected in the number of previous illness episodes (Fig. 4.12a).

In, short, it follows that the effectiveness of maintenance treatment with antidepressants as well as most psychotropic agents remains somewhat uncertain due to risk of exaggerated differences in recurrence risks with medication discontinued vs. continued, especially soon after discontinuation. In addition, the design of many modern trials of psychotropic drugs may add further complexity to the interpretation of long-term trials involving drug discontinuation. Many trials involve selection biases that require initial clinical response to a test treatment of interest, followed by continuation vs. discontinuation of that initially effective treatment, based on random selection. Such so-called "enrichment" trial designs are not necessarily representative of all patients with a particular diagnosis, since the subjects continued for long-term study are selected as having responded to a treatment of interest as well as by the somewhat arbitrary timing of long-term treatment with

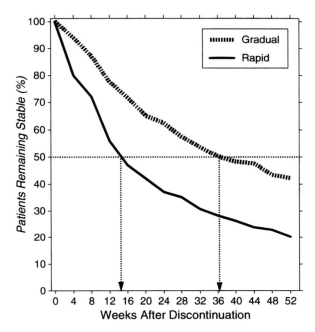

Fig. 4.13 Latency to recurrence of major depression following discontinuation of antidepressants rapidly vs. gradually. Data are for 398 antidepressant-treated patients with bipolar-I, -II, unipolar major depressive, or panic disorder (2/3 were women, mean age=42 years), in whom ongoing antidepressant treatment was discontinued abruptly or rapidly (1–7 days; *n* = 188; *solid line*) or gradually (≥14 days; *n* = 210; *dotted line*), omitting patients who were symptomatic at the time of discontinuing. *Horizontal dotted line* = 50% recurrence risk and *vertical dotted arrows* indicate median latency (15.1 [95%CI: 7.96–22.2] vs. 36.3 [27.1–45.6] weeks) by survival analysis (2.4-fold difference, *p* < 0.0001). Latency was shortest among bipolar-I disorder patients and longest with antidepressants of long elimination half-life (such as fluoxetine). Adapted from Baldessarini et al. [42]

respect to a particular index clinical state associated with an acute episode of illness and the residual vulnerabilities that may be associated with incomplete recovery.

One of the few long-term treatment trials that did not involve enrichment randomized 27 patients diagnosed with unipolar major depressive disorder who had been clinically stable for at least 6 months to either imipramine or placebo for up to 2 years. The resulting placebo vs. drug difference in recurrence risk (61.5% vs. 43.0%, a 1.4-fold difference, with an estimated NNT of 5.4) was not statistically significant, but the comparison was underpowered [281]. In addition to possible biasing induced by enrichment designs for long-term trials, further uncertainty arises regarding the appropriate duration of what is usually termed the "continuation" phase of treatment following initial response in acute illness, and prior to randomization to continue or to discontinue a treatment, hopefully under adequately blinded conditions, to a placebo. It is usually assumed that the continuation phase of treatment is sufficiently prolonged as to assure virtually full recovery from the index acute illness episode, even though this is not always certain. It is also usual currently to discontinue treatment sufficiently gradually and slowly as to limit the impact of discontinuation-associated stress. However, for most psychotropic treatments, neither the duration of continued treatment so as to assure full clinical recovery and substantial stability, nor an appropriate duration of dose reduction to discontinuation is established. That is, there is some possibility that too-rapid discontinuation of a treatment shown to be effective, and in a patient who is barely recovering from an index episode of illness can contribute artifactually to apparently marked differences in recurrence risk over time with treatment continued vs. discontinued. These uncertainties have led some investigators to advise great caution in interpreting apparent recurrence-risk differences with continued vs. discontinued long-term medication as representing compelling evidence of a long-term protective or prophylactic effect [40, 41, 44, 46, 160, 580, 582].

Despite the uncertainties just discussed, some consensus about practical approaches to the use of long-term treatment with antidepressants and other psychotropic drugs has evolved over the years. Usually, three typical phases of treatment are considered: (a) treatment of an acute episode of depression or other illness; (b) continuation treatment to solidify initial therapeutic responses, to avoid early relapses, and to attain clinical remission that is sustained for a substantial period (typically at least 2 or 3 months); and (c) a long-term, recurrence prevention or prophylactic phase [319]. Precise definitions of the duration of these somewhat arbitrary phases remain elusive, but with major depression, acute treatment typically is considered to involve 6–12 weeks, continuation–stabilization another 6–12 months, and recurrence–prevention, times later than 1 year.

Most experts advise continuing tolerated doses of an initially effective antidepressant for at least 6–12 months, and then undertaking critical assessments of individual strengths and vulnerabilities of patients before considering whether to continue or discontinue treatment [59, 197, 232, 283, 319, 495, 548]. If the patient has had relatively few previous episodes of depression and appears to be well recovered from the treated, index episode, and to be both stable and functioning well without unusual stress, with due consideration to the burdens of continued treatment (adverse effects and cost), then a slow and cautious trial of discontinuation can be considered. Ideally, this should involve very slow, stepwise, reduction of drug dose over at least several weeks, and perhaps longer, with close follow-up and communication between patient and clinician to consider evidence of an impending recurrence of illness that would call for a return to adequate, active treatment. Such cautious attempts at treatment discontinuation can be repeated, perhaps once or twice a year to determine the feasibility and safety of drug-free maintenance. Ongoing support and use of psychotherapy may well add a degree of safety and stability, even if medication can be discontinued successfully.

A final consideration, especially for long-term treatment with an antidepressant drug, is acquisition cost. Generic antidepressants that no longer are patent-protected average as much as 75 times less costly than brand-name products, whereas the uncommonly used MAO inhibitors remain 20–50 times more costly than commonly used, generic antidepressants [452].

Broadening Applications of Antidepressants

As was reviewed above, the overall efficacy of antidepressants in the complex and highly heterogeneous group of illnesses included within the diagnostic concept of major depressive disorder is limited. However, the range of other potential applications of antidepressant medicines is substantial and growing. They include relatively well established indications for continued postrecovery treatment of recurrent depression, for depression associated with other psychiatric or general medical disorders, and for applications in various anxiety-related disorders including panic disorder, generalized anxiety and various phobias, as well as obsessive-compulsive disorder and other compulsive conditions or impulse-control disorders. In addition, various other conditions have been treated with antidepressants, with a range of supporting evidence of efficacy and safety.

These applications include posttraumatic stress disorder (PTSD), premenstrual dysphoria, menopausal flushing, enuresis and stress incontinence, attention deficit-hyperactivity disorder (ADHD), conditions in elderly patients that include combinations of affective and cognitive features, chronic and neuropathic pain syndromes, bulimia nervosa, premature ejaculation, migraine, various inflammatory disorders including colitis, dermatitis, and myositis, and possible reduction of risk of recurrences of myocardial infarction. These uses of antidepressants are listed with representative supporting literature in Table 4.13.

Some of these uses call for further comment. For many, if not most of the disorders for which antidepressant treatment is given, clinical responses are typically only partial and usually require additional therapeutic interventions, often with behaviorally oriented psychotherapy. This is usually the case with obsessive-compulsive and severe anxiety disorders, attention disorders, PTSD, and others. For obsessive-compulsive disorder, beneficial effects appear to be limited to serotonin- and serotonin–norepinephrine-reuptake inhibitors given in relatively high doses [3, 182, 198, 308, 339]. It is also important to realize that older antidepressants continue to have utility in some of the indications listed in Table 4.13. These include tricyclic antidepressants and stimulants as well as bupropion in attention disorders [259, 434], and tertiary amine tricyclics as well as modern serotonin–norepinephrine-reuptake inhibitors for neuropathic and other chronic pain syndromes [510].

A particularly interesting application of antidepressants is for depressed, post-myocardial infarction patients. Initially, the relatively medically safe serotonin-reuptake inhibitors were tried in the treatment of depression, not uncommonly associated with heart attacks. Initial findings indicated not only safety and effective therapeutic effects on depression, but also possible reduction of risk of recurrences of myocardial infarction. These were specifically related to thrombotic events, with

Table 4.13 Antidepressants: growing clinical applications

Applications	References
Acute major depression, especially unipolar	[388, 391, 570]
Preventing early (? late) relapse in depression	[580, 581]
Major depression with medical or neurological comorbidity	[441, 445, 613]
Generalized anxiety disorder (SRIs)	[47, 67]
Obsessive-compulsive disorder (SRIs, high doses)	[3, 198, 308]
Impulse control disorders (SRIs)	[4, 74, 265, 374]
Panic disorder (TCAs, MAOIs, SRIs)	[361, 489]
Social phobias (SRIs)	[190, 481]
Posttraumatic stress disorder (SRIs)	[78, 442]
Premenstrual dysphoria (SRIs)	[184, 614]
Menopausal flushing (SRIs)	[511, 533]
Enuresis (pediatric and geriatric; tertiary amine TCAs)	[28, 616]
Stress incontinence (duloxetine)	[57, 450]
Attention deficit-hyperactivity disorder (TCAs, bupropion)	[119, 434, 469]
Geriatric pseudodementia (antidepressants and stimulants)	[230, 247, 306, 519]
Chronic or neuropathic pain (tertiary amine TCAs, duloxetine, venlafaxine)	[88, 510]
Bulimia nervosa (SRIs)	[273]
Tic disorders (TCAs)	[128, 517]
Premature ejaculation (SRIs)	[1, 240]
Migraine	[50, 56]
Inflammatory syndromes (colitis, dermatitis, fibromyositis/ chronic fatigue)	[222, 234, 401, 402, 561]
Myocardial infarction (SRIs)	[210, 606]

Except for depression and some anxiety disorders most applications are off-label; references are representative; for most of these applications, comprehensive treatment programs with a strong emphasis on behavioral methods and rehabilitation are appropriate
TCAs tricyclic antidepressants; *SRIs* serotonin-reuptake inhibitors

less impact on cardiac arrhythmias that can be sensitive to emotional states [210]. Several serotonin-reuptake inhibitors appear to interfere with the functioning of platelets, possibly through serotonin-related mechanisms, and may reduce the risk of adverse outcomes of atherosclerotic disease with clot formation [606]. Similar mechanisms, or effects on gastric acid secretion, have also been associated inconsistently with risk of gastrointestinal bleeding during treatment with such drugs [19, 69, 98].

Specific approved indications for individual modern antidepressants are provided in Table 4.14.

Antidepressants and Suicidal Risk

A pressing question is whether antidepressants have effects on risk of suicide. Research on the effects of psychiatric treatment, particularly pharmacological treatments, in relation to risk of suicidal behaviors was virtually unknown a decade ago.

Table 4.14 FDA-approved indications for serotonin-reuptake inhibitors

Indication	Citalopram	S-Citalopram	Fluoxetine	Fluvoxamine	Paroxetine	Sertraline	Vilazodone	Venlafaxine
Acute major depression (unipolar)	yes	yes	yes	no	yes	yes	yes	yes
Treatment-resistant depression	no	no	with olanzapine	no	no	no	no	no
Acute major depression (bipolar)	no	no	with olanzapine	no	no	no	no	no
Anxiety (generalized)	no	yes	no	no	yes	no	no	yes
Bulimia nervosa	no	no	yes	no	no	no	no	no
Obsessive-compulsive disorder	no	no	yes	yes	yes	yes	no	no
Panic disorder	no	no	yes	no	yes	yes	no	yes
Premenstrual syndrome	no	no	yes	no	yes	yes	no	no
Posttraumatic stress disorder	no	no	no	no	yes	yes	no	no
Social phobia	no	no	no	yes	yes	yes	no	no

Several agents have specific indication for long-term continuation, typically for 6–12 months
Fluoxetine, fluvoxamine, paroxetine, and sertraline are approved for use in juveniles (all for anxiety disorders, only fluoxetine for major depression)
Information is from manufacturers' product information bulletins concerning FDA approval status at the end of 2011

This deficiency is remarkable in view of the very significant risk of suicide and of other causes of excess or early mortality in major affective disorder patients and possibly also in severe anxiety disorders [504, 504]. In addition to suicide, causes of excess mortality associated with mood disorders include apparent accidents, complications of comorbid abuse of alcohol or other substances, and a several-fold increase of mortal risks from intercurrent medical disorders, especially stress-sensitive disorders in elderly patients [399]. Notably, the total numbers of deaths from medical comorbidity and from suicide among mood-disorder patients are similar overall, but risks for suicide relative to the general population are very large, especially in younger patients early in the course of illness [559]. Relative risks of suicide, sometimes reported as standardized mortality ratios (SMRs) average approximately tenfold greater than in age- and sex-matched general populations [231]. Risks are even higher in bipolar I and II disorders than in most other psychiatric disorders, as discussed above in Chapter 3 [559, 560]. Given the broad heterogeneity of unipolar major depressive disorder, however, suicide risks vary markedly with illness severity. SMR may be as low as five among moderately depressed, never-hospitalized patients, and can approach 20 among patients with severe, recurrent depression leading to hospitalization [559].

In addition, as noted above regarding bipolar disorder, the ratio of suicide attempts to fatalities (A/S) in major affective disorder patients is far lower than in the general population, with ratios of 3–6 in mood disorders compared to 20–30 in the general population. The A/S ratio probably reflects greater potential lethality of suicidal behavior in patients with major depressive or bipolar disorders, based on methods employed and intent [559].

Initial interest in possible effects of psychotropic drug treatments and suicidal risk arose with evidence that lithium may reduce risk of suicides and attempts among bipolar disorder patients and perhaps those with other recurrent mood disorders [6, 42, 558). An historically important next step was FDA approval of a specific indication for treatment with clozapine in schizophrenia to reduce suicidal risk in 2003 [244, 353]. In addition, British and American regulatory agencies found evidence from controlled pharmaceutical trials submitted for review to suggest increased risk of suicidal ideation and perhaps suicidal behavior among juvenile patients with depressive or anxiety disorders during treatment with modern antidepressants, compared to placebos [224, 225].

This concern led to the decision to require a strongly emphasized ("black box") warning of this possible risk in informational materials for all drugs with antidepressant effects or used to treat any form of depressive illness, at least in young patients [323, 530]. In turn, this high level warning was followed by a decrease in use of modern antidepressants, particularly serotonin-reuptake inhibitors, in both juveniles and adults, as well as decreased rates of diagnosis of major depression in children, all with uncertain, but probably minor effects on national suicide rates [397, 573].

Whether the finding and concern are valid or not remains controversial for several reasons [206]. One is that the data involved were drawn from relatively short-lasting, controlled trials that usually attempted to exclude potentially suicidal subjects. Also, the indications of suicidal risk were obtained largely passively and incidentally as "adverse event" comments arising from the trials, not based on explicit assessment of suicidal risk as a defined outcome measure. Efforts have, however, been undertaken to include more explicit, planned assessments of suicidal risks and behaviors in trials of virtually all centrally active drug candidates, using validated rating methods [352]. An additional limitation is that most of the FDA reported outcomes involved suicidal thinking, which is very hard to evaluate reliably even with explicit efforts, and bears only an indirect and numerically limited relationship to suicidal behaviors, and very limited predictive relationship to suicide [293]. Moreover, there is growing evidence that the pharmacology of suicidal thinking, attempts, and of suicide may differ [36, 43, 296]. A further limitation is that the overall risk of suicidal thoughts or behavior has been similar with vs. without antidepressant treatment in several reviews and meta-analyses, including in analyses of the FDA database on antidepressant trials [35, 44, 224, 225, 444, 530]. An adverse relationship with antidepressant treatment was found only in post hoc analyses of data stratified by subject age, with evidence of increased risk of suicidal ideation and perhaps behavior only in juveniles and young adults, up to age 24 years [225, 530]. Indeed, at older ages, there is evidence that such risks may be lower with antidepressant treatment [225, 323, 530]. This age-associated phenomenon was later replicated in a meta-analysis of several large clinical cohort or case-control studies of treatment with serotonin-reuptake inhibitors [51]. Both findings are summarized in Fig. 4.14. If this relationship of possible suicidal risk and age to antidepressant treatment is valid, it may suggest greater effectiveness of antidepressants in older adults, an hypothesis that remains to be tested adequately in depressed adults.

Additional uncertainty about the relationship of antidepressant treatment to suicidal risk arises from seemingly inconsistent research, including some indicating beneficial effects [554]. For example, many controlled trials in depressed adults find greater lowering with antidepressants more than placebo of scores on rating-scale items specific for suicide (e.g., item 3 on the Hamilton, and item 10 on the Montgomery-Åsberg depression rating scales) [5, 36, 43]. Such evidence may well be unreliable, however, if overall clinical impressions of improvement affect ratings of individual scale items. Additional studies, usually based on retrospective analysis of large collections of clinical data, have also found evidence of lowered risks of suicidal behavior during treatment with antidepressants [120, 314, 328, 444, 456, 532, 560]. Others have found little association of antidepressant treatment and suicidal risk, particularly of suicide attempts and fatalities; still others have found some evidence of increased risk [43, 170, 218, 444]. Many of these inconsistent findings arise from meta-analyses of incidental findings concerning adverse events in controlled trials for the treatment of depression or from large clinical samples. Notably, however, even the FDA meta-analyses leading to the age-specific "black-box" warning found nothing when data from all trials and all ages were pooled, and evidence of some increased risk of suicidal ideation and possible attempts only at ages below 25 years, based on retrospective stratification by age (Fig. 4.14).

A number of studies also found parallels between enormous increases of sales of modern antidepressants in the 1990s and small decreases in suicide rates in specific countries or regions, mainly in Nordic countries and North America [43, 267]. Such findings are difficult to interpret since they involve "ecological" or correlational analyses based on aggregate trends, which cannot identify relationships between

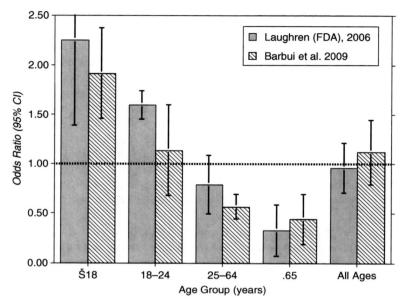

Fig. 4.14 Odds ratio (±95% confidence interval) for risk of suicidal ideation or behavior vs. age groups of patients treated with or without a modern antidepressant. Data are derived from FDA retrospective meta-analyses of data from 396 placebo-controlled trials involving some 113,000 subjects submitted for review (*shaded bars* [323]), or from a meta-analysis of 8 large, case-control or cohort studies (*striped bars* [51]). Note that there is no overall effect of treatment with vs. without antidepressant exposure (all ages), but that differences arise only when specific age groups are considered, with increased risk in juveniles and young adults, and lower risks in older adults

treatments and outcomes at the level of individuals. Moreover, the findings have been highly inconsistent internationally, and even among European countries. Worldwide, the distribution of ratios of suicide rates reported by the World Health Organization before and after the introduction of modern antidepressants in the late 1980s vs. early 2000s was essentially random, with half finding decreases and half finding increases [37, 43]. Moreover, even in countries where decreases in suicide rates were documented, similar trends toward declining rates were noted in the 1970s and 1980s, well before the introduction and explosively increased use of modern antidepressants [37, 43]. One possibly relevant consideration is that suicide rates appear to be strongly associated with indices of what might be termed "access to care," generally, including such factors as population density, numbers of trained clinicians per 100,000 population, individual and public funding for medical and mental health services, income, and others [555]. It seems likely that antidepressant sales are only one indication of increased interest in mood disorders in recent decades, and perhaps generally increased levels of care for mood disordered patients who are at highest risk for suicide [39].

In sum, the evidence concerning suicidal behavior, and life-threatening attempts and suicides specifically, in relation to antidepressant treatment remains conflicted and largely inconclusive. It should be noted that recent regulatory concerns are limited largely to suicidal thoughts and infrequent nonlethal behaviors in young patients, in whom the efficacy of antidepressants for major depression remains even less certain than in adults [565]. A clinically prudent response to these confusing and inconsistent observations is to be aware that older as well as modern antidepressants can have adverse emotional and behavioral effects on depressed persons. These are most likely to emerge early in treatment [277]. The risks may be numerically greater with modern antidepressants, given their extraordinarily widespread use among nonpsychiatric clinicians as well as mental health specialists, encouraged by the reputation of these drugs for relatively safety. The relatively safety of modern antidepressants appears to have encouraged a decline in close supervision of depressed patients newly treated with antidepressants in recent decades [179]. However, uncommon adverse behavioral responses call for particularly close clinical supervision and support early in treatment and with patients who are not yet well known [36, 37, 179, 526, 572]. Adverse responses can include increased agitation and anger, decreased sleep, and suicidal thinking. In most cases, given adequately close clinical supervision, such reactions can be detected as they evolve and appropriate clinical steps taken to limit risk of suicidal behavior. These include reducing or removing antidepressants and using sedative, mood stabilizing, or antipsychotic agents, along with increased frequency and intensity of clinical supervision and support [429].

Adverse Effects

General and Autonomic Effects

Characteristic adverse effects of antidepressants, as a group, include annoying and sometimes dangerous autonomic effects. Some of these reactions (especially xerostomia, impaired visual accommodation, acute glaucoma, mild tachycardia, constipation, urinary retention, and confusion) have been ascribed to the appreciable antimuscarinic actions of most tricyclic antidepressants, especially the tertiary amine subgroup, which also carry elevated risks of sedation and postural hypotension. In contrast, initial nausea and other gastrointestinal effects are routinely associated with serotonin-reuptake inhibitors, especially early in treatment and with rapid dose increases, often with gradual relief with longer use. They also produce high rates of sexual dysfunction—especially ejaculatory incompetence in men and anorgasmia in women—which is routine with most serotonin-reuptake inhibitors and occurs occasionally with tricyclic antidepressants and with agents lacking appreciable anticholinergic-parasympathetic activity including MAO inhibitors [17, 211, 366, 369, 484, 496]. Anorgasmia is almost certainly induced by serotonin- and serotonin–norepinephrine-reuptake inhibitors over and above any sexual dysfunction associated with depression. It is not easily treated, but suggestions have included use of yohimbine, cyproheptadine, amantadine, and agents used for erectile dysfunction such as sildenafil, or changing to bupropion [129, 324, 390, 585]. Autonomic effects are common with use of phenelzine in commonly relatively high, effective, doses above 45 mg/day, possibly due to a shift toward adrenergic-sympathetic dominance, but not antiparasympathetic effects. Priapism is associated with trazodone and may be due to antiadrenergic actions, particularly those mediated by α receptors [118, 552]. Headaches are also a common complaint associated with many dissimilar antidepressants [17].

Another general risk of antidepressant treatment is induction of restlessness and agitation as well as hypomania, mania, or mixed manic-depressive states [506, 556]. Such reactions are not rare among persons considered to have unipolar major depressive illness, especially young patients. They have sometimes been viewed as an adverse effect of treatment with antidepressants and other mood-elevating agents, but reactions that resemble those encountered among bipolar disorder patients should lead to heightened suspicion of previously undiagnosed bipolar disorder [152].

First-generation antidepressants, especially, are associated with weight gain, which can become a significant medical problem and may lead to discontinuation of treatment. Some relatively stimulating antidepressants have a low risk of weight gain but may tend to induce agitation and insomnia as well as anorexia [68]. Weight gain appears to be most common with amitriptyline, mirtazapine, and paroxetine, and lowest, or with at least transient weight loss, with bupropion and fluoxetine [490]. Older antidepressants have even been associated with metabolic syndrome, including hypertension, type-2 diabetes mellitus, and dyslipidemia as well as weight gain, possibly interacting with risks associated with major affective disorders apart from their treatment [349].

Bupropion, clomipramine, and maprotiline carry a dose-dependent risk of inducing epileptic seizures. Tricyclic antidepressants, especially the tertiary amines, are strongly antimuscarinic and can induce confusion and delirium, especially if combined with other drugs with prominent anticholinergic effects, such as antiparkinsonism drugs, as well as clozapine and thioridazine [183, 216, 217]. Such toxic reactions can be reversed by cautions and slow administration of the central acetylcholinesterase inhibitor physostigmine [216]. Nefazodone (as well as carbamazepine, pemoline, and valproate) has a substantial risk of inducing hepatic toxicity, which led to its current disfavor [485]. Serotonin-reuptake inhibitors can also interfere with platelet function to increase risk of bleeding, but may also have a myocardial infarction-preventive effect [19, 210]. Characteristic adverse effects of antidepressant drugs are summarized in Table 4.15.

Cardiac Risks

Risk of life-threatening cardiac arrhythmias with clinical doses of tricyclic antidepressants can be exaggerated and so limit use of these effective agents. Cardiotoxic effects can occur in acute overdoses of tricyclic antidepressants involving 1 or 2 week's supply taken at once; they were a frequent cause of death in acute overdoses with suicidal intent when first-generation antidepressants were more widely employed clinically

[63, 82, 380, 414, 471, 545, 604]. Their risks include liability considerations [61]. The quinidine-like direct cardiac depressant actions of most tricyclic antidepressants contraindicate their use with other cardiac depressants, including such antipsychotics as thioridazine, mesoridiazine, or ziprasidone, and probably present extra risks in depressed cardiac patients. Transient postural hypotension or sustained, usually mild, elevation of diastolic blood pressure and other adverse cardiac effects occurs with tricyclic antidepressants and MAO inhibitors, as well as some serotonin–norepinephrine-reuptake inhibitors including venlafaxine and desvenlafaxine [256, 520]. Most modern antidepressants are much less likely than tricyclics to induce life-threatening toxic cardiovascular effects, or delirium and fatalities on acute overdoses. Indeed, the serotonin- and serotonin–norepinephrine-reuptake inhibitors have displaced tricyclic antidepressants for treating depressed cardiac, poststroke, and other medically ill and elderly patients [513]. Nevertheless, there is evidence of an association of antidepressant treatment and increased risk of stroke, possibly even greater with serotonin-reuptake inhibitors than other types of antidepressants, including tricyclics [607]. It is not clear how many such strokes have been thrombotic or hemorrhagic, but the relationship with serotonin-reuptake inhibitors seems inconsistent with evidence that risk of repeat myocardial infarction may be reduced with such drugs [210, 606].

Withdrawal Reactions

Most types of antidepressant drugs, especially following prolonged administration and abrupt or rapid withdrawal of relatively high doses of short-acting agents, can be followed by discontinuation or withdrawal reactions in a substantial proportion of patients, but have been associated most often with serotonin- and serotonin–norepinephrine-reuptake inhibitors [73, 220, 474, 494, 540, 602, 618]. It has been suggested that such reactions may differ from other forms of drug dependency and withdrawal reactions, and that those associated with rapid discontinuation of serotonin-reuptake inhibitors may have unique features [473, 494]. However, the status of a special type of serotonin-reuptake discontinuation syndrome remains uncertain. Evidence of drug craving has been observed clinically in some patients soon after stopping a serotonin-reuptake inhibitor, and the withdrawal symptoms that occur in such circumstances bear many similarities to those associated with other dependency-producing agents including benzodiazepines [385]. However, the similarities of discontinuation-associated symptoms among pharmacologically dissimilar types of antidepressants are striking and efforts to explain the symptoms physiologically are limited. Systems invoked have included serotonergic, noradrenergic, and muscarinic-acetylcholinergic neurotransmission, with consideration of adjustments required on removal of an agonist or antagonist of such transmission. With the amine-uptake inhibiting and variably anticholinergic tricyclic antidepressants, mixtures of effects on all of these systems may be involved.

Typical symptoms include malaise, gastrointestinal symptoms, shifts in blood pressure and heart rate, strange bodily sensations including symptoms resembling

Table 4.15 Characteristic adverse effects of antidepressants

Agents	Characteristic adverse effects
Tricyclic-like agents	
Amitriptyline, doxepin, imipramine	Anticholinergic effects, cognitive impairment, hypotension, cardiac depressant, leukopenia
Amoxapine	Extrapyramidal effects, tardive dyskinesia, rare seizures
Clomipramine	Sedation, nausea, anticholinergic effects, cardiac depressant, hemolysis
Desipramine, protriptyline	Mildly anticholinergic, cardiotoxic, agitation
Maptrotiline	Seizures (dose-dependent), rashes
Monoamine oxidase inhibitors	
Phenelzine, tranylcypromine	Hypotension, sexual dysfunction, hepatotoxic; severe hypertension with pressor amines in drugs and foods, serotonin syndrome with serotonin-reuptake inhibitors, meperidine, dextromethorphan
Mocobemide	Moderate pressor risk, possible serotonin syndrome
Selegiline	Low pressor risk at low doses, some risk of serotonin syndrome
St. John's wort	Possible serotonin syndrome with serotonin-reuptake inhibitors
Serotonin-reuptake inhibitors	
Any serotonin-reuptake inhibitor	Nausea, anorgasmia, increased levels of other drugs, serotonin syndrome (especially with MAO inhibitors), withdrawl syndrome, variable agitation or insomnia, variable lethargy, bleeding (platelet dysfunction)
Atypical agents	
Bupropion	Seizures (dose-dependent), insomnia, anorexia
Mirtazapine	Sedation, rare leukopenia
Nefazodone	Sedation, hepatotoxicity, minor gastric and sexual dysfunction, rare serotonin syndrome
Trazodone	Sedation, some hypotension, priapism

Based on reports in manufacturers' product information bulletins

electric shocks, a range of emotional responses that can include sadness and tearfulness, irritability, or anxiety and agitation; there may also be motor symptoms, including muscle twitching, myoclonus, or tremor, especially with the discontinuation of serotonin-reuptake inhibitors. Such responses are particularly likely when short-acting antidepressants such as paroxetine or venlafaxine are withdrawn rapidly. Associations of risk with drug dose and duration of treatment have been suggested, but are not certain [474]. Often, such reactions are of moderate intensity and persist for some days, following a latency of several days, and rarely persist for more than 1 or 2 weeks. They can be alarming if patients are not warned in advance of their possibility and can limit acceptance of future treatment. Since fluoxetine discontinuation is unlikely to induce a discontinuation or withdrawal reaction [617], some experts advise substituting it when a serotonin-reuptake inhibitor of shorter half-life is involved. In addition, treatment with an antiparkinson–anticholinergic agent when a tricyclic antidepressant is withdrawn has been reported [602]. However, symptomatic management is usually achieved simply by restarting the discontinued medicine and reducing its doses slowly after achieving control of symptoms. These withdrawal syndromes are to be distinguished from later-appearing increased risk of relapse associated with antidepressant discontinuation as was discussed above (pp 200–204).

Another antidepressant discontinuation phenomenon involves very uncommon case reports of new episodes of mania or hypomania among patients being treated for depression [357, 376]. All types of antidepressants have been implicated, although the majority of cases have involved tricyclics. Curiously, there are more reports of cases involving apparent unipolar major depressive disorder than known cases of bipolar disorder. Successful treatment has included use of antimanic drugs, but again, paradoxically, reinstatement of antidepressant treatment has sometimes been followed by rapid resolution of manic symptoms [376]. It follows that possible mechanisms underlying such reactions are not known.

Drug Interactions

Interactions among drugs used to treat patients with mood disorders are not uncommon and sometimes can be dangerous or life threatening [476]. Most serotonin-reuptake inhibitors interact importantly with hepatic microsomal cytochrome-P450 (CYP) oxidases, acting as substrates for their own metabolic clearance, or as antagonists, or, sometimes, both. As CYP antagonists, they can decrease the clearance and increase circulating concentrations of many other drugs, with an increased risk of toxic effects [518]. Fluoxetine and paroxetine have particularly strong inhibitory effects on the microsomal cytochrome P-450 drug-metabolizing isozyme CYP-2D6.

Due to the prolonged duration of action of fluoxetine and its active metabolite norfluoxetine, inhibition of 2D6 can persist for days following discontinuation of fluoxetine. Sertraline and fluvoxamine have less of this effect, and venlafaxine, desvenlafaxine as well as racemic and both isomers of citalopram, very little. Mirtazapine also appears to have a low risk of interactions with CYP drug-metabolizing enzymes [472]. Bupropion has some antagonistic effect on CYP-2D6 and may potentiate tricyclic antidepressants and cardiac antiarrhythmia drugs [518]. This effect of some serotonin-reuptake inhibitors alters the pharmacokinetics of many drugs and can increase serum concentrations and effects of tricyclic antidepressants, and most phenothiazine and thioxanthene antipsychotic agents as well as clozapine and probably carbamazepine, and can increase the cardiac depressant effects of quinidine-like antiarrhythmia agents, as well as potentiating cimetidine and some beta-adrenergic antagonists [103, 457, 518]. Sertraline, while a less potent inhibitor of CYP-2D6, is highly bound to serum proteins and can compete for such binding with other agents, including warfarin, with increased risk of bleeding [433]. Paroxetine also interacts with warfarin without increasing serum concentrations of free warfarin, but increasing risk of bleeding through an unknown mechanism [49]. An additional risk of bleeding can arise from gastrointestinal ulceration associated with common combinations of serotonin-reuptake inhibitors, which have anticlotting effects, with nonsteroidal anti-inflammatory agents (NSAIDs), especially in elderly patients [135]. Fluvoxamine strongly inhibits CYP isozymes 1A2, 2C19 and

3A4 and can potentiate tricyclic antidepressants, clozapine, propranolol, warfarin, and caffeine [141]. A particularly dangerous practice was to deliberately potentiate clozapine with fluvoxamine so as to reduce the required doses of the relatively expensive antipsychotic, but at the risk of uncontrolled and potentially dangerous increases in its serum concentrations. It is not widely appreciated that MAO inhibitors can also interfere with the functioning of drug-metabolizing CYP enzymes, with a particular risk of inhibition of CYP-2A6 by tranylcypromine, and weak effects on CYP 2C19 and 3A4 by phenelzine [209].

Opposite risks arise when many psychotropic drugs are given with carbamazepine and other potent inducers of CYP enzymes (such as barbiturates and phenytoin, as well as the experimental natural-product antidepressant St. John's wort), so as to lose serum concentrations and effects of the second psychotropic agent [144, 219, 275, 591]. Oxcarbazepine is less likely to have such effects [102]. A further complication among anticonvulsants is that valproate sometimes has drug-potentiating actions, including on serotonin-reuptake inhibitors and the H_2 antihistaminic antiulcer agent cimetidine [381].

Tricyclic antidepressants can interact additively with agents with substantial antimuscarinic-anticholinergic effects, sometimes inducing a state of atropinic toxicity, marked by delirium, tachycardia, mydriatic pupils, reddened and dry skin and mucosae, and loss of bowel activity. Such states have been reversed safely with the centrally active anticholinesterase physostigmine, given slowly. However, this antidote has substantial risks of inducing epileptic seizures if administered rapidly intravenously [183, 216].

Drug interactions of especially great clinical importance and potentially lethal outcomes include those between monoamine oxidase inhibitors and agents of two types. First, MAO inhibitors can interact with norepinephrine-releasing agents (indirect sympathomimetic agents) such as tyramine in some foods, and stimulants including amphetamines and ephedrine, as well as decongestants including phenylephrine, phenylpropanolamine, with less certain effects of bupropion. These interactions lead to the rapid release of increased concentrations of norepinephrine at sympathetic autonomic and central noradrenergic nerve terminals that have built up owing to the anti-MAO-A actions of many MAO inhibitors, including phenelzine and tranylcypromine, to block the MAO in mitochondria of nerve terminals. This interaction, as well as inhibition of the clearance of tyramine and other amines by MAO in the gut and liver, leads to sharp and potentially dangerous or fatal increases in blood pressure, with risk of intracranial bleeding, collapse, and death. Many of the implicated norepinephrine-releasing agents are available in forms that may not be recognized as dangerous, including tyramine-containing fermented or processed foods as well as over-the-counter cold remedies. The "hypertensive crisis" phenomenon has been recognized for several decades, and lists of foods and medicinal agents to be avoided during treatment with MAO inhibitors are readily available; many of these are summarized in Table 4.16.

A second type of toxic interaction that is especially likely to follow combinations of MAO inhibitors with many other types of commonly used drugs are indications of cerebral intoxication, usually with autonomic and motor signs, fever and variable hypertension, often rapidly appearing and progressing potentially to coma and

Table 4.16 Contraindicated substances and conditions during treatment with monoamine oxidase inhibitors

Foods (HBP)

Aged cheeses and meats[a]

Banana peel[a] (especially if over-ripe)

Concentrated yeast extracts (e.g., Marmite)[a]

Draft beer (including nonalcoholic)[a]

Broad beans (fava. snowpea, and pods)[a]

Fermented products (sauerkraut, kimchi)[a]

Concentrated soy sauce,[a] *monosodium glutamate* (MSG)

Spoiled meat or fish, or improperly stored[a] and pickled fish

Tyramine-containing food supplements

Wine (especially red, in large quantity)

Drugs and Other Substances

Analgesics (fentanyl, local anesthetics with epinephrine [HBP], meperidine[a], pentazocine, propoxyphene, tramadol) [DEL]

Antihypertensives (guanethidine, α-methyldopa, reserpine, tetrabenazine) (HBP)

Antiparasitics, insecticides (amitraz) (DEL)

Antitussive (dextromethorphan) (DEL)

Catecholamine-releasing or uptake inhibiting agents (amphetamines[a], bupropion (?), cathinone, cocaine, sMDMA, mescaline, 4-methylaminorex, α-methyldopa; methylphenidate, pemoline, phenethylamine, sibutramine, sympathomimetics[a] [e.g., ephedrine, pseudoephedrine, phenylephrine, phenylpropanolamine], tyramine [including in foods]) (HBP)

Ergolines and lysergamides (ergotamine, bromocriptine, LSD) (DEL)

MAO inhibitors (including nonpsychiatric agents, e.g., furazolidone, hydrazines, linezolid, methylene blue, pargyline, procarbazine, and possibly isoniazid) (DEL)

Monoamine precursors (L-DOPA [HBP], 5-hydroxytryptophan, L-tryptophan, 5-hydroxytryptophan) (DEL)

Natural products (St. John's wort, *Echinops, Ginkgo* [all ?]) (DEL)

Muscle relaxants (cyclobenzaprine) (DEL)

Serotonin agonists (sumatriptan and other triptans [evidence inconsistent]) (DEL)

Serotonin-reuptake inhibitors (SRIs, SNRIs [especially fluoxetine], SNRIs [including milnacipran], mirtazapine)[a] *or releasers* (e.g., fenfluramine)[a] (DEL)

Tricyclic-type antidepressants (amoxapine, maprotiline, nomifensine, TCAs [especially 3-amines]), and possibly carbamazepine (DEL)

Tryptamines or precursors (dimethyltryptamine, psilocin) (DEL)

Medical Conditions

Dental or other *local anesthesia* (with vasoconstrictors such as epinephrine) (HBP)

Elective surgery due to potential drug interactions

Electroconvulsive treatment due to potential drug interactions

Pheochromocytoma (release of endogenous pressor catecholamines) (HBP)

Pregnancy (safety status remains uncertain)

Foods considered dangerous typically contain more than 6 mg of tyramine per serving. Risks of hypertensive crises (with norepinephrine-releasing agents) or cerebral intoxication syndromes (including serotonin syndrome and other forms of delirium) are highly variable with the substances and conditions listed above. Adapted from Gardner et al. [193]; Rapaport [440]; Sun-Edelstein et al. [535]; Ables and Nagubilli [2]; Wimbiscus et al. [598]

[a]Absolutely contraindicated substances with particularly high risk of hypertensive crisis (HBP) or cerebral intoxication with delirium (DEL)

death. It has become popular to term such reactions as manifestations of a relatively specific "serotonin syndrome." This term reflects the clearly identified risks of combining MAO inhibitors with serotonin-reuptake inhibitors, serotonin–norepinephrine-reuptake inhibitors, and other agents that can potentiate the availability or actions of serotonin ([2, 85, 208, 375, 440, 535, 539, 542]; Table 4.16). An additional basis for the term is that motor symptoms are a common component of the reactions, including ataxia, shivering, and initial muscle twitching, increased tendon reflexes and muscle tone, myoclonus, and other forms of motor overactivity, that has been associated theoretically with the effects of the descending serotonergic projections to the spinal motor neurons of the ventral horn of the spinal cord. Moreover, proposed antidotes include centrally active serotonin antagonists such as the antihistamine cyproheptadine, which also has antagonistic (or inverse-agonistic) actions at various $5HT_2$ receptor subtypes 2A, 2B, and 2C [2].

However, similar syndromes can involve agents in which mediation by excessive central serotonergic neurotransmission is less likely. Examples include combinations of an MAO inhibitor with a tricyclic-type antidepressant, analgesics including meperidine, and the antitussive agent dextromethorphan (Table 4.16). Moreover, there are several similarities of serotonin syndrome with the neuroleptic malignant syndrome (NMS) associated with potent antipsychotic agents, including rhabdomyolysis in both syndromes. Similarities are especially prominent in atypical forms of NMS associated with modern antipsychotic agents such as clozapine, which are marked mainly by delirium with minor or negligible motor signs (described above in Chapter 2; [512]). There is also substantial overlap with central anticholinergic (atropine-like) poisoning, sometimes including mydriasis or dilated pupils, and delirium is a prominent feature in anticholinergic intoxication and NMS as well as serotonin syndrome. Also, peripheral autonomic signs associated with loss of parasympathetic activity may not clearly distinguish anticholinergic poisoning with the other cerebral intoxication syndromes [192].

A critical point is that all of these types of cerebral intoxication can occur with clinical use of psychotropic agents, and they may not be recognized immediately as a change in mental status toward confusion, memory loss, and other indications of delirium [26]. All are usually best managed by stopping the suspected cerebral toxins and support of vital functions, as alternative neuromedical causes are considered. A further generalization is that serotonin- and serotonin–norepinephrine-reuptake inhibitors, and meperidine should never be combined with an MAO inhibitor, including agents with anti-MAO activity that are not used to treat psychiatric disorders, such as furazolidone, hydrazines, linezolid, methylene blue, procarbazine, and possibly isoniazid (Table 4.16).

Lethality of Overdoses

Older antidepressants, including MAO inhibitors and tricyclic-type antidepressants are notoriously toxic and potentially lethal on acute overdosage ([312, 471]; Table 4.17). Even with the advent of far less potentially toxic modern agents,

Table 4.17 Lethality on overdose of common psychotropic drugs

Agents	Fatalities [rate (%)]	RR vs. TCAs (95% CI)
Tricyclic antidepressants	89/12,522 (0.711)	1.00 (comparator)
Barbiturates	16/3,274 (0.489)	0.689 (0.405–1.05)
Phenothiazines	22/4,704 (0.486)	0.660 (0.414–1.05)
Benzodiazepines	180/60,014 (0.300)	0.424 (0.329–0.546)[a]
Lithium	16/5,296 (0.245)	0.347 (0.194–0.620)[a]
Modern antipsychotics	72/32,422 (0.222)	0.314 (0.230–0.428)[a]
Valproate	21/9,619 (0.218)	0.309 (0.192–0.496)[a]
Serotonin-reuptake inhibitors	106/55,977 (0.189)	0.268 (0.202–0.355)[a]
Carbamazepine	9/5,144 (0.175)	0.247 (0.125–0.491)[a]

Rates of fatalities/persons exposed to an overdose, regardless of severity or apparent intent. Shown are observed crude rates in descending rank order and relative risk (RR with 95% confidence intervals [CI]) compared to tricyclic antidepressants as a standard class of potentially lethal toxins. Tricyclics with highest risk were long-acting protriptyline (3.70%) and desipramine, which yields highly cardiotoxic hydroxy-metabolites (1.72%); nortriptyline was lowest (0.090%)
[a]Indicates statistically significantly lower risk than tricyclic antidepressants (by χ^2, $p < 0.001$). Adapted from Watson et al. [587]

antidepressants continue to represent a high proportion of agents involved in self-poisoning with suspected suicidal intent, especially among females and young persons. Life threatening or fatal cases often involve combinations of antidepressants with other drugs or alcohol [176]. Overdoses of antidepressants, particularly older agents, involve complex intoxication, usually marked by delirium and unstable vital signs.

Risks of severe, life threatening, or fatal intoxication following acute overdoses of antidepressants are greatest with MAO inhibitors and tricyclic-type agents and lowest with most serotonin-reuptake inhibitors (Table 4.17). Agents with mixed effects on serotonin and norepinephrine reuptake, including duloxetine and venlafaxine, may be somewhat more toxic than most serotonin-reuptake inhibitors. In addition, bupropion and citalopram among modern antidepressants may have greater toxic risks than most serotonin-reuptake inhibitors [106, 367, 592]. Venlafaxine overdoses have been associated with hypertension and cardiac conduction defects, and citalopram with depressed cardiac conduction times; duloxetine has been associated with hepatic toxicity [176, 268, 269, 367]. In contrast, overdoses of mirtazapine, unless associated with other drugs or alcohol, generally have been benign [586].

Tricyclic-type antidepressants, in particular, produce cardiac depressant effects and high risks of potentially fatal ventricular arrhythmias and cardiac arrest, risks of which can be limited by immediate administration of intravenous bicarbonate [426]. Particularly high fatality rates have been associated with acute overdoses of amitriptyline, desipramine, and maprotiline [592]. Tricyclic antidepressants can be removed by hemodialysis, sometimes augmented by extracorporeal sorbent detoxification in addition to orogastric administration of activated charcoal, although such treatment may require several days even after initial recovery of consciousness and stable vital signs [25]. There also have been promising results with the use of intravenous infusions of lipids to sequester high circulating concentrations of the lipophilic tricyclic antidepressants [70, 101]. Although modern antidepressants are much less likely to be lethally toxic in overdose, high doses of serotonin-reuptake inhibitors have been associated with delirium and coma, supposedly of the "serotonin syndrome" type [268]. A summary of recent rates of fatality per overdose of various drugs of interest is provided in Table 4.17 [587].

Antidepressants in Pregnancy

Except for some anticonvulsants, information about teratogenic or other neonatal risks of most psychotropic agents remains incomplete and somewhat ambiguous. Indications for treatment of major depression, or to avoid recurrences, during pregnancy are high owing to high prevalence of recurrences during pregnancy (especially in the first trimester) and particularly in the early postpartum period, as are pressures to avoid potential adverse effects on fetal development of maternal psychiatric illness [172, 175, 340, 516, 581, 590]. Recurrence involves more than two-thirds of pregnant women who elect to discontinue treatment for recurrent major depression during pregnancy, and this risk is reduced by more than half if antidepressants are continued [114]. However, there are concerns about potential adverse effects on fetal and neonatal development in association with use of antidepressants during pregnancy [143]. The effectiveness of psychotherapy as an alternative, though clearly likely to be safer than use of antidepressant drugs, is not extensively evaluated specifically in the perinatal periods. It may be effective in mild-moderate cases but is likely to prove inadequate to prevent postpartum depression [87, 145, 326, 411, 564]. Other alternatives include use of ECT and other physical methods such as rTMS [299, 463].

Specific teratogenic risks with particular agents or classes of antidepressants remain uncertain. Findings include inconsistent, and usually low-level, associations of fetal developmental anomalies with antidepressant use early in pregnancy [143, 567]. These include shortened gestation, low birth-weight, neonatal respiratory distress, and impaired physiological vigor or transient neonatal irritability and adaptational problems, with suspicion of risk of neonatal withdrawal reactions [143, 159, 335, 393, 544, 578]. In addition, there have been reports of neonatal pulmonary hypertension following exposure to serotonin-reuptake inhibitors during pregnancy, and there may or may not be a particular risk of cardiac malformations associated with paroxetine [11, 143, 158, 567]. However, the prevalence of most of the observed associations is low and contributions of antidepressant-exposure specifically vs. maternal or environmental factors have not been adequately distinguished [143, 566, 601].

An additional concern about use of psychotropic drugs during pregnancy is that their metabolic clearance can shift, often toward decreased circulating concentrations in mid-pregancy, and again at delivery, sometimes with sharp increases in blood levels. Moreover, most agents that pass the blood–brain diffusion barrier are also able to access fetal tissue across the placental barrier, and many antidepressants are also found in breast milk [131, 382, 411]. With modern antidepressant drugs, shifts in drug clearance during pregnancy are usually moderate and variable among individual patients, although precise details, by trimester of pregnancy, and postpartum changes are lacking for most agents [185, 395, 507].

Conclusions

Antidepressant drugs are among the most commonly prescribed medicines of all kinds, as well as among the most widely used psychotropic agents. They include older tricyclic-type agents and monoamine oxidase (MAO) inhibitors. From the end of the 1980s, there was a decade-long explosive increase and broadening of the use of newer, second-generation antidepressants (notably, serotonin- and serotonin–norepinephrine-reuptake inhibitors, bupropion, and mirtazapine), not only for major depressive disorder, but also for a range of anxiety-related and other disorders. Since their introduction in the late 1950s, antidepressants have always had evidence of quite limited overall efficacy, particularly for acute major depression, sometimes barely surpassing the effects of placebo treatment, and with a substantial proportion of failed trials that did not separate statistically from placebo treatment. In addition, drug-placebo differences (effect-sizes) have gradually been declining over the past three decades, owing almost entirely to increased rates of responses during treatment with a placebo. In turn, this decline has stimulated ever-larger trials with more patients and more collaborating sites, often in cultures based in regions other than North America or Western Europe, probably with a paradoxical tendency to make placebo-associated responses even larger. Although the efficacy of this class of agents for acute major depression has been marginal and has not improved with the development of second-generation or modern antidepressants, their range of clinical applications to other disorders has risen steadily, contributing to their broad popularity among a wide range of clinicians in general medicine as well as psychiatry.

Efforts have been made to identify characteristics of patients who respond to antidepressants and to improve chances for successful treatment. There is currently considerable uncertainty about the relative efficacy of antidepressants for bipolar depressive episodes. Initially unsatisfactory responses in unipolar depression can sometimes be overcome with increased doses of a standard agent, longer treatment, or changing to an alternative type of antidepressant, or use of adjunctive treatments ranging from lithium carbonate to psychotherapy.

Long-term, prophylactic effects of antidepressants have been proposed, and typical practice is to recommend their continuation for at least several months or longer after apparent clinical recovery from an acute episode of major depression. However, the sustained effectiveness of antidepressants for long-term prevention of recurrences by treatment over several years remains uncertain. In large part, this uncertainty rests on study of highly selected patients defined by initial short-term responses to treatment, as well as by high rates of recurrences within the weeks or few months following antidepressant discontinuation that appear to be in excess of expectations from the natural history of the untreated illness or the patient's own previous pattern of morbidity. Such later drug discontinuation-associated reactions are different from early, short-term physiological withdrawal reactions to discontinuing antidepressants, and particularly, short-acting serotonin- or serotonin–norepinephrine-reuptake inhibitors.

Adverse effects of antidepressants are many and complex, and can sometimes be lethal. Overdoses of tricyclic-type antidepressants carry high risks of cardiac dysfunction, coma, and death. Overdoses of serotonin- and serotonin–norepinephrine-reuptake inhibitors, or their combination or administration of many other substances with MAO inhibitors can lead to acute and potentially fatal forms of acute, delirious cerebral intoxication. Although most modern antidepressants are safer on overdose than first-generation agents, rates of overdoses with suicidal intent have declined only to a limited extent, and suicidal outcomes have also changed little over the past several decades, and quite variably among regions and between countries.

Important questions concerning antidepressants include: (a) how to optimize their efficacy by identifying subgroups within the broad spectrum of major depressive disorders, (b) clarifying their relative value for bipolar depression, (c) resolving the ambiguity concerning their possible long-term prophylactic value for depressive and anxiety disorders, and (d) continuing to search for generally more effective treatments for severe depression.

References

1. Abdel-Hamid IA. Pharmacologic treatment of rapid ejaculation: levels of evidence-based review. Curr Clin Pharmacol. 2006;1:243–54.
2. Ables AZ, Nagubilli R. Prevention, recognition, and management of serotonin syndrome. Am Fam Physician. 2010;81:1139–42.
3. Abudy A, Juven-Wetzler A, Zohar J. Pharmacological management of treatment-resistant obsessive-compulsive disorder. CNS Drugs. 2011;25:585–96.
4. Achab S, Khazaal Y. Psychopharmacological treatment in pathological gambling: critical review. Curr Pharm Des. 2011;17:1389–95.
5. Acharya N, Rosen AS, Polzer JP, D'Souza DN, Perahia DG, Cavazzoni PA, et al. Duloxetine: meta-analyses of suicidal behaviors and ideation in clinical trials for major depressive disorder. J Clin Psychopharmacol. 2006;26:587–94.
6. Ahrens B, Müller-Oerlinghausen B. Does lithium exert an independent antisuicidal effect? Pharmacopsychiatry. 2001;34:132–6.
7. Aiken CB. Pramipexole in psychiatry: systematic review of the literature. J Clin Psychiatry. 2007;68:1230–6.
8. Alexopoulos GS, Kiosses DN, Heo M, Murphy CF, Shanmugham B, Gunning-Dixon F. Executive dysfunction and the course of geriatric depression. Biol Psychiatry. 2005;58:204–10.
9. Allen C, Kalu U-G, Sexton CE, Ebmeier KP. Transcranial stimulation in depression. Br J Psychiatry. 2012;200:10–1.
10. Alonso J, Angermeyer MC, Bernert S, Bruffaerts R, Brugha TS, Bryson H. Use of mental health services in Europe: results from the European Study of the Epidemiology of Mental Disorders (ESEMeD) project. Acta Psychiatr Scand. 2004;(Suppl 420):47–54.
11. Alwan S, Reefhuis J, Rasmussen SA, Olney RS, Friedman JM. Use of selective serotonin-reuptake inhibitors in pregnancy and the risk of birth defects. N Engl J Med. 2007;356:2684–92.

12. American Psychiatric Association (APA). Diagnostic and Statistical Manual (DSM-III). 3rd ed. Washington, DC: American Psychiatric Press; 1980.
13. Ampuero E, Rubio FJ, Falcon R, Sandoval M, Diaz-Veliz G, Gonzalez RE, et al. Chronic fluoxetine treatment induces structural plasticity and selective changes in glutamate receptor subunits in the rat cerebral cortex. Neuroscience. 2010;169:98–108.
14. Amsterdam JD, Williams D, Michelson D, Adler LA, Dunner DL, Nierenberg AA, et al. Tachyphylaxis after repeated antidepressant drug exposure in patients with recurrent major depressive disorder. Neuropsychobiology. 2009;59:227–33.
15. Anacker C, Zunszain PA, Cattaneo A, Carvalho LA, Garabedian MJ, Thuret S, et al. Antidepressants increase human hippocampal neurogenesis by activating the glucocorticoid receptor. Mol Psychiatry. 2011;16:738–50.
16. Anderson IM. Selective serotonin reuptake inhibitors vs. tricyclic antidepressans: meta-analysis of efficacy and tolerability. J Affect Disord. 2000;58:19–36.
17. Anderson HD, Pace WD, Libby AM, West DR, Valuck RJ. Rates of five common antidepressant side effects among new adult and adolescent cases of depression: retrospective US claims study. Clin Ther. 2012;34:113–23.
18. Andrade C, Rao NS. How antidepressant drugs act: primer on neuroplasticity as the eventual mediator of antidepressant efficacy. Indian J Psychiatry. 2010;52:378–86.
19. Andrade C, Sandarsh S, Chethan KB, Nagesh KS. Serotonin reuptake inhibitor antidepressants and abnormal bleeding: review for clinicians and reconsideration of mechanisms. J Clin Psychiatry. 2010;71:1565–75.
20. Angell M. editor. The illusions of psychiatry: an exchange. 2011. NY Rev Books 58. p. 11 http://www.nybooks.com/articles/archives/2011/aug/18.
21. Angelucci F, Croce N, Spalletta G, Dinallo V, Gravina P, Bossù P, et al. Paroxetine rapidly modulates the expression of brain-derived neurotrophic factor mRNA and protein in a human glioblastoma-astrocytoma cell line. Pharmacology. 2011;87:5–10.
22. Ansseau M, Kupfer DJ, Reynolds III CF, Coble PA. "Paradoxical" shortening of REM latency on first recording night in major depressive disorder: clinical and polysomnographic correlates. Biol Psychiatry. 1985;20:135–45.
23. Anthony J, Sexton T, Neumaier J. Antidepressant-induced regulation of 5-HT-1B mRNA in rat dorsal raphe nucleus reverses rapidly after drug discontinuation. J Neurosci Res. 2000;61:82–7.
24. Arana GW, Baldessarini RJ, Ornsteen M. The dexamethasone suppression test for diagnosis and prognosis in psychiatry: commentary and review. Arch Gen Psychiatry. 1985;42: 1193–204.
25. Ash SR, Levy H, Akmal M, Mankus RA, Sutton JM, Emery DR, et al. Treatment of severe tricyclic antidepressant overdose with extracorporeal sorbent detoxification. Adv Ren Replace Ther. 2002;9:31–41.
26. Attar-Herzberg D, Apel A, Gang N, Dvir D, Mayan H. Serotonin syndrome: initial misdiagnosis. Isr Med Assoc J. 2009;11:367–70.
27. Axelrod J, Whitby LG, Hertting G. Effect of psychotropic drugs on the uptake of ^{3}H-norepinephrine by tissues. Science. 1961;133:383–4.
28. Aubert D, Berard E, Blanc JP, Lenoir G, Liard F, Lottmann H. [Isolated primary nocturnal enuresis: international evidence based management. Consensus recommendations by French expert group (French)]. Prog Urol. 2010;20:343–9.
29. Bakker A, van Balkom AJ, Spinhoven P. SSRIs vs TCAs in the treatment of panic disorder: meta-analysis. Acta Psychiatr Scand. 2002;106:163–7.
30. Baldessarini RJ. Neuropharmacology of S-adenosyl-L-methionine. Am J Med. 1987; 83(Suppl 5A):95–103.
31. Baldessarini RJ. Update on pharmacology and treatment with antidepressants. J Clin Psychiatry. 1989;50:117–26.
32. Baldessarini RJ. American biological psychiatry and psychopharmacology 1944–1994. Chapt 16. In: Menninger RW, Nemiah JC, editors. American Psychiatry after World War II (1944–1994). Washington DC: APA Press; 2000. p. 371–412.

33. Baldessarini RJ. Drug therapy of depression and anxiety disorders. Chapt 17. In: Brunton LL, Lazo JS, Parker KL, editors. Goodman and Gilman's The Pharmacological Basis of Therapeutics. 11th ed. New York: McGraw-Hill; 2005. p. 429–59.

34. Baldessarini RJ, Henk HJ, Sklar AR, Chang J, Leahy LF. Psychotropic medications for bipolar disorder patients in the United States: polytherapy and adherence. Psychiatr Serv. 2008;59:1175–83.

35. Baldessarini RJ, Leahy LF, Arcona S, Gause D, Zhang W, Hennen J. Prescribing patterns of psychotropic medicines in the United States for patients diagnosed with bipolar disorders. Psychiatr Serv. 2007;58:85–91.

36. Baldessarini RJ, Pompili M, Tondo L. Suicidal risk in antidepressant trials. Arch Gen Psychiatry. 2006;63:246–8.

37. Baldessarini RJ, Pompili M, Tondo L, Tsapakis E, Soldani F, Faedda GL, et al. Antidepressants and suicidal behavior: are we hurting or helping? Clin Neuropsychiatry. 2005;2:73–5.

38. Baldessarini RJ, Salvatore P, Khalsa HM, Gebre-Medhin P, Imaz H, González-Pinto A, et al. Morbidity in 303 first-episode bipolar I disorder patients. Bipolar Disord. 2010;12:264–70.

39. Baldessarini RJ, Suppes T, Tondo L. Lithium withdrawal in bipolar disorder: implications for clinical practice and experimental therapeutics research. Am J Ther. 1996;3:492–6.

40. Baldessarini RJ, Tondo L. Psychopharmacology for suicide prevention. Chapt 9. In: Pompili M, editor. Suicide: A Global Perspective. Sarjah, UAE: Bentham Science Publishers; 2006. p. 114–27.

41. Baldessarini RJ, Tondo L, Faedda GL, Viguera AC, Baethge C, Bratti I, et al. Latency, discontinuation, and re-use of lithium treatment. Chapter 38. In: Bauer M, Grof P, Müller-Oerlinghausen B, editors. Lithium in Neuropsychiatry: The Comprehensive Guide. London: Taylor & Francis; 2006. p. 465–81.

42. Baldessarini RJ, Tondo L, Ghiani C, Lepri B. Illness risk following rapid versus gradual discontinuation of antidepressants. Am J Psychiatry. 2010;167:934–41.

43. Baldessarini RJ, Tondo L, Hennen J. Reduced suicide risk during long-term treatment with lithium. Ann NY Acad Sci. 2001;932:24–43.

44. Baldessarini RJ, Tondo L, Strombom I, Dominguez S, Fawcett J, Oquendo M, et al. Analysis of ecological studies of relationships between antidepressant utilization and suicidal risk. Harv Rev Psychiatry. 2007;15:133–45.

45. Baldessarini RJ, Tondo L, Viguera AC. Discontinuing psychotropic agents. J Psychopharmacol. 1999;13:292–3.

46. Baldessarini RJ, Vieta E, Calabrese JR, Tohen M, Bowden C. Bipolar depression: overview and commentary. Harv Rev Psychiatry. 2010;18:143–57.

47. Baldwin DS, Ajel KI, Garner M. Pharmacological treatment of generalized anxiety disorder. Curr Top Behav Neurosci. 2010;2:453–67.

48. Bandelow B, Rüther E. Treatment-resistant panic disorder. CNS Spectr. 2004;9:725–39.

49. Bannister SJ, Houser VP, Hulse JD, Kisicki JC, Rasmussen JG. Evaluation of the potential for interaction of paroxetine with diazepam, cimetidine, warfarin, and digoxin. Acta Psychiatr Scand Suppl. 1989;350:102–6.

50. Barbanti P, Aurilia C, Egeo G, Fofi L. Migraine prophylaxis: what is new and what we need? Neurol Sci. 2011;32 (Suppl 1):S111–5.

51. Barbui C, Esposito E, Cipriani A. Selective serotonin reuptake inhibitors and risk of suicide: systematic review of observational studies. Can Med Assoc J (CMAJ). 2009;180:291–7.

52. Bares M, Novak T, Brunovsky M, Kopecek M, Stopkova P, Krajca V, et al. Change of QEEG prefrontal cordance as a response predictor to antidepressive intervention in bipolar depression: pilot study. J Psychiatr Res. 2012;46:219–25.

53. Bares M, Novak T, Kopecek M, Stopkova P, Kozeny J, Höschl C. Early improvement of depressive symptoms as a potential predictor of response to antidepressants in depressive patients who failed to respond to previous antidepressant treatments. Analysis of naturalistic data. Eur Psychiatry. 2011 [Epub ahead of print, 28 Nov].

54. Barnes TR. Evidence-based guidelines for the pharmacological treatment of schizophrenia: recommendations from the British Association for Psychopharmacology. J Psychopharmacol. 2011;25:567–620.

55. Bartova L, Berger A, Pezawas L. Is there a personalized medicine for mood disorders? Eur Arch Psychiatry Clin Neurosci. 2010;260 (Suppl 2):S121–6.
56. Baskin SM, Smitherman TA. Comorbidity between migraine and depression: update on traditional and alternative treatments. Neurol Sci 2011;32 (Suppl 1):S9–13.
57. Basu M, Duckett JR. Update on duloxetine for the management of stress urinary incontinence. Clin Interv Aging. 2009;4:25–30.
58. Baudry A, Mouillet-Richard S, Schneider B, Launay JM, Kellermann O. MiR-16 targets the serotonin transporter: new facet for adaptive responses to antidepressants. Science. 2010;329:1537–41.
59. Bauer M, Bschor T, Pfennig A, Whybrow PC, Angst J, Versiani M, et al. World Federation of Societies of Biological Psychiatry (WFSBP) guidelines for biological treatment of unipolar depressive disorders in primary care. World J Biol Psychiatry. 2007;8:67–104.
60. Baumann P, Nil R, Souche A, Montaldi S, Baettig D, Lambert S, et al. A double-blind, placebo-controlled study of citalopram with and without lithium in the treatment of therapy-resistant depressive patients: a clinical, pharmacokinetic, and pharmacogenetic investigation. J Clin Psychopharmacol. 1996;16:307–14.
61. Bech P. Is the antidepressive effect of second-generation antidepressants a myth? Psychol Med. 2010;40:181–6.
62. Beekwilder JP, Beems T. Overview of the clinical applications of vagus nerve stimulation. J Clin Neurophysiol. 2010;27:130–8.
63. Beerworth EE, Tiller JW. Liability in prescribing choice: the example of the antidepressants. Aust N Z J Psychiatry. 1998;32:560–6.
64. Benazzi F. Reviewing the diagnostic validity and utility of mixed depression (depressive mixed states). Eur Psychiatry. 2008;23:40–8.
65. Benmansour S, Altamirano AV, Jones DJ, Sanchez TA, Gould GG, Pardon MC, et al. Regulation of the norepinephrine transporter by chronic administration of antidepressants. Biol Psychiatry. 2004;55:313–6.
66. Berntson GG, Uchino BN, Cacioppo JT. Origins of baseline variance and the Law of Initial Values. Psychophysiology. 1994;31:204–10.
67. Bespalov AY, van Gaalen MM, Gross G. Antidepressant treatment in anxiety disorders. Curr Top Behav Neurosci. 2010;2:361–90.
68. Bhuvaneswar CG, Baldessarini RJ, Harsh VL, Alpert JE. Adverse endocrine and metabolic effects of psychotropic drugs: selective clinical review. CNS Drugs. 2009;23:1003–21.
69. Bismuth-Evenzal Y, Gonopolsky Y, Gurwitz D, Iancu I, Weizman A, Rehavi M. Decreased serotonin content and reduced agonist-induced aggregation in platelets of patients chronically medicated with SSRI drugs. J Affect Disord. 2012;136:99–103.
70. Blaber MS, Khan JN, Brebner JA, McColm R. "Lipid rescue" for tricyclic antidepressant cardiotoxicity. J Emerg Med. 2012 [Epub ahead of print, 11 Jan].
71. Blier P. Pharmacology of rapid-onset antidepressant treatment strategies. J Clin Psychiatry. 2001;62:12–7.
72. Blier P, de Montigny C, Chaput Y. Modification of the serotonin system by antidepressants: implications for the therapeutic response in major depression. J Clin Psychopharmacol. 1987;7:24S–35.
73. Blier P, Tremblay P. Physiological mechanisms underlying the antidepressant discontinuation syndrome. J Clin Psychiatry. 2006;67 (Suppl 4):8–13.
74. Bloch MH, Landeros-Weisenberger A, Dombrowski P, Kelmendi B, Wegner R, Nudel J, et al. Systematic review: pharmacological and behavioral treatment for trichotillomania. Biol Psychiatry. 2007;62:839–46.
75. Bloch MH, McGuire J, Landeros-Weisenberger A, Leckman JF, Pittenger C. Meta-analysis of the dose-response relationship of SSRI in obsessive-compulsive disorder. Mol Psychiatry. 2010;15:850–5.
76. Bobo WV, Chen H, Trivedi MH, Stewart JW, Nierenberg AA, Fava M, et al. Randomized comparison of selective serotonin reuptake inhibitor (escitalopram) monotherapy and antidepressant combination pharmacotherapy for major depressive disorder with melancholic features. J Affect Disord. 2011;133:467–76.

77. Bodkin JA, Zornberg GL, Lukas SE, Cole JO. Buprenorphine treatment of refractory depression. J Clin Psychopharmacol. 1995;15:49–57.
78. Baker DG, Nievergelt CM, Risbrough VB. Post-traumatic stress disorder: emerging concepts of pharmacotherapy. Expert Opin Emerg Drugs. 2009;14:251–72.
79. Bolasco A, Carradori S, Fioravanti R. Focusing on new monoamine oxidase inhibitors. Expert Opin Ther Pat. 2010;20:909–39.
80. Bollini P, Pampallona S, Tibaldi G, Kupelnick B, Munizza C. Effectiveness of antidepressants. Meta-analysis of dose-effect relationships in randomised clinical trials. Br J Psychiatry. 1999;174:297–303.
81. Bonanno G, Giambelli R, Raiteri L, Tiraboschi E, Zappettini S, Musazzi L, et al. Chronic antidepressants reduce depolarization-evoked glutamate release and protein interactions favoring formation of SNARE complex in hippocampus. J Neurosci. 2005;25:3270–9.
82. Bodmer M. [Poisoning with antidepressants (German)]. Ther Umsch. 2009;66:335–41.
83. Booij L, Van der Does AJ, Riedel WJ. Monoamine depletion in psychiatric and healthy populations: review. Mol Psychiatry. 2003;8:951–73.
84. Borges R, Pereda D, Beltrá n B, Prunell M, Rodríguez M, Machado JD. Intravesicular factors controlling exocytosis in chromaffin cells. Cell Mol Neurobiol. 2010;30:1359–64.
85. Boyer EW, Shannon M. The serotonin syndrome. N Engl J Med. 2005;352:1112–20.
86. Bradley SR, Uslaner JM, Flick RB, Lee A, Groover KM, Hutson PH. The mGluR7 allosteric agonist AMN082 produces antidepressant-like effects by modulating glutamatergic signaling. Pharmacol Biochem Behav. 2012;10:35–40.
87. Brandon AR, Freeman MP. When she says "no" to medication: psychotherapy for antepartum depression. Curr Psychiatry Rep. 2011;13:459–666.
88. Brinkers M, Petz T, Hoffmeyer D. [General importance of antidepressants in treatment of patients with chronic pain (German)]. Anasthesiol Intensivmed Notfallmed Schmerzther. 2010;45:712–6.
89. Brooks WH, Guida WC, Daniel KG. Significance of chirality in drug design and development. Curr Top Med Chem. 2011;11:760–70.
90. Brown WA. Treatment response in melancholia. Acta Psychiatr Scand. 2007;116 (Suppl 433):125–9.
91. Burke WM, Gergel I, Bost A. Fixed-dose trial of the single-isomer SSRI escitalopram in depressed outpatients. J Clin Psychiatry. 2003;63:331–6.
92. Bymaster FP, Katner JS, Nelson DL, Hemrick-Luecke SK, Threlkeld PG, Heiligenstein JH, et al. Atomoxetine increases extracellular levels of norepinephrine and dopamine in prefrontal cortex of rat: a potential mechanism for efficacy in attention deficit/hyperactivity disorder. Neuropsychopharmacology. 2002;27:699–711.
93. Byrne SE, Rothchild AJ. Loss of antidepressant efficacy during maintenance therapy: possible mechanisms and treatments. J Clin Psychiatry. 1998;59:279–88.
94. Cape J, Whittington C, Buszewicz M, Wallace P, Underwood L. Brief psychological therapies for anxiety and depression in primary care: meta-analysis and meta-regression. BMC Med. 2010;8:38–51.
95. Carlsson A, Kehr W, Lindqvist M, Magnusson T, Atack CV. Regulation of monoamine metabolism in the central nervous system. Pharmacol Rev. 1972;24:371–84.
96. Carlsson A, Lindqvist M. Effects of antidepressant agents on the synthesis of brain monoamines. J Neural Transm. 1978;43:73–91.
97. Carpenter DJ. St. John's wort and S-adenosyl methionine as "natural" alternatives to conventional antidepressants in the era of the suicidality boxed warning: what is the evidence for clinically relevant benefit? Altern Med Rev. 2011;16:17–39.
98. Carvajal A, Ortega S, Del Olmo L, Vidal X, Aguirre C, Ruiz B, et al. Selective serotonin reuptake inhibitors and gastrointestinal bleeding: case-control study. PLoS One. 2011;6:e19819.
99. Carvalho AF, Cavalcante JL, Castelo MS, Lima MC. Augmentation strategies for treatment-resistant depression: literature review. J Clin Pharm Ther. 2007;32:415–28.

100. Catterson ML, Preskorn SH. Pharmacokinetics of selective serotonin reuptake inhibitors: clinical relevance. Pharmacol Toxicol. 1996;78:203–38.
101. Cave G, Harvey M, Graudins A. Intravenous lipid emulsion as antidote: summary of published human experience. Emerg Med Australas. 2011;23:123–41.
102. Centorrino F, Albert MJ, Berry JM, Kelleher JP, Fellman V, Line G, et al. Oxcarbazepine: clinical experience with hospitalized psychiatric patients. Bipolar Disord. 2003;5:370–4.
103. Centorrino F, Baldessarini RJ, Frankenburg FR, Kando J, Volpicelli SA, Flood JG. Serum levels of clozapine and norclozapine in patients treated with selective serotonin reuptake inhibitors. Am J Psychiatry. 1996;153:820–2.
104. Cerletti U. Old and new information about electroshock. Am J Psychiatry. 1950;107:87–94.
105. Chakos M, Patel JK, Rosenheck R, Glick ID, Hammer MB, Tapp A, et al. Concomitant psychotropic medication use during treatment of schizophrenia patients: longitudinal results from the CATIE study. Clin Schizophr Relat Psychoses. 2011;5:124–34.
106. Chan AN, Gunja N, Ryan CJ. Comparison of venlafaxine and SSRIs in deliberate self-poisoning. J Med Toxicol. 2010;6:116–21.
107. Chang T, Fava M. The future of psychopharmacology of depression. J Clin Psychiatry. 2010;71:971–5.
108. Chaput Y, de Montigny C, Blier P. Presynaptic and postsynaptic modification of the serotonin system by long-term administration of antidepressant treatments. Neuropsychopharmacology. 1991;5:219–29.
109. Charney DS. Monoamine dysfunction and the pathophysiology and treatment of depression. J Clin Psychiatry. 1998;59 (Suppl 14):11–4.
110. Chen F, Lawrence AJ. Effects of antidepressant treatment on serotonergic and dopaminergic systems in Fawn-Hooded rats: a quantitative autoradiography study. Brain Res. 2003;976:22–9.
111. Cipriani A, Furukawa TA, Salanti G, Geddes JR, Higgins JP, Churchill R, et al. Comparative efficacy and acceptability of 12 new-generation antidepressants: a multiple-treatments meta-analysis. Lancet. 2009;373:746–58.
112. Clausius N, Born C, Grunze H. [Relevance of dopamine agonists in the treatment of depression (German)]. Neuropsychiatr. 2009;23:15–25.
113. Coenen VA, Schlaepfer TE, Maedler B, Panksepp J. Cross-species affective functions of the medial forebrain bundle-implications for the treatment of affective pain and depression in humans. Neurosci Biobehav Rev. 2011;35:1971–81.
114. Cohen LS, Altshuler LL, Harlow BL, Nonacs R, Newport DJ, Viguera AC, et al. Relapse of major depression during pregnancy in women who maintain or discontinue antidepressant treatment. JAMA. 2006;295:499–507.
115. Cole JO. Therapeutic efficacy of antidepressant drugs: review. JAMA. 1964;190:448–55.
116. Connolly KR, Thase ME. If at first you don't succeed: review of the evidence for antidepressant augmentation, combination and switching strategies. Drugs. 2011;71:43–64.
117. Cooper TB, Simpson GM. Prediction of individual dosage of nortriptyline. Am J Psychiatry. 1978;135:333–5.
118. Correas-Gómez MA, Portillo Martín JA, Martín García B, Hernández Rodríguez R, Gutiérrez Baños JL, del Valle Schaan JI, et al. [Trazodone-induced priapism (Spanish)]. Actas Urol Esp. 2000;24:840–2.
119. Correll CU, Kratochvil CJ, March JS. Developments in pediatric psychopharmacology: focus on stimulants, antidepressants, and antipsychotics. J Clin Psychiatry. 2011;72:655–70.
120. Courtet P. Suicidality: risk factors and the effects of antidepressants: parallel reduction of suicidality and other depressive symptoms during treatment with the SNRI, milnacipran. Neuropsychiatr Dis Treat. 2010;6:3–8.
121. Covington III HE, Vialou V, Nestler EJ. From synapse to nucleus: novel targets for treating depression. Neuropharmacology. 2010;58:683–93.
122. Crane GE. The psychiatric side effects of iproniazid. Am J Psychiatry. 1956;112:494–501.

123. Cremers TI, Spoelstra EN, de Boer O, Bosker FJ, Mørk A, den Boer JA, et al. Desensitization of 5-HT autoreceptors upon pharmacokinetically monitored chronic treatment with citalopram. Eur J Pharmacol. 2000;397:351–7.

124. Crews FT, Scott JA, Shorsetin NH. Rapid down-regulation of serotonin2 receptor binding during combined administration of tricyclic antidepressant drugs and alpha-2 antagonists. Neuropharmacology. 1983;22:1203–9.

125. Crossley NA, Bauer M. Acceleration and augmentation of antidepressants with lithium for depressive disorders: two meta-analyses of randomized, placebo-controlled trials. J Clin Psychiatry. 2007;68:935–40.

126. Croxtall JD, Scott LJ. Olanzapine-fluoxetine: review of its use in patients with treatment-resistant major depressive disorder. CNS Drugs. 2010;24:245–62.

127. Cuijpers P, Geraedts AS, van Oppen P, Andersson G, Markowitz JC, van Straten A. Interpersonal psychotherapy for depression: meta-analysis. Am J Psychiatry. 2011;168:581–92.

128. Daly JM, Wilens T. Use of tricyclic antidepressants in children and adolescents. Pediatr Clin North Am. 1998;45:1123–35.

129. Damis M, Patel Y, Simpson GM. Sildenafil in the treatment of SSRI-induced sexual dysfunction: pilot study. Prim Care Companion J Clin Psychiatry. 1999;1:184–7.

130. Damluji NF, Fergusson JM. Paradoxical worsening of depressive symptomatology caused by antidepressants. J Clin Psychopharmacol. 1988;8:347–9.

131. Davanzo R, Copertino M, De Cunto A, Minen F, Amaddeo A. Antidepressant drugs and breastfeeding: review of the literature. Breastfeed Med. 2011;6:89–98.

132. Davids E, Zhang K, Baldessarini RJ. Stereoselective effects of methylphenidate on hyperactivity in juvenile rats induced by neonatal 6-hydroxydopamine lesioning. Psychopharmacology. 2002;160:92–8.

133. Davis JM, Giakas WJ, Qu J, Prasad P, Leucht S (2011). Should we treat depression with drugs or psychological interventions? Philos Ethics Humanit Med. 2011;10:8–21.

134. Davis JM, Glassman AH. Antidepressant drugs. In: Kaplan HI, Sadock BJ, editors. Comprehensive Textbook of Psychiatry. 5th ed. Baltimore: Williams & Wilkins; 1989. p. 1627–55.

135. de Abajo FJ. Effects of selective serotonin reuptake inhibitors on platelet function: mechanisms, clinical outcomes and implications for use in elderly patients. Drugs Aging. 2011;28:345–67.

136. Delay J, Laine R, Bouisson J-F. Note concernant l'action de l'isonicotinyl-hydrazide dans le traitement des états dépressifs. Ann Méd-Psychol (Paris). 1952;110:689–92.

137. Delgado PL, Miller HL, Salomon RM, Licinio J, Heninger GR, Gelenberg AJ, et al. Monoamines and the mechanism of antidepressant action: effects of catecholamine depletion on mood of patients treated with antidepressants. Psychopharmacol Bull. 1993;29:389–96.

138. de Leon J, Armstrong SC, Cozza KL. Clinical guidelines for psychiatrists for the use of pharmacogenetic testing for CYP450 2D6 and CYP450 2 C19. Psychosomatics. 2006;47:75–85.

139. Demyttenaere K. Agomelatine: narrative review. Eur Neuropsychopharmacol. 2011;21 (Suppl 4):S703–9.

140. Demyttenaere K, Bonnewyn A, Bruffaerts R, De Girolamo G, Gasquet I, Kovess V, et al. Clinical factors influencing the prescription of antidepressants and benzodiazepines: results from the European study of the epidemiology of mental disorders (ESEMeD). J Affect Disord. 2008;110:84–93.

141. DeVane CL, Gill HS. Clinical pharmacokinetics of fluvoxamine: applications to dosage regimen design. J Clin Psychiatry. 1997;58 (Suppl 5):7–14.

142. DeVane CL, Pollock BG. Pharmacokinetic considerations of antidepressant use in the elderly. J Clin Psychiatry. 1999;60 (Suppl 20):38–44.

143. Diav-Citrin O, Ornoy A. Selective serotonin reuptake inhibitors in human pregnancy: to treat or not to treat? Obstet Gynecol Int. 2011. [Epub before print, 10 Dec].

144. Díaz RA, Sancho J, Serratosa J. Antiepileptic drug interactions. Neurologist. 2008;14 (Suppl 6):S55–65.

145. Dimidjian S, Goodman S. Nonpharmacologic intervention and prevention strategies for depression during pregnancy and the postpartum. Clin Obstet Gynecol. 2009;52:498–515.
146. Diniz JB, Shavitt RG, Pereira CA, Hounie AG, Pimentel I, Koran LM, et al. Quetiapine versus clomipramine in the augmentation of selective serotonin reuptake inhibitors for the treatment of obsessive-compulsive disorder: randomized, open-label trial. J Psychopharmacol. 2010;24:297–307.
147. Dorfman W. Treatment of depression. Psychosomatics. 1964;5:7–13.
148. Dowson JH. MAO inhibitors in mental disease: current status. J Neural Transm (Suppl. 1987;23):121–38.
149. Doze VA, Handel EM, Jensen KA, Darsie B, Luger EJ, Haselton JR, et al. Alpha(1A)- and alpha(1B)-adrenergic receptors differentially modulate antidepressant-like behavior in the mouse. Brain Res. 2009;1285:148–57.
150. Druss BG, Hoff RA, Rosenheck RA. Underuse of antidepressants in major depression: prevalence and correlates in a national sample of young adults. J Clin Psychiatry. 2000;61:234–7.
151. Dudek D, Rybakowski JK, Siwek M, Pawlowski T, Lojko D, Roczen R, et al. Risk factors of treatment-resistance in major depression: association with bipolarity. J Affect Disord. 2010;126:268–71.
152. Dumlu K, Orhon Z, Özerdem A, Tural Ü, Ulas H, Tunca Z. Treatment-induced manic switch in the course of unipolar depression can predict bipolarity: cluster analysis based evidence. J Affect Disord. 2011;134:91–101.
153. Dunner DL, Amsterdam JD, Shelton RC. Loebel, Romano SJ. Efficacy and tolerability of adjunctive ziprasidone in treatment-resistant depression: randomized, open-label pilot study. J Clin Psychiatry. 2007;68:1071–7.
154. Dupuy JM, Ostacher MJ, Huffman J, Perlis RH, Nierenberg AA. Critical review of pharmacotherapy for major depressie disorder. Int J Neuropsychopharmacol. 2011;14:1417–31.
155. Dwyer AV, Whitten DL, Hawrelak JA. Herbal medicines, other than St. John's Wort, in the treatment of depression: systematic review. Altern Med Rev. 2011;16:40–9.
156. Eddy DM. Evidence-based medicine: unified approach. Health Aff. 2005;24:9–17.
157. Edeleanu L. Über einige Derivate der Phenylmethacrylsäure und der Phenylisobuttersäure. Berichte Deutsch Chem Ges. 1887;20:616–22.
158. Einarson A, Pistelli A, DeSantis M, Malm H, Paulus WD, Panchaud A, et al. Evaluation of the risk of congenital cardiovascular defects associated with use of paroxetine during pregnancy. Am J Psychiatry. 2008;165:749–52.
159. Ellfolk M, Malm H. Risks associated with in utero and lactation exposure to selective serotonin reuptake inhibitors (SSRIs). Reprod Toxicol. 2010;30:249–60.
160. El-Mallakh RD, Briscoe B. Studies of long-term use of antidepressants. CNS Drugs. 2012;26:97–109.
161. El-Mallakh RS, Gao Y, Jeannie-Roberts R. Tardive dysphoria: the role of long term antidepressant use in-inducing chronic depression. Med Hypotheses. 2011;6:769–73.
162. Erbe S, Gutwinski S, Bschor T. [Augmentation of antidepressants with atypical antipsychotics in non-delusional unipolar depressed patients with non-response to antidepressant-monotherapy (German)]. Psychiatr Prax. 2012;39:57–63.
163. Extein I, Pottash ALC, Gold MS. Thyrotropin-releasing hormone test in the diagnosis of unipolar depression. Psychiatry Res. 1981;5:311–6.
164. Fairchild CJ, Rush AJ, Vasavada N, Giles DE, Khatami M. Which depressions respond to placebo? Psychiatry Res. 1986;18:217–26.
165. Fava M, Davidson KG. Definition and epidemiology of treatment-resistant depression. Psychiatr Clin North Am. 1996;19:179–200.
166. Fava M, Rosenbaum JF, McGrath PJ, Stewart JW, Amsterdam JD, Quitkin FM. Lithium and tricyclic augmentation of fluoxetine treatment for resistant major depression: a double-blind, controlled study. Am J Psychiatry. 1994;151:1372–4.
167. Fawcett J, Barkin RL. Efficacy issues with antidepressants. J Clin Psychiatry. 1997;58 (Suppl 6):32–9.

168. Feighner JP, Boyer WF, Merideth CH, Hendrickson GG. Double-blind comparison of fluoxetine, imipramine and placebo in outpatients with major depression. Int Clin Psychopharmacol. 1989;4:127–34.

169. Fekadu A, Wooderson SC, Markopoulo K, Donaldson C, Papadopoulos A, Cleare AJ. What happens to patients with treatment-resistant depression? Systematic review of medium to long term outcome studies. J Affect Disord. 2009;116:4–11.

170. Fergusson D, Doucette S, Glass JC, Shapiro S, Healy D, Hebert P, et al. Association between suicide attempts and selective serotonin reuptake inhibitors: systematic review of randomized controlled trials. BMJ. 2005;330:396–9.

171. Ferrand Y, Kelton CM, Guo JJ, Levy MS, Yu Y. Using time-series intervention analysis to understand U.S. Medicaid expenditures on antidepressant agents. Res Social Adm Pharm. 2011;7:64–80.

172. Field T, Diego M, Hernandez-Reif M. Prenatal depression effects and interventions: review. Infant Behav Dev. 2010;33:409–18.

173. Fink M. Meduna and the origins of convulsive therapy. Am J Psychiatry. 1984;141:1034–41.

174. Fink KB, Göthert M. 5-HT receptor regulation of neurotransmitter release. Pharmacol Rev. 2007;59:360–417.

175. Fishell A. Depression and anxiety in pregnancy. J Popul Ther Clin Pharmacol. 2010;17:e363–9.

176. Flanagan RJ. Fatal toxicity of drugs used in psychiatry. Hum Psychopharmacol. 2008;23 (Suppl 1):43–51.

177. Foley KF, DeSanty KP, Kast RE. Bupropion: pharmacology and therapeutic applications. Expert Rev Neurother. 2006;6:1249–65.

178. Fox HH, Gibas JT. Synthetic tuberculostats: monoaklyl derivatives of isonicotinylhydrazine. J Org Chem. 1953;18:994–1002.

179. Food and Drug Administration (FDA) of the United States (2009). Public health advisory: suicidality in adults being treated with antidepressant medications. http://www.fda. gov/Drugs/DrugSafety/PostmarketDrugSafetyInformationforPatientsandProviders/ DrugSafetyInformationforHeathcareProfessionals/PublicHealthAdvisories/ucm053169. htm.

180. Fournier JC, DeRubeis RJ, Hollon SD, Dimidjian S, Amsterdam JD, Shelton RC, et al. Antidepressant drug effects and depression severity: patient-level meta-analysis. JAMA. 2010;303:47–53.

181. Fox AW. Subcutaneous sumatriptan pharmacokinetics: delimiting the monoamine oxidase inhibitor effect. Headache. 2010;50:249–55.

182. Franklin ME, Foa EB. Treatment of obsessive compulsive disorder. Annu Rev Clin Psychol. 2011;7:229–43.

183. Frascogna N. Physostigmine: is there a role for this antidote in pediatric poisonings? Curr Opin Pediatr. 2007;19:201–5.

184. Freeman EW. Therapeutic management of premenstrual syndrome. Expert Opin Pharmacother. 2010;11:2879–89.

185. Freeman MP, Nolan Jr PE, Davis MF, Anthony M, Fried K, Fankhauser M, et al. Pharmacokinetics of sertraline across pregnancy and postpartum. J Clin Psychopharmacol. 2008;28:646–53.

186. Freire RC, Hallak JE, Crippa JA, Nardi AE. New treatment options for panic disorder: clinical trials from 2000 to 2010. Expert Opin Pharmacother. 2011;12:1419–28.

187. Furukawa TA, Cipriani A, Barbui C, Geddes JR. Long-term treatment of depression with antidepressants: systematic narrative review. Can J Psychiatry. 2007;52:545–52.

188. Galbaud du Fort G. [Hematologic toxicity of antidepressive agents (French)]. Encéphale. 1988;14:307–18.

189. Galton F. Regression towards mediocrity in hereditary stature. J Anthropol Inst Gr Brit Ireland. 1886;15:246–63.

190. Ganasen KA, Stein DJ. Pharmacotherapy of social anxiety disorder. Curr Top Behav Neurosci. 2010;2:487–503.

191. Garattini S, Bonaccorsi A, Jori A, Samanin R. [Monoamines and the pharmacological effects of tricyclic antidepressive agents (French)]. Rev Neurol (Paris). 1972;127:265–92.
192. Gardner DM. Serotonin Syndrome. Halifax, MS: Dalhousie University Medical Center; 2011.
193. Gardner DM, Shulman KI, Walker SE, Tailor SA. Making of a user-friendly MAOI diet. J Clin Psychiatry. 1996;57:99–104.
194. Garver DL, Davis JM. Biogenic amine hypotheses of affective disorders. Life Sci. 1979;24: 383–94.
195. Gaynes BN, Dusetzina SB, Ellis AR, Hansen RA, Farley JF, Miller WC, et al. Treating depression after initial treatment failure: directly comparing switch and augmenting strategies in STAR*D. J Clin Psychopharmacol. 2012;32:114–9.
196. Gaynes BN, Warden D, Trivedi MH, Wisniewski SR, Fava M, Rush AJ. What did STAR*D teach us? Results from a large-scale, practical, clinical trial for patients with depression. Psychiatr Serv. 2009;60:1439–45.
197. Gelenberg AJ. Review of the current guidelines for depression treatment. J Clin Psychiatry. 2011;71:e15–20.
198. Gentile S. Efficacy of antidepressant medications in children and adolescents with obsessive-compulsive disorder: systematic appraisal. J Clin Psychopharmacol. 2011;31:625–32.
199. George MS. Transcranial magnetic stimulation for the treatment of depression. Expert Rev Neurother. 2010;10:1761–2.
200. George MS, Lisanby SH, Avery D, McDonald WM, Durkalski V, Pavlicova M, et al. Daily left prefrontal transcranial magnetic stimulation therapy for major depressive disorder: sham-controlled randomized trial. Arch Gen Psychiatry. 2010;67:507–16.
201. George MS, Ward Jr HE, Ninan PT, Pollack M, Nahas Z, Anderson B, et al. Pilot study of vagus nerve stimulation (VNS) for treatment-resistant anxiety disorders. Brain Stimul. 2008;1:112–21.
202. Geschwind N, Nicolson NA, Peeters F, van Os J, Barge-Schaapveld D, Wichers M. Early improvement in positive rather than negative emotion predicts remission from depression after pharmacotherapy. Eur Neuropsychopharmacol. 2011;21:241–7.
203. Ghaemi SN, Ostacher MM, El-Mallakh RS, Borrelli D, Baldassano CF, Kelley ME, et al. Antidepressant discontinuation in bipolar depression: randomized clinical trial of long-term effectiveness and safety. J Clin Psychiatry. 2010;71:372–80.
204. Ghaemi SN, Wingo AP, Filkowski MA, Goodwin FK, Baldessarini RJ. Effectiveness of long-term antidepressant treatment in bipolar disorder: meta-analysis. Acta Psychiatr Scand. 2008;118:347–56.
205. Ghanbari R, El Mansari M, Blier P. Enhancement of serotonergic and noradrenergic neurotransmission in the rat hippocampus by sustained administration of bupropion. Psychopharmacology. 2011;217:61–73.
206. Gibbons RD, Mann JJ. Strategies for quantifying the relationship between medications and suicidal behavior: what has been learned? Drug Saf. 2011;34:375–95.
207. Gijsman HJ, Geddes JR, Rendell JM, Nolen WA, Goodwin GM. Antidepressants for bipolar depression: systematic review of randomized, controlled trials. Am J Psychiatry. 2004;161: 1537–47.
208. Gillman PK. Tripans: serotonin agonists and serotonin syndrome (serotonin toxicity): review. Headache. 2010;50:264–72.
209. Gillman PK. Advances pertaining to the pharmacology and interactions of irreversible, nonselective monoamine oxidase imhibitors. J Clin Psychopharmacol. 2011;31:66–74.
210. Glassman AH, O'Connor CM, Califf RM, Swedberg K, Schwartz P, Bigger Jr JT, et al. Sertraline treatment of major depression in patients with acute MI or unstable angina. JAMA. 2002;288:701–9.
211. Glowinski J, Baldessarini RJ. Metabolism of norepinephrine in the central nervous system. Pharmacol Rev. 1966;18:1201–38.
212. Glue P, Donovan MR, Kolluri S, Emir B. Meta-analysis of relapse prevention antidepressant trials in depressive disorders. Aust N Z J Psychiatry. 2010;44:697–705.

213. Goodwin GM, Green AR, Johnson P. 5-HT2 receptor characteristics in frontal cortex and 5-HT2 receptor-mediated head-twitch behaviour following antidepressant treatment to mice. Br J Pharmacol. 1984;83:225–42.
214. Gourion D. [Antidepressants and their onset of action: a major clinical, methodological and prognostical issue (French)]. Encéphale. 2008;34:73–81.
215. Gram LF. Dose-effect relationships for tricyclic antidepressants: basis for rational clinical testing of new antidepressants. Psychopharmacol Ser. 1993;10:163–73.
216. Granacher RP, Baldessarini RJ. Physostigmine: use in acute anticholinergic syndrome with antidepressant and antiparkinson drugs. Arch Gen Psychiatry. 1975;32:375–80.
217. Güloglu C, Orak M, Üstündag M, Altunci YA. Analysis of amitriptyline overdose in emergency medicine. Emerg Med J. 2011;28:296–309.
218. Gunnell D, Saperia J, Ashby D. Selective serotonin reuptake inhibitors (SSRIs) and suicide in adults: meta-analysis of drug company data from randomized, placebo-controlled trials submitted to the MHRA's safety review. BMJ. 2005;330:385–8.
219. Gurley BJ, Swain A, Williams DK, Barone G, Battu SK. Gauging the clinical significance of P-glycoprotein-mediated herb-drug interactions: comparative effects of St. John's wort, echinacea, clarithromycin, and rifampin on digoxin pharmacokinetics. Mol Nutr Food Res. 2008;52:772–9.
220. Haddad PM. Antidepressant discontinuation syndromes. Drug Saf. 2001;24:183–97.
221. Haddjeri N, Blier P, de Montigny C. Acute and long-term actions of the antidepressant drug mirtazapine on central 5-HT neurotransmission. J Affect Disord. 1998;51:255–66.
222. Hadley SK, Gaarder SM. Treatment of irritable bowel syndrome. Am Fam Physician. 2005;72:2501–6.
223. Hajos-Korcsok E, McTavish SF, Sharp T. Effect of selective 5-hydroxytryptamine reuptake inhibitors on brain extracellular noradrenaline: microdialysis studies using paroxetine. Eur J Pharmacol. 2000;407:101–7.
224. Hammad TA, Laughren TP, Racoosin JA. Suicide rates in short-term, randomized controlled trials of newer antidepressants. J Clin Psychopharmacol. 2006;26:203–7.
225. Hammad TA, Laughren TP, Racoosin JA. Suicidality in pediatric patients treated with antidepressant drugs. Arch Gen Psychiatry. 2006;63:332–9.
226. Hamoda HM, Osser DN. Psychopharmacology Algorithm Project at the Harvard South Shore Program: update on psychotic depression. Harv Rev Psychiatry. 2008;16:235–47.
227. Hansen RA, Gartlehner G, Lohr KN, Gaynes BN, Carey TS. Efficacy and safety of second-generation antidepressants in the treatment of major depressive disorder. Ann Intern Med. 2005;143:415–26.
228. Hansen RA, Gaynes BN, Thieda P, Gartlehner G, Deveaugh-Geiss A, Krebs E, et al. Meta-analysis of major depressive disorder relapse and recurrence with second-generation antidepressants. Psychiatr Serv. 2008;59:1121–30.
229. Hardeland R, Poeggeler B, Srinivasan V, Trakht I, Pandi-Perumal SR, Cardinali DP. Melatonergic drugs in clinical pracitice. Arzneimittelforschung. 2008;58:1–10.
230. Hardy SE. Methylphenidate for the treatment of depressive symptoms, including fatigue and apathy, in medically ill older adults and terminally ill adults. Am J Geriatr Pharmacother. 2009;7:34–59.
231. Harris EC, Barraclough B. Suicide as an outcome for mental disorders: meta-analysis. Br J Psychiatry. 1997;170:205–28.
232. Härter M, Klesse C, Berger M, Bermejo I, Bschor T, Gensichen J, et al. [Evidence-based treatment of depression (German)]. Z Psychosom Med Psychother. 2010;56:334–42.
233. Harvard Medical School. Cytochrome P450 enzymes and psychiatric drugs. In: Harvard Mental Health Letter. 2009. http://harvardpartnersinternational.staywellsolutionsonline.com/HealthNewsLetters/69,M0209b. Accessed 15 Dec 2011.
234. Häuser W, Bernardy K, Uçeyler N, Sommer C. Treatment of fibromyalgia syndrome with antidepressants: meta-analysis. JAMA. 2009;301:198–209.
235. Hausmann A, Hörtnagl C, Walpoth M, Fuchs M, Conca A. [Are there substantial reasons for contraindicating antidepressants in bipolar disorder: facts or artifacts? (German)]. Neuropsychiatr. 2007;21:131–58.

236. Hayes E Nonpsychiatric prescribing fuels rise in antidepressant use. 2011. The Pink Sheet, August 9. http://survivingantidepressants.org/index.php?/topic/1025-us-antidepressant-sales-rise-to-11b-due-to-primary-care-docs/.

237. Hayes DJ, Greenshaw AJ. 5-HT receptors and reward-related behavior: review. Neurosci Biobehav Rev. 2011;35:1419–149.

238. Healy D. The Antidepressant Era. Cambridge, MA: Harvard University Press; 1997.

239. Heijnen WT, Birkenhäger TK, Wierdsma AI, van den Broek WW. Antidepressant pharmacotherapy failure and response to subsequent electroconvulsive therapy: meta-analysis. J Clin Psychopharmacol. 2010;30:616–9.

240. Hellstrom WJ. Update on treatments for premature ejaculation. Int J Clin Pract. 2011;65: 16–26.

241. Heninger GR, Charney DS, Sternberg DE. Lithium carbonate augmentation of antidepressant treatment: effective prescription for treatment-refractory depression. Arch Gen Psychiatry. 1983;40:1335–42.

242. Henkel V, Mergl R, Allgaier AK, Kohnen R, Möller HJ, Hegerl U. Treatment of depression with atypical features: meta-analytic approach. Psychiatry Res. 2006;141:89–101.

243. Henkel V, Seemüller F, Obermeier M, Adli M, Bauer M, Mundt C, et al. Does early improvement triggered by antidepressants predict response/remission? Analysis of data from a naturalistic study on a large sample of inpatients with major depression. J Affect Disord. 2009;115:439–49.

244. Hennen J, Baldessarini RJ. Reduced suicidal risk during treatment with clozapine: meta-analysis. Schizophr Res. 2005;73:139–45.

245. Hennings JM, Owashi T, Binder EB, Horstmann S, Menke A, Kloiber S, et al. Clinical characteristics and treatment outcome in a representative sample of depressed inpatients. J Psychiatr Res. 2009;43:215–29.

246. Henry JA, Martin AJ. Risk-benefit assessment of antidepressant drugs. Med Toxicol Adverse Drug Exp. 1987;2:445–62.

247. Henry G, Williamson D, Tampi RR. Efficacy and tolerability of antidepressants in the treatment of behavioral and psychological symptoms of dementia: literature review of evidence. Am J Alzheimers Dis Other Dement. 2011;26:169–83.

248. Hermann EK, Munsch S, Biedert E, Lang W. [Psychotherapy for depression (German)]. Ther Umsch. 2010;67:581–4.

249. Hickie IB, Rogers NL. Novel melatonin-based therapies: potential advances in the treatment of major depression. Lancet. 2011;378:621–31.

250. Hjorth S, Bengtsson HJ, Kullberg A, Carlzon D, Peilot H. Serotonin autoreceptor function and antidepressant drug action. J Psychopharmacol. 2000;14:177–85.

251. Hollon SD, Ponniah K. Review of empirically supported psychological therapies for mood disorders in adults. Depress Anxiety. 2010;27:891–932.

252. Holsboer F, Ising M. Central CRH system in depression and anxiety–evidence from clinical studies with CRH1 receptor antagonists. Eur J Pharmacol. 2008;583:350–7.

253. Holzheimer PE, Kelley ME, Gross RE, Filkowski MM, Garlow SJ, Barrocas A, et al. Subcallosal cingulate deep brain stimuation for treatment-resistant unipolar and bipolar depression. Arch Gen Psychiatry. 2012;69:150–8.

254. Hordern A. The antidepressant drugs. N Engl J Med. 1965;272:1159–69.

255. Hornykiewicz O. Dopamine (3-hydroxytyramine) and brain function. Pharmacol Rev. 1966;18:925–64.

256. Howell C, Wilson AD, Waring WS. Cardiovascular toxicity due to venlafaxine poisoning in adults: review of 235 consecutive cases. Br J Clin Pharmacol. 2007;64:192–7.

257. Howland RH, Shutt LS, Berman SR, Spotts CR, Denko T. Emerging use of technology for the treatment of depression and other neuropsychiatric disorders. Ann Clin Psychiatry. 2011;23:48–62.

258. Howland RH, Wilson MG, Kornstein SG, Clayton AH, Trivedi MH, Wohlreich MM, et al. Factors predicting reduced antidepressant response: experience with the SNRI duloxetine in patients with major depression. Ann Clin Psychiatry. 2008;20:209–18.

259. Huang YS, Tsai MH. Long-term outcomes with medications for attention-deficit hyperactivity disorder: current status of knowledge. CNS Drugs. 2011;25:539–54.
260. Hughes S, Cohen D. Systematic review of long-term studies of drug treated and non-drug treated depression. J Affect Disord. 2009;118:9–18.
261. Hunter DG. Do we need evidence for everything? Am Orthopt J. 2010;60:59–62.
262. Ioannidis JPA. Effectiveness of antidepressants: an evidence myth constructed from a thousand randomized trials? Philos Ethics Humanit Med. 2008;3:14–27.
263. Iosifescu DV, Bankier B, Fava M. Impact of medical comorbid disease on antidepressant treatment of major depressive disorder. Curr Psychiatry Rep. 2004;6:193–201.
264. Iovieno N, Tedeschini E, Bentley KH, Evins AE, Papakostas GI. Antidepressants for major depressive disorder and dysthymic disorder in patients with comorbid alcohol use disorders: meta-analysis of placebo-controlled randomized trials. J Clin Psychiatry. 2011;72: 1144–51.
265. Ipser JC, Sander C, Stein DJ. Pharmacotherapy and psychotherapy for body dysmorphic disorder. Cochrane Database Syst Rev. 2009;1:CD005332.
266. Invernizzi RW, Garattini S. Role of presynaptic alpha$_2$-adrenoreceptors in antidepressant action: recent findings from microdialysis studies. Prog Neuropsychopharmacol Biol Psychiatry. 2004;28:819–27.
267. Isacsson G, Rich CL, Jureidini J, Raven M. Increased use of antidepressants has contributed to the worldwide reduction in suicide rates. Br J Psychiatry. 2010;196:429–33.
268. Isbister GK, Bowe SJ, Dawson A, Whyte IM. Relative toxicity of selective serotonin reuptake inhibitors (SSRIs) in overdose. J Toxicol Clin Toxicol. 2004;42:277–85.
269. Isbister GK, Buckley NA, Whyte IM. Serotonin toxicity: a practical approach to diagnosis and treatment. Med J Aust. 2007;187:361–5.
270. Iversen LL. Speed, Ecstacy, Ritalin: The Science of Amphetamines. New York: Oxford University Press; 2006.
271. Iversen LL, Iversen SD, Bloom FE, Roth RH. Introduction to Neuropsychopharmacology. New York: Oxford University Press; 2009.
272. Jakobsen JC, Hansen JL, Simonsen S, Simonsen E, Gluud C. Effects of cognitive therapy versus interpersonal psychotherapy in patients with major depressive disorder: systematic review of randomized clinical trials with meta-analyses and trial sequential analyses. Psychol Med. 2011. [Epub ahead of print 3 Nov].
273. Jackson CW, Cates M, Lorenz R. Pharmacotherapy of eating disorders. Nutr Clin Pract. 2010;25:143–59.
274. Jakobsen JC, Hansen JL, Simonsen E, Gluud C. Effect of adding psychodynamic therapy to antidepressants in patients with major depressive disorder: systematic review of randomized clinical trials with meta-analyses and trial sequential analyses. J Affect Disord. 2012;137: 4–14.
275. Jankovic SM, Dostic M. Choice of antiepileptic drugs for the elderly: possible drug interactions and adverse effects. Expert Opin Drug Metab Toxicol. 2012;8:81–91.
276. Jefferson JW. Review of the cardiovascular effects and toxicity of tricyclic antidepressants. Psychosom Med. 1975;37:160–79.
277. Jick H, Kaye JA, Jick SS. Antidepressants and the risk of suicidal behaviors. JAMA. 2004;292:338–43.
278. Joffe RT, Singer W, Levitt AJ, MacDonald C. A placebo-controlled comparison of lithium and triiodothyronine augmentation of tricyclic antidepressants in unipolar refractory depression. Arch Gen Psychiatry. 1993;50:387–93.
279. Joyce PR, Mulder RT, Luty SE, Sullivan PF, McKenzie JM, Abbott RM, et al. Patterns and predictors of remission, response and recovery in major depression treated with fluoxetine or nortriptyline. Aust N Z J Psychiatry. 2002;36:384–91.
280. Kalinowsky LB. History of convulsive therapy. Ann N Y Acad Sci. 1986;462:1–4.
281. Kane JM, Quitkin FM, Rifkin A, Ramos-Lorenzi JR, Nayak DD, Howard A. Lithium carbonate and imipramine in the prophylaxis of unipolar and bipolar II illness: prospective, placebo-controlled comparison. Arch Gen Psychiatry. 1982;39:1065–9.

282. Kantor D, McNevin S, Leichner P, Harper D, Krenn M. The benefit of lithium carbonate adjunct in refractory depression–fact or fiction? Can J Psychiatry. 1986;31:416–8.
283. Kashner MT, Rush JA, Altshuler KZ. Measuring costs of guideline-driven mental health care: the Texas Medication Algorithm Project. J Ment Health Policy Econ. 1999;2:111–21.
284. Katona CL. Refractory depression: a review with particular reference to the use of lithium augmentation. Eur Neuropsychopharmacol. 1995;5:109–13.
285. Katz G. Tachyphylaxis/tolerance to antidepressants in treatment of dysthymia: results of a retrospective naturalistic chart review study. Psychiatry Clin Neurosci. 2011;65:499–504.
286. Katz G. Tachyphylaxis/tolerance to antidepressive medications: review. Isr J Psychiatry Relat Sci. 2011;48:129–35.
287. Keers R, Aitchison KJ. Pharmacogenetics of antidepressant response. Expert Rev Neurother. 2011;11:101–25.
288. Keller-Ashton A, Hamer R, Rosen RC. Serotonin reuptake inhibitor-induced sexual dysfunction and its treatment: a large-scale retrospective study of 596 psychiatric outpatients. J Sex Marital Ther. 1997;23:165–75.
289. Kemp DE, Ganocy SJ, Brecher M, Carlson BX, Edwards S, Eudicone JM, et al. Clinical value of early partial symptomatic improvement in the prediction of response and remission during short-term treatment trials in 3369 subjects with bipolar I or II depression. J Affect Disord. 2011;130:171–9.
290. Kennedy SH, Giacobbe P, Tizvi SJ, Placenza FM, Nishikawa Y, Mayberg HS, et al. Deep brain stimulation for treatment-resistnat depression: follow-up after 3–6 years. Am J Psychiatry. 2011;168:502–10.
291. Kennedy SH, Lam RW, Cohen NL, Ravindran AV. Clinical guidelines for the treatment of depressive disorders: Medications and other biological treatments. Can J Psychiatry. 2001;46 (Suppl 1):38S–58.
292. Kennedy GJ, Marcus P. Use of antidepressants in older patients with co-morbid medical conditions: guidance from studies of depression in somatic illness. Drugs Aging. 2005;22:273–87.
293. Kessler RC, Berglund P, Borges G, Nock M, Wang PS. Trends in suicide ideation, plans, gestures, and attempts in the United States, 1990–1992 to 2001–2003. JAMA. 2005;293: 2487–95.
294. Ketter TA, Citrome L, Wang PW, Culver JL, Srivastava S. Treatments for bipolar disorder: can number needed to treat / harm help inform clinical decisions? Acta Psychiatr Scand. 2011;123:175–89.
295. Khan A, Bhat A, Kolts R, Thase ME, Brown W. Why has the antidepressant–placebo difference in antidepressant clinical trials diminished over the past three decades? CNS Neurosci Ther. 2010;16:217–26.
296. Khan A, Khan S, Brown WA. Suicide rates in clinical trials of SSRIs, other antidepressants, and placebo: analysis of FDA reports. Am J Psychiatry. 2003;160:790–2.
297. Khan A, Leventhal RM, Khan SR, Brown WA. Severity of depression and response to antidepressants and placebo: analysis of the Food and Drug Adinistration database. J Clin Psychopharmacol. 2002;22:40–5.
298. Kim Y, Kim SH, Kim YS, Lee YH, Ha K, Shin SY. Imipramine activates glial cell line-derived neurotrophic factor via early growth response gene 1 in astrocytes. Prog Neuropsychopharmacol Biol Psychiatry. 2011;35:1026–32.
299. Kim DR, Sockol L, Barber JP, Moseley M, Lamprou L, Rickels K, et al. Survey of patient acceptability of repetitive transcranial magnetic stimulation (TMS) during pregnancy. J Affect Disord. 2011;129:385–90.
300. Kirsch I, Deacon BJ, Huedo-Medina TB, Scoboria A, Moore TJ, Johnson BT. Initial severity and antidepressant benefits: meta-analysis of data submitted to the Food and Drug Administration. PLoS Med. 2008;5:e45–60.
301. Kittler K, Lau T, Schloss P. Antagonists and substrates differentially regulate serotonin transporter cell surface expression in serotonergic neurons. Eur J Pharmacol. 2010;629:63–7.
302. Klerman GL, Cole JO. Clinical pharmacology of imipramine and related antdepressant compounds. Pharmacol Rev. 1965;17:101–41.
303. Kline NS. Depression: diagnosis and treatment. Med Clin North Am. 1961;35:1041–53.

304. Kocsis JH, Croughan JL, Katz MM, Butler TP, Secunda S, Bowden CL, et al. Response to treatment with antidepressants of patients with severe or moderate nonpsychotic depression and of patients with psychotic depression. Am J Psychiatry. 1990;147:621–4.
305. Koda K, Ago Y, Cong Y, Kita Y, Takuma K, Matsuda T. Effects of acute and chronic administration of atomoxetine and methylphenidate on extracellular levels of noradrenaline, dopamine and serotonin in the prefrontal cortex and striatum of mice. J Neurochem. 2010;114: 259–70.
306. Kok RM, Heeren TJ, Nolen WA. Continuing treatment of depression in the elderly: systematic review and meta-analysis of double-blinded randomized controlled trials with antidepressants. Am J Geriatr Psychiatry. 2011;19:249–55.
307. Komossa K, Depping AM, Gaudchau A, Kissling W, Leucht S. Second-generation antipsychotics for major depressive disorder and dysthymia. Cochrane Database Syst Rev. 2010;12:CD008121.
308. Kordon A, Zurowski B, Wahl K, Hohagen F. [Evidence-based pharmacotherapy and other somatic treatment approaches for obsessive-compulsive disorder: state of the art (German)]. Nervenarzt. 2011;82:319–24.
309. Kornstein SG. Maintenance therapy to prevent recurrence of depression: summary and implications of the PREVENT study. Expert Rev Neurother. 2008;8:737–42.
310. Kranz GS, Kasper S, Lanzenberger R. Reward and the serotonergic system. Neuroscience. 2010;166:1023–35.
311. Kravitz HM, Edwards JH, Fawcett J, Fogg L. Challenging the amphetamine challenge test: report of an antidepressant treatment study. J Affect Disord. 1990;20:121–8.
312. Krishel S, Jackimczyk K. Cyclic antidepressants, lithium, and neuroleptic agents. Pharmacology and toxicology. Emerg Med Clin North Am. 1991;9:53–86.
313. Krishnan KR. Comorbidity and depression treatment. Biol Psychiatry. 2003;53:701–6.
314. Kuba T, Yakushi T, Fukuhara H, Nakamoto Y, Singeo Jr ST, Tanaka O, et al. Suicide-related events among child and adolescent patients during short-term antidepressant therapy. Psychiatry Clin Neurosci. 2011;65:239–45.
315. Kuhn R. Über die behandlung de depressiver zusstände mit einem iminodibenzyl derivative (G22355). Schweiz Med Wochenschr. 1957;87:1135–40.
316. Kuhn R. Treatment of depressive states with G22355 (imipramine hydrochloride). Am J Psychiatry. 1958;115:459–64.
317. Kuhn R. The discovery of imipramine. In: Ayd F, Blackwell B, editors. Discoveries in Biological Psychiatry. Philadelphia: Lippincott; 1970.
318. Kuk AY, Li J, Rush AJ. Recursive subsetting to identify patients in the STAR*D: method to enhance the accuracy of early prediction of treatment outcome and to inform personalized care. J Clin Psychiatry. 2010;71:1502–8.
319. Kupfer DJ. Pharmacological management of depression. Dialogues Clin Neurosci. 2005;7: 191–205.
320. Lam RW, Chan P, Wilkins-Ho M, Yatham LN. Repetitive transcranial magnetic stimulation for treatment-resistant depression: a systematic review and metaanalysis. Can J Psychiatry. 2008;53:621–31.
321. Lapin IP, Oxenkrug GF. Intensification of the central serotoninergic processes as a possible determinant of the thymoleptic effect. Lancet. 1969;1:132–6.
322. Lasser RA, Baldessarini RJ. Thyroid hormones in depressive disorders: reappraisal of clinical utility. Harv Rev Psychiatry. 1997;4:291–305.
323. Laughren TP. Overview of a meeting of the psychopharmacology drug advisory committee (PDAC) concerning suicidal risk in trials of antidepressant drugs in juvenile and adult patients (p. 27ff). 2006. http://www.fda.gov/ohrms/dockets/ax/06/briefing/2006-4272b1-index.htm.
324. Lauerma H. Successful treatment of citalopram-induced anorgasmia by cyproheptadine. Acta Psychiatr Scand. 1996;93:69–70.
325. Lavretsky H, Kim MD, Kumar A, Reynolds III CF. Combined treatment with methylphenidate and citalopram for accelerated response in the elderly: open trial. J Clin Psychiatry. 2003;64:1410–4.

326. Le HN, Perry DF, Stuart EA. Randomized controlled trial of a preventive intervention for perinatal depression in high-risk Latinas. J Consult Clin Psychol. 2011;79:135–41.
327. Lejoyeux M, Adès J. Antidepressant discontinuation: review of the literature. J Clin Psychiatry. 1997;58 (Suppl 7):11–6.
328. Leon AC, Solomon DA, Li C, Fiedorowicz JG, Coryell WH, Endicott J, et al. Antidepressants and risks of suicide and suicide attempts: 27-year observational study. J Clin Psychiatry. 2011;72:580–6.
329. Leuchter AF, Cook IA, Hunter AM, Korb AS. A new paradigm for the prediction of antidepressant treatment response. Dialogues Clin Neurosci. 2009;11:435–46.
330. Li X, Frye MA, Shelton RC. Review of pharmacological treatment in mood disorders and future directions for drug development. Neuropsychopharmacology. 2012;37: 77–101.
331. Limosin F, Mekaoui L, Hautecouverture S. [Prophylactic treatment for recurrent major depression (French)]. Presse Med. 2007;36:1627–33.
332. Ljung R, Lu Y, Lagergren J. High concomitant use of interacting drugs and low use of gastroprotective drugs among NSAID users in an unselected elderly population: a nationwide register-based study. Drugs Aging. 2011;28:469–76.
333. Long TD, Kathol RG. Critical review of data supporting affective disorder caused by nonpsychotropic medication. Ann Clin Psychiatry. 1993;5:259–70.
334. Lotufo-Neto F, Trivedi M, Thase ME. Meta-analysis of the reversible inhibitors of monomaine oxidase type A, moclobemide and brofaromine, for the treatment of depression. Neuropsychopharmacology. 1999;20:226–47.
335. Lund N, Pedersen LH, Henriksen TB. Selective serotonin reuptake inhibitor exposure in utero and pregnancy outcomes. Arch Pediatr Adolesc Med. 2009;163:949–54.
336. Macaluso M, Zackula R, D'Empaire I, Baker B, Liow K, Preskorn SH. Twenty percent of a representative sample of patients taking bupropion have abnormal, asymptomatic electroencephalographic findings. J Clin Psychopharmacol. 2010;30:312–7.
337. MacGillivray A, Hatcher S, Ogston S, Reid I, Sullivan F, Williams B, et al. Efficacy and tolerability of selective serotonin reuptake inhibitors compared with tricyclic antidepressants in depression treated in primary care: systematic review and meta-analysis. BMJ. 2003;326: 1014–7.
338. Marangell LB, Martinez M, Jurdi RA, Zboyan H. Neurostimulation therapies in depression: review of new modalities. Acta Psychiatr Scand. 2007;116:174–81.
339. Marazziti D, Carlini M, Dell'osso L. Treatment strategies of obsessive-compulsive disorder and panic disorder/agoraphobia. Curr Top Med Chem. 2012;12:238–53.
340. Marcus S, Lopez JF, McDonough S, Mackenzie MJ, Flynn H, Neal Jr CR, et al. Depressive symptoms during pregnancy: impact on neuroendocrine and neonatal outcomes. Infant Behav Dev. 2011;34:26–34.
341. Masi G, Brovedani P. The hippocampus, neurotrophic factors and depression: possible implications for the pharmacotherapy of depression. CNS Drugs. 2011;25:913–31.
342. Mathew SJ, Manji HK, Charney DS. Novel drugs and therapeutic targets for severe mood disorders. Neuropsychopharmacology. 2008;33:2080–92.
343. Mavissakalian MR. Imipramine vs. sertraline in panic disorder: 24-week treatment completers. Ann Clin Psychiatry. 2003;15:171–80.
344. Mavissakalian MR, Perel JM. Imipramine dose-response relationship in panic disorder with agoraphobia. Arch Gen Psychiatry. 1989;46:127–31.
345. Mayberg HS. Targeted electrode-based modulation of neural circuits for depression. J Clin Invest. 2009;119:717–25.
346. Marek GJ. Cortical 5-hydroxytryptamine-2A-receptor mediated excitatory synaptic currents in the rat following repeated daily fluoxetine administration. Neurosci Lett. 2008;438: 312–6.
347. Mayes R, Horwitz AV. DSM-III and the revolution in the classification of mental illness. J Hist Behav Sci. 2005;41:249–67.

348. McGonigle P. Peptide therapeutics for CNS indications. Biochem Pharmacol. 2012;83: 559–66.
349. McIntyre RS, Park KY, Law CW, Sultan F, Adams A, Lourenco MT, et al. Association between conventional antidepressants and the metabolic syndrome: review of the evidence and clinical implications. CNS Drugs. 2010;24:741–53.
350. Meltzer HY, Alphs L, Green AI, Altamura AC, Anand R, Bertoldi A, et al. Clozapine treatment for suicidality in schizophrenia: International Suicide Prevention Trial (InterSePT). Arch Gen Psychiatry. 2003;60:82–91.
351. Mendels J. Role of lithium as an antidepressant. Mod Probl Pharmacopsychiatry. 1982;18:138–44.
352. Meyer RE, Salzman C, Youngstrom EA, Clayton PJ, Goodwin FK, Mann JJ, et al. Suicidality and risk of suicide: definition, drug safety concerns, and a necessary target for drug development: consensus statement. J Clin Psychiatry. 2010;71:e1–21.
353. Meyer RE, Salzman C, Youngstrom EA, Clayton PJ, Goodwin FK, Mann JJ, et al. Suicidality and risk of suicide—definition, drug safety concerns, and a necessary target for drug development. J Clin Psychiatry. 2010;71:1040–6.
354. Micallef J, Fakra E, Blin O. [Use of antidepressant drugs in schizophrenic patients with depression (French)]. Encéphale. 2006;32:263–9.
355. Millan MJ, Gobert A, Rivet JM, Adhumeau-Auclair A, Cussac D, Newman-Tancredi A, et al. Mirtazapine enhances frontocortical dopaminergic and corticolimbic adrenergic, but not serotonergic transmission by blockade of alpha-2-adrenergic and serotonin-2C receptors: comparison with citalopram. Eur J Neurosci. 2000;12:1079–95.
356. Miller HL, Delgado PL, Salomon RM, Berman R, Krystal JH, Heninge GR, Charney DS. Clinical and biochemical effects of catecholamine depletion on antidepressant-induced remission of depression. Arch Gen Psychiatry. 1996;53:117–28.
357. Mirin SM, Schatzberg AF, Creasey DE. Hypomania and mania after withdrawal of tricyclic antidepressants. Am J Psychiatry. 1981;138:87–99.
358. Mischoulon D, Best-Popescu C, Laposata M, Merens W, Murakami JL, Wu SL, et al. Double-blind dose-finding pilot study of docosahexaenoic acid (DHA) for major depressive disorder. Eur Neuropsychopharmacol. 2008;18:639–45.
359. Mitchell PB. Therapeutic drug monitoring of non-tricylcic antidepressant drugs. Clin Chem Lab Med. 2004;42:1212–8.
360. Mitchell PB, Frankland A, Hadzi-Pavlovic D, Roberts G, Corry J, Wright A, et al. Comparison of depressive episodes in bipolar disorder and in major depressive disorder within bipolar disorder pedigrees. Br J Psychiatry. 2011;199:303–9.
361. Mochcovitch MD, Nardi AE. Selective serotonin-reuptake inhibitors in the treatment of panic disorder: systematic review of placebo-controlled studies. Expert Rev Neurother. 2010;10: 1285–93.
362. Moeller O, Norra C, Gründer G. [Monoaminergic function in major depression: possibly helpful tool for choosing treatment strategy (German)]. Nervenarzt. 2006;77:800–8.
363. Mohr P, Rodriguez M, Slavíčková A, Hanka J. Application of vagus nerve stimulation and deep brain stimulation in depression. Neuropsychobiology. 2011;64:170–81.
364. Mojtabai R. Does depression screening have an effect on the diagnosis and treatment of mood disorders in general medical settings: instrumental variable analysis of the national ambulatory medical care survey? Med Care Res Rev. 2011;68:462–89.
365. Moncrieff J, Wessely S, Hardy R. Active placebos versus antidepressants for depression. Cochrane Database Syst Rev. 2004;CD003012.
366. Montejo-González AL, Llorca G, Izquierdo JA, Ledesma A, Bousoño M, Calcedo A, et al. SSRI-induced sexual dysfunction: fluoxetine, paroxetine, sertraline, and fluvoxamine in a prospective, multicenter, and descriptive clinical study of 344 patients. J Sex Marital Ther. 1997;23:176–94.
367. Montgomery SA. Tolerability of serotonin norepinephrine reuptake inhibitor antidepressants. CNS Spectr. 2008;13(7 Suppl 11):27–33.

368. Mooney JJ, Schatzberg AF, Cole JO, Samson JA, Waternaux C, Gerson B, et al. Urinary 3-methoxy-4-hydroxyphenylglycol and the depression-type score as predictors of differential responses to antidepressants. J Clin Psychopharmacol. 1991;11:339–43.
369. Moret C, Isaac M, Briley M. Problems associated with long-term treatment with selective serotonin reuptake inhnibitors. J Psychopharmacol. 2009;23:967974.
370. Morishita S, Kinoshita T. Predictors of response to sertraline in patients with major depression. Hum Psychopharmacol. 2008;23:647–51.
371. Morita M, Nakayama K. Mirtazapine in combination with perospirone synergistically enhances dopamine release in the rat prefrontal cortex via 5-HT1A receptor activation. Psychiatry Clin Neurosci. 2011;65:246–53.
372. Mouchabac S. [Severe depression: pharmacological treatments (French)]. Encéphale. 2009;35 (Suppl 7):S319–24.
373. Mucci A, Volpe U, Merlotti E, Bucci P, Galderisi S. Pharmaco-EEG in psychiatry. Clin EEG Neurosci. 2006;37:81–98.
374. Mueller A, de Zwaan M. [Treatment of compulsive buying (German)]. Fortschr Neurol Psychiatr. 2008;76:478–83.
375. Muzyk AJ, Jakel RJ, Preud'homme X. Serotonin syndrome after a massive overdose of controlled-release paroxetine. Psychosomatics. 2010;51:437–42.
376. Narayan V, Haddad PM. Antidepressant discontinuation manic states: a critical review of the literature and suggested diagnostic criteria. J Psychopharmacol. 2011;25:306–13.
377. National Center for Health Statistics (NCHS). Health, United States, 2007. Washington, DC: US Department of Health and Human Services; 2007.
378. National Institutes of Health (NIH) of the United States. Drug trials. 2011. http://clinicaltrials.gov/.
379. Naudet F, Maria AS, Falissard B. Antidepressant response in major depressive disorder: meta-regression comparison of randomized controlled trials and observational studies. PLoS One. 2011;6:e20811.
380. Nelson LS, Erdman AR, Booze LL, Cobaugh DJ, Chyka PA, Woolf AD, et al. Selective serotonin reuptake inhibitor poisoning: evidence-based consensus guideline for out-of-hospital management. Clin Toxicol. 2007;45:315–32.
381. Nemeroff CB, DeVane CL, Pollock BG. Newer antidepressants and the cytochrome P450 system. Am J Psychiatry. 1996;153:311–20.
382. Newport DJ, Wilcox MM, Stowe ZN. Antidepressants during pregnancy and lactation: defining exposure and treatment issues. Semin Perinatol. 2001;25:177–90.
383. Nibuya M, Nestler EJ, Duman RS. Chronic antidepressant administration increases the expression of cAMP response element binding protein (CREB) in rat hippocampus. J Neurosci. 1996;16:2365–72.
384. Nielsen M, Gøtzsche P. An analysis of psychotropic drug sales. Increasing sales of selective serotonin reuptake inhibitors are closely related to number of products. Int J Risk Saf Med. 2011;23:125–32.
385. Nielsen M, Hansen EH, Gøtzsche PC. What is the difference between dependence and withdrawal reactions? Comparison of benzodiazepines and selective serotonin reuptake inhibitors. Addiction. 2012;107:900–8.
386. Nierenberg AA. Predictors of response to antidepressants: general principles and clinical implications. Psychiatr Clin North Am. 2003;26:345–52.
387. Nierenberg AA, Dougherty D, Rosenbaum JF. Dopaminergic agents and stimulants as antidepressant augmentation strategies. J Clin Psychiatry. 1998;59 (Suppl 5):60–4.
388. Nierenberg AA, Leon AC, Price LH, Shelton RC, Trivedi MH. Crisis of confidence: antidepressant risk vs. benefit. J Clin Psychiatry. 2011;72:e11–5.
389. Nunes EV, Levin FR. Treatment of depression in patients with alcohol or other drug dependence: a meta-analysis. JAMA. 2004;291:1887–96.
390. Nurnberg HG, Hensley PL, Lauriello J, Parker LM, Keith SJ. Sildenafil for women patients with antidepressant-induced sexual dysfunction. Psychiatr Serv. 1999;50:1076–8.
391. Nutt DJ. Highlights of the international consensus statement on major depressive disorder. J Clin Psychiatry. 2011;72:e21–2.

392. Nyhuis PW, Gastpar M, Scherbaum N. Opiate treatment in depression refractory to antidepressants and electroconvulsive therapy. J Clin Psychopharmacol. 2005;28:593–5.
393. Oberlander TF, Papsdorf M, Brain UM, Misri S, Ross C, Grunau RE. Prenatal effects of selective serotonin reuptake inhibitor antidepressants, serotonin transporter promoter genotype (SLC6A4), and maternal mood on child behavior at 3 years of age. Arch Pediatr Adolesc Med. 2010;164:444–51.
394. O'Donnell JM, Zhang HT. Antidepressant effects of inhibitors of cAMP phosphodiesterases (PDE4). Trends Pharmacol Sci. 2004;25:158–63.
395. O'Brien L, Baumer C, Thieme D, Sachs H, Koren G. Changes in antidepressant metabolism in pregnancy evidenced by metabolic ratios in hair: novel approach. Forensic Sci Int. 2010;196: 93–6.
396. Olds J, Milner P. Positive reinforcement produced by electrical stimulation of septal area and other regions of rat brain. J Comp Physiol Psychol. 1954;47:419–327.
397. Olfson M, Marcus SC, Druss BG. Effects of Food and Drug Administration warnings on antidepressant use in a national sample. Arch Gen Psychiatry. 2008;65:94–101.
398. Olsen LR, Mortensen EL, Bech P. Prevalence of major depression and stress indicators in the Danish general population. Acta Psychiatr Scand. 2004;109:96–103.
399. Ösby U, Brandt L, Correia N, Ekbom A, Sparén P. Excess mortality in bipolar and unipolar disorder in Sweden. Arch Gen Psychiatry. 2001;58:844–50.
400. Owens MJ, Morgan WN, Plott SJ, Nemeroff CB. Neurotransmitter receptor and transporter binding profile of antidepressants and their metabolites. J Pharmacol Exp Ther. 1997;283: 1305–22.
401. Pae CU, Marks DM, Patkar AA, Masand PS, Luyten P, Serretti A. Pharmacological treatment of chronic fatigue syndrome: focusing on the role of antidepressants. Expert Opin Pharmacother. 2009;10:1561–70.
402. Pae CU, Tharwani H, Marks DM, Masand PS, Patkar AA. Atypical depression: comprehensive review. CNS Drugs. 2009;23:1023–37.
403. Paez-Pereda M, Hausch F, Holsboer F. Corticotropin releasing factor receptor antagonists for major depressive disorder. Expert Opin Investig Drugs. 2011;20:519–35.
404. Papakostas GI, Charles D, Fava M. Are typical starting doses of the selective serotonin reuptake inhibitors sub-optimal? Meta-analysis of randomized, double-blind, placebo-controlled, dose-finding studies in major depressive disorder. World J Biol Psychiatry. 2010;11:300–7.
405. Papakostas GI, Mischoulon D, Shyu I, Alpert JE, Fava M. S-adenosyl methionine (SAMe) augmentation of serotonin reuptake inhibitors for antidepressant nonresponders with major depressive disorder: double-blind, randomized clinical trial. Am J Psychiatry. 2010;167: 942–8.
406. Papakostas GI, Perlis RH, Seifert C, Fava M. Antidepressant dose reduction and the risk of relapse in major depressive disorder. Psychother Psychosom. 2007;76:266–70.
407. Parikh SV, Segal ZV, Grigoriadis S, Ravindran AV, Kennedy SH, Lam RW, et al. CANMAT clinical guidelines for the management of major depressive disorder in adults: psychotherapy alone or in combination with antidepressant medication. J Affect Disord. 2009;117 (Suppl 1):S15–25.
408. Park J, Aragona BJ, Kile BM, Carelli RM, Wightman RM. In vivo voltammetric monitoring of catecholamine release in subterritories of the nucleus accumbens shell. Neuroscience. 2010;169:132–42.
409. Parker G, Fink M, Shorter E, Taylor MA, Akiskal H, Berrios G, et al. Issues for DSM-5: whither melancholia? The case for its classification as a distinct mood disorder. Am J Psychiatry. 2010;167:745–7.
410. Parker G, Roy K, Wilhelm K, Mitchell P. Assessing the comparative effectiveness of antidepressant therapies: prospective clinical practice study. J Clin Psychiatry. 2001;62:117–25.
411. Payne JL. Antidepressant use in the postpartum period: practical considerations. Am J Psychiatry. 2007;164:1329–32.
412. Pearce JMS. Leopold Auenbrugger: camphor-induced epilepsy: remedy for manic psychosis. Eur Neurol. 2008;59:105–7.

413. Pedrelli P, Iovieno N, Vitali M, Tedeschini E, Bentley KH, Papakostas GI. Treatment of major depressive disorder and dysthymic disorder with antidepressants in patients with comorbid opiate use disorders enrolled in methadone maintenance therapy: meta-analysis. J Clin Psychopharmacol. 2011;31:582–6.

414. Peretti S, Judge R, Hindmarch I. Safety and tolerability considerations: tricyclic antidepressants vs. selective serotonin reuptake inhibitors. Acta Psychiatr Scand Suppl. 2000;403:17–25.

415. Pérez V, Soler J, Puigdemont D, Alvarez E, Artigas F. Double-blind, placebo-controlled trial of pindolol augmentation of depressive patients resistant to serotonin reuptake inhibitors. Arch Gen Psychiatry. 1999;56:375–9.

416. Perlis RH, Alpert J, Nierenberg AA, Mischoulon D, Yeung A, Rosenbaum JF, et al. Clinical and sociodemographic predictors of response to augmentation, or dose increase among depressed outpatients resistant to fluoxetine 20 mg/day. Acta Psychiatr Scand. 2003;108: 432–8.

417. Peroutka SJ, Snyder SH. Regulation of serotonin-2 (5-HT$_2$) receptors labeled with [^{3}H] spiroperidol by chronic treatment with the antidepressant amitriptyline. J Pharmacol Exp Ther. 1980;215:582–7.

418. Perry PJ, Zeilmann C, Arndt S. Tricyclic antidepressant concentrations in plasma: an estimate of their sensitivity and specificity as a predictor of response. J Clin Psychopharmacol. 1994;14:230–40.

419. Perugi G, Fornaro M, Akiskal HS. Are atypical depression, borderline personality disorder and bipolar II disorder overlapping manifestations of a common cyclothymic diathesis? World Psychiatry. 2011;10:45–51.

420. Peselow ED, Sanfilipo MP, Difiglia C, Fieve RR. Melancholic/endogenous depression and response to somatic treatment and placebo. Am J Psychiatry. 1992;149:1324–34.

421. Petit M, Dollfus S. [Biochemical factors of resistance to antidepressants (French)]. Encéphale. 1986;12(Suppl):207–15.

422. Petrides G, Tobias KG, Kellner CH, Rudorfer MV. Continuation and maintenance electroconvulsive therapy for mood disorders: review of the literature. Neuropsychobiology. 2011;64:129–40.

423. Pfeiffer PN, Ganoczy D, Zivin K, Valenstein M. Benzodiazepines and adequacy of initial antidepressant treatment for depression. J Clin Psychopharmacol. 2011;31:360–4.

424. Pfeiffer PN, Szymanski BR, Valenstein M, McCarthy JF, Zivin K. Trends in antidepressant prescribing for new episodes of depression and implications for health system quality measures. Med Care. 2012;50:86–90.

425. Piek E, van der Meer K, Nolen WA. Guideline recommendations for long-term treatment of depression with antidepressants in primary care: critical review. Eur J Gen Pract. 2010;16: 106–12.

426. Pierog JE, Kane KE, Kane BG, Donovan JW, Helmick T. Tricyclic antidepressant toxicity treated with massive sodium bicarbonate. Am J Emerg Med. 2009;27:e3–7.

427. Piet J, Hougaard E. Effect of mindfulness-based cognitive therapy for prevention of relapse in recurrent major depressive disorder: systematic review and meta-analysis. Clin Psychol Rev. 2011;31:1032–40.

428. Pigott HE, Leventhal AM, Alter GS, Boren JJ. Efficacy and effectiveness of antidepressants: current status of research. Psychother Psychosom. 2010;79:267–79.

429. Pompili M, Tondo L, Baldessarini RJ. Suicidal risk emerging during antidepressant treatment: recognition and intervention. Clin Neuropsychiatry. 2005;2:66–72.

430. Porcelli S, Drago A, Fabbri C, Gibiino S, Calati R, Serretti A. Pharmacogenetics of antidepressant response. J Psychiatry Neurosci. 2011;36:87–113.

431. Porcelli S, Fabbri C, Spina E, Serretti A, De Ronchi D. Genetic polymorphisms of cytochrome P450 enzymes and antidepressant metabolism. Expert Opin Drug Metab Toxicol. 2011;7: 1101–15.

432. Preskorn SH. Dose-effect and concentration-effect relationships with new antidepressants. Psychopharmacol Ser. 1993;10:174–89.

433. Preskorn SH, Shah R, Neff M, Golbeck AL, Choi J. Potential for clinically significant drug-drug interactions involving the CTP-2D6 system: effects with fluoxetine and paroxetine vs. sertraline. J Psychiatr Pract. 2007;13:5–12.
434. Pringsheim T, Steeves T. Pharmacological treatment for attention deficit hyperactivity disorder (ADHD) in children with comorbid tic disorders. Cochrane Database Syst Rev. 2011; CD007990.
435. Rakovsky JJ, Holtzheimer PE, Nemeroff CB. Emerging targets for antidepressant therapeies. Curr Opin Chem Biol. 2009;13:291–302.
436. Ramakrishna D, Subhash MN. Differential modulation of α-1 adrenoceptor subtypes by antidepressants in the rat brain. J Neural Transm. 2010;117:1423–30.
437. Ramirez OA, Wang RY. Electrophysiological evidence for locus coeruleus norepinephrine autoreceptor subsensitivity following subchronic administration of d-amphetamine. Brain Res. 1986;385:415–9.
438. Rao S, Zisook S. Anxious depression: clinical features and treatment. Curr Psychiatry Rep. 2009;11:429–36.
439. Rapaport MH. Distary restrictions and drug interactions with monoamine oxidase inhibitors: the state of the art. J Clin Psychiatry. 2007;68 (Suppl 8):42–6.
440. Rasmussen KG. Considerations in choosing electroconvulsive therapy versus transcranial magnetic stimulation for depression. J ECT. 2011;27:51–4.
441. Ramasubbu R. Therapy for prevention of post-stroke depression. Expert Opin Pharmacother. 2011;12:2177–787.
442. Ravindran LN, Stein MB. Pharmacotherapy of post-traumatic stress disorder. Curr Top Behav Neurosci. 2010;2:505–25.
443. Rea K, Folgering J, Westerink BH, Cremers TI. Alpha$_1$-adrenoceptors modulate citalopram-induced serotonin release. Neuropharmacology. 2010;58:962–71.
444. Reeves RR, Ladner ME. Antidepressant-induced suicidality: an update. CNS Neurosci Ther. 2010;16:227–34.
445. Ried LD, Jia H, Feng H, Cameon R, Wang X, Tueth M, et al. Selective serotonin reuptake inhibitor treatment and depression are associated with poststroke mortality. Ann Pharmacother. 2011;45:888–97.
446. Riedel M, Möller HJ, Obermeier M, Adli M, Bauer M, Kronmüller K, et al. Clinical predictors of response and remission in inpatients with depressive syndromes. J Affect Disord. 2011;133:137–49.
447. Richelson E. Pharmacology of antidepressants. Mayo Clin Proc. 2001;76:511–27.
448. Richelson E. Interactions of antidepressants with neurotransmitter transporters and receptors and their clinical relevance. J Clin Psychiatry. 2003;64 (Suppl 13):5–12.
449. Rihmer Z, Gonda X, Döme P, Erdős P, Ormos M, Pani L. Novel approaches to drug-placebo difference calculation: evidence from short-term antidepressant drug-trials. Hum Psychopharmacol. 2011. [Epub ahead of print, 14 July].
450. Robinson D, Cardozo L. New drug treatments for urinary incontinence. Maturitas. 2010;65:340–7.
451. Rogers SC, Clay PM. Statistical review of controlled trials of imipramine and placebo in the treatment of depressive illnesses. Br J Psychiatry. 1975;127:599–603.
452. Rosen S, Baldessarini RJ. Acquisition costs of psychopharmacological agents (internal report). Belmont, MA: McLean Hospital Pharmacy Department; 2011.
453. Rosenthal LJ, Goldner WS, O'Reardon JP. T$_3$ augmentation in major depressive isorder: safety considerations. Am J Psychiatry. 2011;168:1035–40.
454. Rothschild AJ, Dunlop BW, Dunner DL, Friedman ES, Gelenberg A, Holland P, et al. Assessing rates and predictors of tachyphylaxis during the prevention of recurrent episodes of depression with venlafaxine-ER for two years (PREVENT) study. Psychopharmacol Bull. 2009;42:5–20.
455. Rouhi M. Chirality at work: drug developers can learn much from recent successful and failed chiral switches. Chem Eng News. 2003;81:56–61.

456. Rucci P, Frank E, Scocco P, Calugi S, Miniatri M, Fagiolini A, et al. Treatment-emergent suicidal ideation during 4 months of acute management of unipolar major depression with SSRI pharmacotherapy or interpersonal psychotherapy in a randomized clinical trial. Depress Anxiety. 2011;28:303–9.

457. Rudorfer MV, Potter WZ. Combined fluoxetine and tricyclic antidepressants. Am J Psychiatry. 1989;146:562–4.

458. Rudorfer MV, Potter WZ. Metabolism of tricyclic antidepressants. Cell Mol Neurobiol. 1999;19:373–409.

459. Ruhé HG, Mason NS, Schene AH. Mood is indirectly related to serotonin, norepinephrine and dopamine levels in humans: meta-analysis of monoamine depletion studies. Mol Psychiatry. 2007;12:331–59.

460. Rush AJ, Siefert SE. Clinical issues in considering vagus nerve stimulation for treatment-resistant depression. Exp Neurol. 2009;219:36–43.

461. Rush AJ, Warden D, Wisniewski SR, Fava M, Trivedi MH, Gaynes BN, et al. STAR*D: revising conventional wisdom. CNS Drugs. 2009;23:627–47.

462. Rybakowski JK. Bipolarity and inadequate response to antidepressant drugs: clinical and psychopharmacological perspective. J Affect Disord. 2012;136:e13–9.

463. Saatcioglu O, Tomruk NB. Use of electroconvulsive therapy in pregnancy: review. Isr J Psychiatry Relat Sci. 2011;48:6–11.

464. Salomon RM, Miller HL, Delgado PL, Charney D. Use of tryptophan depletion to evaluate central serotonin function in depression and other neuropsychiatric disorders. Int Clin Psychopharmacol. 1993;8 (Suppl 2):41–6.

465. Salzer HM, Lurie ML. Anxiety and depressive states treated with isonicotinyl hydrazide (Isoniazid). Arch Neurol Psychiatry. 1953;70:317–24.

466. Sandler M. Monoamine oxidase inhibitors in depression: history and mythology. J Psychopharmacol. 1990;4:136–9.

467. Sanford M. Quetiapine extended release: adjunctive treatment in major depressive disorder. CNS Drugs. 2011;25:803–13.

468. Sansone RA, Sansone LA. Agomelatine: novel antidepressant. Innov Clin Neurosci. 2011;8: 10–4.

469. Santosh PJ, Sattar S, Canagaratnam M. Efficacy and tolerability of pharmacotherapies for attention-deficit hyperactivity disorder in adults. CNS Drugs. 2011;25:737–63.

470. Sanyal C, Asbridge M, Kisely S, Sketris I, Andreou P. Utilization of antidepressants and benzodiazepines among people with major depression in Canada. Can J Psychiatry. 2011;56:667–76.

471. Sarko J. Antidepressants, old and new: review of their adverse effects and toxicity in overdose. Emerg Med Clin North Am. 2000;18:637–54.

472. Schatzberg AF. Mirtazapine. Chapt 21. In: Schatzberg AF, Nemeroff CB, editors. Textbook of Psychopharmacology. 4th ed. Washington, DC: American Psychiatric Publishing, Inc; 2009. p. 429–37.

473. Schatzberg AF, Blier P, Delgado PL, Fava M, Haddad PM, Shelton RC. Antidepressant discontinuation syndrome: consensus panel recommendations for clinical management and additional research. J Clin Psychiatry. 2006;67 (Suppl 4):27–30.

474. Schatzberg AF, Haddad P, Kaplan EM, Lejoyeux M, Rosenbaum JF, Young AH, et al. Serotonin reuptake inhibitor discontinuation syndrome: hypothetical definition. J Clin Psychiatry. 1997;58 (Suppl 7):5–10.

475. Schatzberg AF, Lindley S. Glucocorticoid antagonists in neuropsychiatric disorders. Eur J Pharmacol. 2008;583:358–64.

476. Schellander R, Donnerer J. Antidepressants: clinically relevant drug interactions. Pharmacology. 2010;86:203–15.

477. Schläpfer TE, Kayser S. [Development of deep brain stimulation as a putative treatment for resistant psychiatric disorders (German)]. Nervenärzt. 2010;81:696–701.

478. Schildkraut JJ. Neuropsychopharmacology and the affective disorders. N Engl J Med. 1969;281:302–8.

479. Schindler W, Haefliger F. Derivatives of iminodibenzyl. Helv Chim Acta. 1954;37:4–5.
480. Schmidt HD, Duman RS. Role of neurotrophic factors in adult hippocampal neurogenesis, antidepressant treatments, and animal models of depressive-like behavior. Behav Pharmacol. 2007;18:391–418.
481. Schneier FR. Pharmacotherapy of social anxiety disorder. Expert Opin Pharmacother. 2011;12:615–25.
482. Schultes RE, editor. Ethnobotany: The Evolution of a Discipline. Portland, OR: Timber Press; 2005.
483. Schwaninger M, Schöfl C, Blume R, Rössig L, Knepel W. Inhibition by antidepressant drugs of cyclic AMP response element-binding protein/cyclic AMP response element-directed gene transcription. Mol Pharmacol. 1995;47:1112–8.
484. Schweitzer I, Maguire K, Ng C. Sexual side-effects of contemporary antidepressants: review. Aust N Z J Psychiatry. 2009;43:795–808.
485. Sedky K, Nazir R, Joshi A, Kaur G, Lippmann S. Which psychotropic medications induce heptotoxicity? Gen Hosp Psychiatry. 2012;34:53–61.
486. Selikoff IJ, Robitzek EH, Ornstein GG. Treatment of pulmonary tuberculosis with hydrazide derivatives of isonicotinic acid. JAMA. 1952;150:973–80.
487. Sen S, Duman RS, Sanacora G. Serum brain-derived neurotrophic factor, depression, and antidepressant medications: meta-analyses and implications. Biol Psychiatry. 2008;64: 527–32.
488. Sepede G, De Berardis D, Gambi F, Campanella D, La Rovere R, D'Amico M, et al. Olanzapine augmentation in treatment-resistant panic disorder: 12-week, fixed-dose, open-label trial. J Clin Psychopharmacol. 2006;26:45–9.
489. Serretti A, Chiesa A, Calati R, Perna G, Bellodi L, De Ronchi D. Novel antidepressants and panic disorder. Neuropsychobiology. 2011;63:1–7.
490. Serretti A, Mandelli L. Antidepressants and body weight: comprehensive review and meta-analysis. J Clin Psychiatry. 2010;71:1259–72.
491. Sharp T, Umbers V, Gartside SE. Effect of a selective 5-HT reuptake inhibitor in combination with 5-HT$_{1A}$ and 5-HT$_{1B}$ receptor antagonists on extracellular 5-HT in rat frontal cortex in vivo. Br J Pharmacol. 1997;121:941–6.
492. Sharpley CF, Bitsika V. Four principal criteria for deciding when to use antidepressants or psychotherapy for unipolar depression: literature review. Int J Psychiatry Clin Pract. 2011;15:2–11.
493. Shaskan EG, Snyder SH. Kinetics of serotonin accumulation into slices from rat brain: relationship to catecholamine uptake. J Pharmacol Exp Ther. 1970;175:404–18.
494. Shelton RC. Nature of the discontinuation syndrome associated with antidepressant drugs. J Clin Psychiatry. 2006;67 (Suppl 4):3–7.
495. Shelton RC, Trivedi MH. Moderators of antidepressant response in major depression. J Clin Psychiatry. 2011;72:e32–7.
496. Shen WW, Hsu JH. Female sexual side effects associated with selective serotonin reuptake inhibitors: descriptive clinical study of 33 patients. Int J Psychiatry Med. 1995;25:239–48.
497. Shopsin B, Friedman E, Gershon S. Parachlorophenylalanine reversal of tranylcypromine effects in depressed patients. Arch Gen Psychiatry. 1976;33:811–9.
498. Shopsin B, Gershon S, Goldstein M, Friedman E, Wilk S. Use of synthesis inhibitors in defining a role for biogenic amines during imipramine treatment in depressed patients. Psychopharmacol Commun. 1975;1:239–49.
499. Shorter E, Fink M. Endocrine Psychiatry: the Dexamethasone Suppression Test and Electroconvulsive Therapy. New York: Oxford University Press; 2010.
500. Shorter E, Healy D. Shock Therapy: A History of Electroconvulsive Treatment in Mental Illness. New Brunswick, NJ: Rutgers University Press; 2007.
501. Sidor MM, MacQueen GM. Antidepressants for the acute treatment of bipolar depression: a systematic review and meta-analysis. J Clin Psychiatry. 2011;72:156–67.
502. Simon RI, Hales RE, editors. Textbook of Suicide Assessment and Management. 2nd ed. Washington, DC: American Psychiatric Publishing Co.; 2012.

503. Simon GE, Perlis RH. Personalized medicine for depression: can we match patients with treatments? Am J Psychiatry. 2010;167:1445–55.
504. Simon GE, Savarino J. Suicide attempts among patients starting depression treatment with medications or psychotherapy. Am J Psychiatry. 2007;164:1029–34.
505. Simon GE, Savarino J, Operkalski B, Wang PS. Suicide risk during antidepressant treatment. Am J Psychiatry. 2006;163:41–7.
506. Sinclair LI, Christmas DM, Hood SD, Potokar JP, Robertson A, Isaac A. Antidepressant-induced jitteriness/anxiety syndrome: systematic review. Br J Psychiatry. 2009;194: 483–90.
507. Sit DK, Perel JM, Helsel JC, Wisner KL. Changes in antidepressant metabolism and dosing across pregnancy and early postpartum. J Clin Psychiatry. 2008;69:652–8.
508. Slotema CW, Blom JD, Hoek HW, Sommer IE. Should we expand the toolbox of psychiatric treatment methods to include Repetitive Transcranial Magnetic Stimulation (rTMS)? A meta-analysis of the efficacy of rTMS in psychiatric disorders. J Clin Psychiatry. 2010;71:873–384.
509. Smith JA. The use of the isopropyl derivative of isonicotinyl hydrazide (Marsalid) in the treatment of mental disease: preliminary report. Am Pract. 1953;4:519–20.
510. Smith HS, Argoff CE. Pharmacological treatment of diabetic neuropathic pain. Drugs. 2011;71:557–89.
511. Soares CN, Arsenio H, Joffe H, Bankier B, Cassano P, Petrillo LF, et al. Escitalopram vs. ethinyl estradiol and norethindrone acetate for symptomatic peri- and postmenopausal women: impact on depression, vasomotor symptoms, sleep, and quality of life. Menopause. 2006;13:780–6.
512. Sokoro AA, Zivot J, Ariano RE. Neuroleptic malignant syndrome versus serotonin syndrome: search for a diagnostic tool. Ann Pharmacother. 2011;45:e50.
513. Somberg TC, Arora RR. Depression and heart disease: therapeutic implications. Cardiology. 2008;111:75–81.
514. Souery D, Oswald P, Massat I, Bailer U, Bollen J, Demyttenaere K, et al. Clinical factors associated with treatment resistance in major depressive disorder. J Clin Psychiatry. 2007;68: 1062–70.
515. Souery D, Papakostas GI, Trivedi MH. Treatment-resistant depression. J Clin Psychiatry. 2006;67 (Suppl 6):16–22.
516. Soufia M, Aoun J, Gorsane MA, Krebs MO. [SSRIs and pregnancy: review of the literature (French)]. Encéphale. 2010;36:513–6.
517. Spencer T, Biederman J, Wilens T, Steingard R, Geist D. Nortriptyline treatment of children with attention-deficit hyperactivity disorder and tic disorder or Tourette's syndrome. J Am Acad Child Adolesc Psychiatry. 1993;32:205–10.
518. Spina E, Trifirò G, Caraci F. Clinically significant drug interactions with newer antidepressants. CNS Drugs. 2012;26:39–67.
519. Spoelhof GD, Davis GL, Licari A. Clinical vignettes in geriatric depression. Am Fam Physician. 2011;84:1149–54.
520. Sproule BA, Hazra M, Pollock BG. Desvenlafaxine succinate for major depressive disorder. Drugs Today (Barc). 2008;44:475–87.
521. Stage KB, Bech P, Gram LF, Kragh-Sørensen P, Rosenberg C, Ohrberg S. Are in-patient depressives more often of the melancholic subtype? Danish University Antidepressant Group. Acta Psychiatr Scand. 1998;98:432–26.
522. Stahl SM. Placebo-controlled comparison of the selective serotonin reuptake inhibitors citalopram and sertraline. Biol Psychiatry. 2000;48:894–901.
523. Starr P, Klein-Schwartz W, Spiller H, Kern P, Ekleberry SE, Kunkel S. Incidence and onset of delayed seizures after overdoses of extended-release bupropion. Am J Emerg Med. 2009;27:911–5.
524. Steffens DC, Krishnan KR, Helms MJ. Are SSRIs better than TCAs? Comparison of SSRIs and TCAs: meta-analysis. Depress Anxiety. 1997;6:10–8.
525. Stein G, Bernadt M. Lithium augmentation therapy in tricyclic-resistant depression: controlled trial using lithium in low and normal doses. Br J Psychiatry. 1993;162:634–40.

526. Stettin GD, Yao J, Verbrugge RR, Aubert RE. Frequency of follow-up care for adult and pediatric patients during initiation of antidepressant therapy. Am J Manag Care. 2006;12: 453–61.
527. Stewart JW, Tricamo E, McGrath PJ, Quitkin FM. Prophylactic efficacy of phenelzine and imipramine in chronic atypical depression: likelihood of recurrence on discontinuation after 6 months' remission. Am J Psychiatry. 1997;154:31–6.
528. Stigler SM. Regression toward the mean, historically considered. Stat Methods Med Res. 1997;6:103–14.
529. Stoll AL, Mayer PV, Kolbrener M, Goldstein E, Suplit B, Lucier J, et al. Antidepressant-associated mania: controlled comparison with spontaneous mania. Am J Psychiatry. 1994;151: 1642–5.
530. Stone M, Laughren TP, Jones ML, Levenson M, Holland PC, Hughes A, et al. Risk of suicidality in clinical trials of antidepressants in adults: analysis of proprietary data submitted to US Food and Drug Administration. BMJ. 2009;339:b2880–9.
531. Straus SE, Richardson WS, Glasziou P, Haynes RB. Evidence-Based Medicine. 3rd ed. Edinburgh: Churchill-Livingstone; 2005.
532. Stübner S, Grohmann R, von Straendorff I, Rüther E, Möller HJ, Müller-Oerlinghausen B, et al. Suicidality as rare adverse event of antidepressant medication: report from the AMSP multicenter drug safety surveillance project. J Clin Psychiatry. 2010;71:1293–307.
533. Sturdee DW. Menopausal hot flush–anything new? Maturitas. 2008;60:42–9.
534. Sulser F, Vetulani J, Mobley PL. Mode of action of antidepressant drugs. Biochem Pharmacol. 1978;27:257–61.
535. Sun-Edelstein C, Tepper SJ, Shapiro RE. Drug-induced serotonin syndrome: review. Expert Opin Drug Saf. 2008;7:587–96.
536. Svensson TH. Brain noradrenaline and the mechanisms of action of antidepressant drugs. Acta Psychiatr Scand Suppl. 2000;402:18–27.
537. Szegedi A, Jansen WT, van Willigenburg AP, van der Meulen E, Stassen HH, Thase ME. Early improvement in the first 2 weeks as a predictor of treatment outcome in patients with major depressive disorder: a meta-analysis including 6562 patients. J Clin Psychiatry. 2009;70: 344–53.
538. Tajalizadekhoob Y, Sharifi F, Fakhrzadeh H, Mirarefin M, Ghaderpanahi M, Badamchizade Z, et al. Effect of low-dose omega 3 fatty acids on the treatment of mild to moderate depression in the elderly: double-blind, randomized, placebo-controlled study. Eur Arch Psychiatry Clin Neurosci. 2011;261:539–49.
539. Talarico G, Tosto G, Pietracupa S, Piacentini E, Canevelli M, Lenzi GL, et al. Serotonin toxicity; short review. Neurol Sci. 2011;32:507–9.
540. Tamam L, Ozpoyraz N. Selective serotonin reuptake inhibitor discontinuation syndrome: review. Adv Ther. 2002;19:17–26.
541. Tatsumi M, Groshan K, Blakely RD, Richelson E. Pharmacological profile of antidepressants and related compounds at human monoamine transporters. Eur J Pharmacol. 1997;340: 249–58.
542. Tepper SJ. Serotonin syndrome: SSRIs, SNRIs, triptans, and current clinical practice. Headache. 2012;52:195–7.
543. Terevnikov V, Stenberg JH, Tiihonen J, Joffe M, Burkin M, Tchoukhine E, et al. Add-on mirtazapine improves depressive symptoms in schizophrenia: double-blind randomized placebo-controlled study with an open-label extension phase. Hum Psychopharmacol. 2011;26:188–93.
544. Ter Horst PG, Jansman FG, van Lingen RA, Smit JP, de Jong-van den Berg LT, Brouwers JR. Pharmacological aspects of neonatal antidepressant withdrawal. Obstet Gynecol Surv. 2008;63:267–79.
545. Thanacoody HK, Thomas SH. Tricyclic antidepressant poisoning: cardiovascular toxicity. Toxicol Rev. 2005;24:205–14.
546. Thase ME. Preventing relapse and recurrence of depression: abrief review of therapeutic options. CNS Spectr. 2006;11(12 Suppl 15):12–21.

547. Thase ME. New directions in the treatment of atypical depression. J Clin Psychiatry. 2007;68 (Suppl 3):4–9.
548. Thase ME. Treating major depression: antidepressant algorithms. J Clin Psychiatry. 2009;70:e46–50.
549. Thase ME. Antidepressant combinations: widely used, but far from empirically validated. Can J Psychiatry. 2011;56:317–23.
550. Thase ME, Larsen KG, Kennedy SH. Assessing the 'true' effect of active antidepressant therapy vs. placebo in major depressive disorder: use of a mixture model. Br J Psychiatry. 2011;199:501–7.
551. Thiele J, Holzinger O. Über O-diaminodibenzyl. Ann Chem Liebigs. 1899;305:96–102.
552. Thompson Jr JW, Ware MR, Blashfield RK. Psychotropic medication and priapism: comprehensive review. J Clin Psychiatry. 1990;51:430–3.
553. Thuile J, Even C, Roullon F. Long-term outcome in anxiety disorders: review of double-blind studies. Curr Opin Psychiatry. 2009;22:84–9.
554. Tiihonen J, Lonnqvist J, Wahlbeck K, Klaukka T, Tanskanen A, Haukka J. Antidepressants and the risk of suicide, attempted suicide, and overall mortality in a nationwide cohort. Arch Gen Psychiatry. 2006;63:1358–67.
555. Tondo L, Albert M, Baldessarini RJ. Suicide rates In relation to health-care access in the United States. J Clin Psychiatry. 2006;67:517–23.
556. Tondo L, Baldessarini RJ, Vázquez G, Lepri B, Visioli C. Clinical responses to treatment in acutely depressed patients with bipolar versus unipolar major affective disorders. Acta Psychiatr Scand. 2012; in press.
557. Tondo L, Hennen J, Baldessarini RJ. Reduced suicide risk with long-term lithium treatment in major affective Illness: A meta-analysis. Acta Psychiatr Scand. 2001;104:163–72.
558. Tondo L, Lepri B, Baldessarini RJ. Risks of suicidal ideation, attempts and suicides among 2826 men and women with types I and II bipolar, and recurrent major depressive disorders. Acta Psychiatr Scand. 2007;116:419–28.
559. Tondo L, Lepri B, Baldessarini RJ. Suicidal status during antidepressant treatment in 789 Sardinian patients with major affective disorder. Acta Psychiatr Scand. 2008;118:106–15.
560. Tondo L, Vázquez G, Baldessarini RJ. Mania associated with antidepressant-treatment: comprehensive meta-analysis. Acta Psychiatr Scand. 2010;121:404–14.
561. Traynor LM, Thiessen CN, Traynor AP. Pharmacotherapy of fibromyalgia. Am J Health Syst Pharm. 2011;68:1307–19.
562. Tremblay B, Blier P. Catecholaminergic strategies for the treatment of major depression. Curr Drug Targets. 2006;7:149–58.
563. Trivedi MH, Fava M, Wisniewski SR, Thase ME, Quitkin F, Warden D, et al. Medication augmentation after the failure of SSRIs for depression. N Engl J Med. 2006;354:1243–52.
564. Trivedi RB, Nieuwsma JA, Williams Jr JW. Examination of the utility of psychotherapy for patients with treatment resistant depression: systematic review. J Gen Intern Med. 2011;26:643–50.
565. Tsapakis E, Soldani F, Tondo L, Baldessarini RJ. Efficacy of antidepressants in depressed children and adolescents: meta-analysis. Br J Psychiatry. 2008;193:10–7.
566. Tuccori M, Montagnani S, Testi A, Ruggiero E, Mantarro S, Scollo C, et al. Use of selective serotonin reuptake inhibitors during pregnancy and risk of major and cardiovascular malformations: update. Postgrad Med. 2010;122:49–65.
567. Tuccori M, Testi A, Antonioli L, Fornai M, Montagnani S, Ghisu N, et al. Safety concerns associated with use of serotonin reuptake inhibitors and other serotonergic/noradrenergic antidepressants during pregnancy: review. Clin Ther. 2009;31:1426–53.
568. Turner EH, Matthews AM, Linardatos E, Tell RA, Rosenthal R. Selective publication of antidepressant trials and its influence on apparent efficacy. N Engl J Med. 2008;358:252–60.

569. Uher R, Perlis RH, Henigsberg N, Zobel A, Rietschel M, Mors O, et al. Depression symptom dimensions as predictors of antidepressant treatment outcome: replicable evidence for interest-activity symptoms. Psychol Med. 2012;42:967–80.
570. Undurraga J, Baldessarini RJ. Randomized, placebo-controlled trials of antidepressants for acute major depression: thirty-year meta-analytic review. Neuropsychopharmacology. 2012;37:851–64.
571. Uzan A, Kabouche M, Rataud J, Le Fur G. Pharmacological evidence of a possible tryptaminergic regulation of opiate receptors by using indalpine, a selective 5-HT uptake inhibitor. Neuropharmacology. 1980;19:1075–9.
572. Valenstein M, Eisenberg D, McCarthy JF, Austin KL, Ganoczy D, Kim HM, et al. Service implications of providing intensive monitoring during high-risk periods for suicide among VA patients with depression. Psychiatr Serv. 2009;60:439–44.
573. Valuck RJ, Libby AM, Orton HD, Morrato EH, Allen R, Baldessarini RJ. Spillover effects on treatment of adult depression in primary care after FDA advisory on risk of pediatric suicidality with SSRIs. Am J Psychiatry. 2007;164:1198–205.
574. Vázquez G, Baldessarini RJ, Yildiz A, Tamayo J, Tondo L, Salvatore P. Multi-site international collaborative clinical trials in mania. Int J Neuropsychopharmacol. 2011;14:1013–6.
575. Vázquez G, Tondo L, Baldessarini RJ. Comparison of antidepressant responses in patients with bipolar vs. unipolar depression: meta-analytic review. Pharmacopsychiatry. 2011;44:21–6.
576. Ventimiglia J, Kalali AH. Generic penetration in the retail antidepressant market. Psychiatry (Edgmont). 2010;7:9–11.
577. Vermeiden M, van den Broek WW, Mulder PG, Birkenhäger TK. Influence of gender and menopausal status on antidepressant treatment response in depressed inpatients. J Psychopharmacol. 2010;24:497–502.
578. Ververs TF, van Wensen K, Freund MW, van der Heide M, Visser GH, Schobben AF, et al. Association between antidepressant drug use during pregnancy and child healthcare utilization. BJOG. 2009;116:1568–77.
579. Vieta E, Colom F. Therapeutic options in treatment-resistant depression. Ann Med. 2011;43:512–30.
580. Viguera AC, Baldessarini RJ, Friedberg J. Discontinuing antidepressant treatment in major depression. Harv Rev Psychiatry. 1998;5:293–306.
581. Viguera AC, Tondo L, Koukopoulos AE, Reginaldi D, Lepri B, Baldessarini RJ. Episodes of mood disorders in 2,252 pregnancies and postpartum periods. Am J Psychiatry. 2011;168:1179–85.
582. Vöhringer PA, Ghaemi SN. Solving the antidepressant efficacy question: effect sizes in major depressive disorder. Clin Ther. 2011;33:B49–61.
583. Wagner KD, Jonas J, Findling RL, Ventura D, Saikali K. Double-blind, randomized, placebo-controlled trial of escitalopram in the treatment of pediatric depression. J Am Acad Child Adolesc Psychiatry. 2006;45:280–8.
584. Wagstaff AJ, Cheer SM, Matheson AJ, Ormrod D, Goa KL. Paroxetine: an update of its use in psychiatric disorders in adults. Drugs. 2002;62:655–703.
585. Walker PW, Cole JO, Gardner EA, Hughes AR, Johnston JA, Batey SR, et al. Improvement in fluoxetine-associated sexual dysfunction in patients switched to bupropion. J Clin Psychiatry. 1993;54:459–65.
586. Waring WS, Good AM, Bateman DN. Lack of significant toxicity after mirtazapine overdose: five-year review of cases admitted to a regional toxicology unit. Clin Toxicol. 2007;45:45–50.
587. Watson WA, Litovitz TL, Klein-Schwartz W, Rodgers Jr GC, Youniss J, Reid N, et al. Annual report of the American Association of Poison Control Centers Toxic Exposure Surveillance System. Am J Emerg Med. 2004;22:335–404.
588. Weihs K, Wert JM. Primary care focus on the treatment of patients with major depressive disorder. Am J Med Sci. 2011;342:324–230.
589. Weil-Malherbe H, Whitby LG, Axelrod J. Uptake of circulating [³H]norepinephrine by the pituitary gland and various areas of the brain. J Neurochem. 1961;8:55–64.

590. Weissman MM, Pilowsky DJ, Wickramaratne PJ, Talati A, Wisniewski SR, Fava M, et al. Remissions in maternal depression and child psychopathology. JAMA. 2006;295:1389–98.

591. Wenk M, Liliane Todesco L, Krähenbühl S. Effect of St John's wort on activities of CYP1A2, CYP3A4, CYP2D6, N-acetyltransferase 2, and xanthine oxidase in healthy males and females. Br J Clin Pharmacol. 2004;57:495–9.

592. White N, Litovitz T, Clancy C. Suicidal antidepressant overdoses: comparative analysis by antidepressant type. J Med Toxicol. 2008;4:238–50.

593. Wijeratne C, Sachdev P. Treatment-resistant depression: critique of current approaches. Aust N Z J Psychiatry. 2008;42:751–62.

594. Wijkstra J, Lijmer J, Balk FJ, Geddes JR, Nolen WA. Pharmacological treatment for unipolar psychotic depression: systematic review and meta-analysis. Br J Psychiatry. 2006;188:410–5.

595. Wilens TE, Biederman J, Baldessarini RJ, Puopolo P, Flood JG. Developmental changes in serum concentrations of desipramine and 2-hydroxydesipramine during treatment with desipramine. J Am Acad Child Adolesc Psychiatry. 1992;31:691–8.

596. Wilder J. Basimetric approach (law of initial value) to biological rhythms. Ann N Y Acad Sci. 1962;98:1211–28.

597. Wilson DL, Ried LD. Identifying iatrogenic depression using confirmatory factor analysis of the Center for Epidemiologic Studies Depression Scale in patients prescribed a verapamil-sustained-release-led or atenolol-led hypertension treatment strategy. Res Social Adm Pharm. 2011. [Epub ahead of print, 29 Nov].

598. Wimbiscus M, Kostenko O, Malone D. MAO inhibitors: risks, benefits, and lore. Cleve Clin J Med. 2010;77:859–82.

599. Witkin JM, Marek GJ, Johnson BG, Schoepp DD. Metabotropic glutamate receptors in the control of mood disorders. CNS Neurol Disord Drug Targets. 2007;6:87–100.

600. Wittchen HU, Jacobi F, Rehm J, Gustavsson A, Svensson M, Jönsson B, et al. Size and burden of mental disorders and other disorders of the brain in Europe 2010. Eur Neuropsychopharmacol. 2011;21:655–79.

601. Wogelius P, Nørgaard M, Gislum M, Pedersen L, Munk E, Mortensen PB, et al. Maternal use of selective serotonin reuptake inhibitors and risk of congenital malformations. Epidemiology. 2006;17:701–4.

602. Wolfe RM. Antidepressant withdrawal reactions. Am Fam Physician. 1997;56:455–62.

603. Wong D, Horng J, Bymaster F, Hauser K, Molloy B. A selective inhibitor of serotonin uptake: Lilly-110150, 3-(p-trifluoromethylphenoxy)-N-methyl-3-phenylpropylamine. Life Sci. 1974;15:471–9.

604. Woolf AD, Erdman AR, Nelson LS, Caravati EM, Cobaugh DJ, Booze LL, et al. Tricyclic antidepressant poisoning: evidence-based consensus guideline for out-of-hospital management. Clin Toxicol. 2007;45:203–33.

605. World Health Organization [WHO]. International Classification of Diseases, ninth revision. Geneva: World Health Organization; 1979.

606. Wozniak G, Toska A, Saridi M, Mouzas O. Serotonin reuptake inhibitor antidepressants (SSRIs) against atherosclerosis. Med Sci Monit. 2011;17:205–14.

607. Wu C-S, Wang S-C, Cheng Y-C, Gau SS-F. Association of cerebrovascular events with antidepressant use: case-crossover study. Am J Psychiatry. 2011;168:511–21.

608. Yamamura S, Abe M, Nakagawa M, Ochi S, Ueno S, Okada M. Different actions for acute and chronic administration of mirtazapine on serotonergic transmission associated with raphe nuclei and their innervation cortical regions. Neuropharmacology. 2011;60:550–60.

609. Yang SJ, Kim SY, Stewart R, Kim JM, Shin IS, Jung SW, et al. Gender differences in 12-week antidepressant treatment outcomes for a naturalistic secondary care cohort. Psychiatry Res. 2011;189:82–90.

610. Yeung AS, Ameral VE, Chuzi SE, Fava M, Mischoulon D. Pilot study of acupuncture augmentation therapy in antidepressant partial and non-responders with major depressive disorder. J Affect Disord. 2011;130:285–9.

611. Yildiz A, Vieta E, Leucht S, Baldessarini RJ. Efficacy of antimanic treatments; meta-analysis of randomized, controlled trials. Neuropsychopharmacology. 2011;36:375–89.

612. Yildiz A, Vieta E, Tohen M, Baldessarini RJ. Factors modifying drug and placebo responses in randomized trials for bipolar mania. Int J Neuropsychopharmacol. 2011;14:863–75.
613. Yohannes AM, Hann M, Sibbald B. Management of depressive symptoms in patients with COPD: a postal survey of general practitioners. Prim Health Care Res Dev. 2011;12: 237–44.
614. Yonkers KA. Management strategies for PMS/PMDD. J Fam Pract. 2004;53 (Suppl 9):S15–20.
615. Young EA, Kornstein SG, Marcus SM, Harvey AT, Warden D, Wisniewski SR, et al. Sex differences in response to citalopram. J Psychiatr Res. 2009;43:503–11.
616. Zaffanello M, Giacomello L, Brugnara M, Fanos V. Therapeutic options in childhood nocturnal enuresis. Minerva Urol Nefrol. 2007;59:199–205.
617. Zajecka J, Fawcett J, Amsterdam J, Quitkin F, Reimherr F, Rosenbaum J, et al. Safety of abrupt discontinuation of fluoxetine: randomized, placebo-controlled study. J Clin Psychopharmacol. 1998;18:193–7.
618. Zajecka J, Tracy KA, Mitchell S. Discontinuation symptoms after treatment with serotonin reuptake inhibitors: literature review. J Clin Psychiatry. 1997;58:291–7.
619. Zarate Jr C, Machado-Vieira R, Henter I, Ibrahim L, Diazgranados N, Salvadore G. Glutamatergic modulators: the future of treating mood disorders? Harv Rev Psychiatry. 2010;18:293–303.
620. Zeller EA, Barsky J, Berman ER, Fouls JR. Actions of isonicotinic acid hydrazide and related compounds on enzymes of brain and other tissues. J Lab Clin Med. 1952;14:965–6.
621. Zhang HT, Whisler LR, Huang Y, Xiang Y, O'Donnell JM. Postsynaptic alpha-2 adrenergic receptors are critical for the antidepressant-like effects of desipramine on behavior. Neuropsychopharmacology. 2009;34:1067–77.
622. Zhao Z, Zhang HT, Bootzin E, Millan MJ, O'Donnell JM. Association of changes in norepinephrine and serotonin transporter expression with the long-term behavioral effects of antidepressant drugs. Neuropsychopharmacology. 2009;34:1467–81.
623. Zhou D, Guo J, Linnenbach AJ, Booth-Genthe CL, Grimm SW. Role of human UGT2B10 in N-glucuronidation of tricyclic antidepressants, amitriptyline, imipramine, clomipramine, and trimipramine. Drug Metab Dispos. 2010;38:863–70.
624. Zhu MY, Kim CH, Hwang DY, Baldessarini RJ, Kim KS. Effects of desipramine treatment on norepinephrine transporter gene expression in the cultured SK-N-BE(2)M17 cells and rat brain tissue. J Neurochem. 2002;82:146–53.
625. Zimmerman M, Thongy T. How often do SSRIs and other new-generation antidepressants lose their effect during continuation treatment? Evidence suggesting the rate of true tachyphylaxis during continuation treatment is low. J Clin Psychiatry. 2007;68:1271–6.

Chapter 5
Status and Prospects for Psychopharmacology

...Where is the wisdom we have lost in knowledge?

Where is the knowledge we have lost in information?...

TS Eliot, The Rock [9]

Overview of Contemporary Efforts to Develop Psychotropic Drugs

Modern psychopharmacology has been evolving since the 1950s. Its contributions have been enormous and of fundamental importance to modern psychiatric thera-peutics, encouraging and enabling the shift away from long-term institutional care, to brief inpatient interventions for crises, and reliance on ambulatory care. In addi-tion, the clinical benefits provided by much of psychiatric chemotherapy, though typically incomplete, have stimulated decades of renewed efforts to develop a more biomedically based theoretical and academic psychiatry. In addition, the psychop-harmaceutical industry has been especially successful in blending science and business to provide products of considerable social value.

Despite these striking and laudable advances, many problems and challenges remain. Notable among these is the evident slowing of progress in the development of novel, let alone superior, psychotropic (and other) drugs in recent years, despite impressive advances in many technical aspects of drug discovery and development that were summarized in Chapter 1. Following years of expanded industrial effort yielding psychotropic drugs with multi-billion-dollar annual markets, many standard agents are losing patent protection, and the rate of innovation has slowed greatly [7, 8, 18, 33]. As chemical congeners, metabolites, isolated isomers, or functional homologs of known drugs, most recently introduced psychotropic drug products are conceptually and therapeutically very similar to older agents [17, 32, 37]. Save for the hope of eventually favorable positioning in competitive markets, there is little incen-tive to encourage the very difficult and uncertain development of truly innovative and possibly clinically superior new agents. Notably, most national systems that regulate

R.J. Baldessarini, *Chemotherapy in Psychiatry: Pharmacologic Basis of Treatments for Major Mental Illness*, DOI 10.1007/978-1-4614-3710-9_5, © Springer Science+Business Media New York 2013 patenting and licensing of new drugs do not allow for more prolonged patent protection or higher prices for new drugs that are proved to outperform existing agents.

Pharmaceutical corporations are showing less interest in pursuing psychotropic or other drugs that act on the central nervous system than formerly [8, 18]. Indeed, one wonders whether modern psychopharmacology is passing through a life-cycle, with periods of birth, development, mature functioning, and risk of decline into senescence and death. A particularly curious turn in the industry has been away from formerly vigorous and competitive investment in scientifically based drug-development toward functioning more and more as marketing organizations. To some extent, the gap in efforts aimed at scientific drug-development can be covered by discoveries in university and other basic scientific laboratories and by acquisition of small start-up companies with innovative products.

There is a fundamental irony underlying the entire history of development of medicinal psychotropic treatments. That is, most of the innovations in the field, as outlined in Chapter 1, have arisen not from rational, scientifically based principles or knowledge of biological characteristics of psychiatric disorders, but from the important but often undervalued element of serendipity [21]. Most of the founding discoveries in modern psychopharmacology between 1949 and 1960 were largely serendipitous, and have been followed by more than a half-century of relatively minor refinements and rather obvious repetitions. Even more ironic is that, at the same time, there has arisen an enormously large, complex, well-funded, and well-staffed program of basic and applied research based not only in the pharmaceutical industry, but also in university laboratories, governmental and private research organizations, and start-up biotechnology companies. Nevertheless the disparity between investment in expensive and seemingly powerful new scientific methods and the production of commercially viable and clinically useful new products is large and growing (see Chap. 1, Fig. 1.2; [12, 33]).

The incongruence between investment in new scientific methods and the discovery of new drugs may in part reflect the fundamentally conservative nature of the work of both the pharmaceutical and academic-scientific industries, in both of which daringly innovative ideas typically are considered risky and not much encouraged—perhaps limiting innovation and discoveries of therapeutic value. However, other tends may be at least as important or more important contributors to the current state of the therapeutic art. A fundamental one is the lack of knowledge of plausible and operationally effective pathophysiology for major mental illnesses and many neuropsychiatric disorders—let along etiology. Even conditions in which there are both an undoubted tissue pathology and known molecular derangements—such as Alzheimer's dementia or Parkinson's disease—etiologies or fundamental causes remain strikingly elusive, as do effective treatments. Oddly, however, even in disorders with an established biology to guide therapeutic developments—in infectious diseases, cardiovascular and metabolic disorders, and cancer research—innovation moves ahead very slowly [20, 24, 33]. A biologically based, rational therapeutics for major psychiatric disorders, including the psychoses, severe mood disorders, and anxiety-related conditions, remains even more elusive.

In addition, both corporate and scientific conservatism may have limited the possibility of advances based on what might be termed "managed serendipity." The approach would acknowledge that an entirely theory-based, rational therapeutics has proved highly elusive with respect to brain and behavioral disorders, and that most advances in the past half-century have come about through unexpected, empirical, observations, albeit by trained and experienced observers who could recognize something of potential value. However, scientists cannot develop programs and careers based on the hope of serendipitous discovery, and industry cannot manage its business obligations on such an unpredictable basis. A notable aspect of industrial drug-development programs is the extent to which marketing plans come into the picture early in the development of a new-drug candidate, including very specific and often narrowly targeted indications for eventual clinical application. As an alternative, there might be value in encouraging very small, even uncontrolled and unblinded, early trials in a range of disorders for virtually any agent what can exert effects in the central nervous system. Such an approach might have identified much earlier the value of antipsychotic agents for mania and severe depression, of the anti-estrogen compound tamoxifen for mania, of antidepressants for severe anxiety disorders, and anticonvulsants for mania, among others [6, 19, 25, 29, 31, 36].

A possibly related, progress-limiting, factor may be a tendency toward compartmentalization of efforts in research and development, with far too little cross-talk among highly specialized approaches to both basic and clinical sciences. As an example, current discussions of updating of both the DSM and ICD as standard, international diagnostic systems for psychiatry appear to be driven as much by economic and guild-protective considerations as by scientific advances, resulting in the proliferation of ever-larger numbers of strikingly insecure diagnostic entities [1, 27, 28, 35]. Indeed, there has been remarkably little advance in psychiatric nosology beyond clinical-descriptive systems that have been evolving for more than two centuries. Psychiatric nosology, despite its severe shortcomings, is fundamental to progress in psychobiological research. It is essential to support the refinement of phenotypes on which more rational and possibly productive efforts in genetics, brain imaging, molecular studies, and other aspects of biomedical research in psychiatry, as well as in experimental therapeutics. Nevertheless, despite many unrealistic promises, there have been remarkably few contributions to the refinement of clinical nosology by biomedical psychiatric research, and the research and nosological activities remain largely separate.

Highlights of the current status of psychopharmacology are summarized in Table 5.1.

Overview of Clinical Therapeutics in Contemporary Psychiatry

Drug-Associated Changes in Clinical Care

The development of clinical psychopharmacology over the past half-century, as noted, has had an enormously salutary impact on clinical therapeutics in psychiatry and, indeed, led to or supported a revolution in the organization of psychiatric services, institutions, research agendas, and training programs. Nevertheless, some less-than-ideal aspects of the process are worth considering. The attraction of rela-

Table 5.1 Current status of psychopharmacology

Many agents are available to treat most major psychiatric disorders; most are reasonably effective and acceptably safe

Effect sizes for most psychotropic drugs are moderate and agents within a class are not distinguishable by efficacy, leaving rational, evidence-based therapeutics an elusive hope

Incentives to develop agents of superior efficacy or tolerability are limited, save for hopes of superior market-positioning

There remain major limitations in short-term efficacy, long-term prophylactic effectiveness, and tolerability

Limited efficacy encourages more nonrational and largely untested polytherapy

Narrow over-reliance on medication and the allopathic compulsion to use more drugs more aggressively risk disappointments concerning treatment outcomes, and may encourage poor treatment-adherence

Characteristics of patients who respond best/worst to particular treatments or tolerate them most/least are not well defined

Indications, diagnoses, and clinical subgroups tend to blur as drugs developed initially as anticonvulsants, antidepressants, antipsychotics, or anxiolytics find wider applications among broad and heterogeneous diagnostic categories

Innovation in all classes of psychotropics is limited and probably slowing, owing largely to a lack of secure psychopathophysiology or etiology of most psychiatric disorders, and all but a descriptive nosology

Reliance on pharmaceutical research aimed primarily at licensing and marketing leaves severe shortcomings in research on clinically relevant questions, with increasingly limited resources

Designs of therapeutic trials and assessment measures require improvement (e.g., functional as well as symptomatic measures, dealing with risks of drug carryover and discontinuation effects, adequate testing for prophylaxis vs. "relapse prevention," accurate and early assessment of risks of adverse effects, adequate testing of dose-effect and dose-risk relationships)

Many plausible and important therapeutic challenges and opportunities are overlooked or avoided (e.g., mortality, suicide, degenerative brain disorders, very young and elderly patients, syndromes of great severity, co-morbidity and complexity)

Combinations of pharmacological and psychosocial treatment methods require testing and optimization

Expectations for the effectiveness of pharmacotherapy in the face of fragmented, nonsystematic, and narrow treatment are greatly exaggerated and driven largely by economic rather than scientific or clinical considerations

tively simple, efficient, and inexpensive approaches to the management of complex human problems identified as psychiatric disorders may have encouraged overvalued and unrealistic expectations of psychotropic medicines, encouraged strongly by the quest for efficiency and cost-containment. However, heavy and unbalanced reliance on psychotropic drug-treatment may risk virtual abandonment of advances of the previous century in clinical, psychopathological, psychological, and social understanding of psychiatrically ill persons and disorders. A realistic appraisal of the power of psychotropic medicines is that they can go far toward alleviating suffering and disability, but that their effects are largely nonspecific and, at best, only partial.

During the era in which medicinal and technical treatments throughout medicine have come to the fore, the evolution of modern clinical services has been grappling—largely unsuccessfully—to limit costs and to improve "efficiency." The very real and substantial value of modern drugs combined with pressures to limit time and costs of care have led to a radically changed system of clinical care in psychiatry. Currently, clinical psychiatry, perhaps particularly in the minimally socialized systems in the United States, is dominated by essentially technical, time-limited, and cost-containment oriented approaches to diagnosis, clinical assessment, and treatment. Prominent now are checklists, algorithms, and brief, infrequent interactions with patients; often sacrificed are detailed knowledge of patients as persons, and individualized understanding of their progress. Treatment based on ever-greater use of psychotropic medicines surely has had a powerful enabling effect on the striking changes in contemporary systems of psychiatric care, and a gradual shift from identification of psychiatrists as comprehensive mental health medical specialists to "psychopharmacologists."

Throughout this book, there are repeated indications of the limitations and shortcomings of pharmacologic treatments for psychiatric illnesses. Indeed, their limitations and risks represent the basis for the title of this book. Most psychiatric disorders are not adequately treated by drugs alone. Most involve highly individual and complex manifestations of hypothetical "disorders" of unknown cause, modified by individual reactions, strengths, vulnerabilities, and circumstances that include social and cultural factors. Gradually, programs marked by limitations of time and of administrative and financial support have evolved as a current standard of care in many clinical programs in the United States and internationally. It is unlikely that most services are adequate to assure consistent, comprehensive assessment and understanding of the complex, individual problems that are identified clinically as psychiatric disorders. To some extent, psychiatric disorders may be more at risk of narrow reductionism than many other medical disorders, and at a disadvantage in competition for resources. Nevertheless, despite traditionally limited interest in the topic, many psychiatric disorders are disabling, and therefore very costly, and often fatal, not only through suicide, but also by their interactions with other intercurrent medical illnesses, complications of substance-abuse, accidents, and other causes [15, 23, 34]. Despite the need for balanced and comprehensive assessment and treatment for patients with psychiatric illnesses, current conditions, and expectations, as well as the highly fragmented and nonsystematic care provided in many sites, severely limit the breadth, comprehensiveness, and integration of psychiatric care that realistically can be provided.

The growing predominance of psychopharmacological treatment of psychiatric illnesses also may have had other adverse effects. In addition to simplified, narrow, and rather technical approaches to complex problems, along with increased time-pressure aimed at containing costs of clinical care, there also has been a highly *pharmacocentric* approach to psychiatric theory, as was discussed in Chapter 1 (see Fig. 1.3), with a decline in research-based clinical innovation. A large proportion of what serves as clinical psychiatric research is currently based on therapeutic trials of new psychotropic drugs, or new applications of established drugs, and a large proportion of that work is directly or indirectly industry-sponsored. Such studies, appropriately, are aimed at problems and questions relevant to the pharmaceutical industry, including their need for licensing and marketing in order to remain in business, rather than questions of broader clinical importance. Nevertheless, alternative sources of support and encouragement of clinical research in psychiatry, notably by private or governmental programs and funding sources, has failed to keep pace with modern needs for clinical psychiatric research. Much of the non-therapeutic clinical research that is done reflects interest in applying current technological developments, as has been traditionally characteristic of biomedically oriented academic psychiatry over the past century [3]. No doubt, interest in applied technology is motivated by hopes of achieving a better understanding of presumed biomedical aspects of psychiatric disorders, perchance leading to improved and more rational treatments. However, as valuable as modern biomedical research in psychiatry may be, it is only one approach and, as a source of innovative treatments based on theories of disease, it has been notably unsuccessful.

Given the clinical value of treatments reviewed in this book, faith in their effectiveness has encouraged manifestations of what may be termed an "allopathic compulsion" to treat, and treat vigorously, whenever possible, with or, more often, without scientific evidence to back-up treatment choices and combinations. Confidence in psychotropic drugs appears to have become so dominant (or alternative clinical skills so atrophied) that it is commonplace to see multiple trials of increasing numbers of drugs, at higher doses, and unusual combinations employed as a strategy-of-choice in dealing with partial or nonresponses. For example, use of three or more psychotropic drugs simultaneously among psychiatric outpatients doubled between 1996 and 2006 to involve one-third of patients, with two or more agents used by 60% of outpatients—most notably, involving antipsychotics and antidepressants [22]. However, such practices involving "polytherapy" are almost entirely very poorly supported by credible experimental evidence of safety and added efficacy compared to more conservative and selective approaches, including non-pharmacological methods [4, 5, 10, 22].

Improving Clinical Care

Many factors impact on the success or failure of modern psychiatric treatment based on the use of psychotropic drugs. These include what might be considered "patient factors" and "clinician factors" (Table 5.2). Patient factors include interactions between particular aspects of psychopathology that are characteristic of particular disorders. For example, psychotic disorder patients who are confused or suspicious may refuse treatment, often surreptitiously; anxious patients may be fearful of adverse effects of treatment; obsessional patients may debate the pros and cons of treatment and fail to take it; very depressed or otherwise pessimistic patients may feel unworthy of treatment or hopeless; grandiose patients may deny illness and the need for treatment.

Given enormous growth in reliance on ambulatory rather than hospital-based care of even severe psychiatric illnesses in recent decades, a fundamental fact of clinical life is that the locus of control over the treatment process has shifted from the institutional staff to the patient. It is therefore important to try to perceive the treatment process from the patient's perspective. Indeed, a central challenge for successful use of psychotropic medicines is to seek to increase the overlap or shared

Table 5.2 Limitations to treatment effectiveness in clinical psychopharmacology

Patient factors

Aspects of psychopathology can limit acceptance of treatment (e.g., psychosis, denial, grandiosity, pessimism, obsessionality, doubt)

Adverse effects often precede benefits and may persist

Benefits are often slow and not obvious to patients

Paradoxically, occasional lapses in treatment-adherence typically lead to relief from the burdens of adverse effects rather than rapid and obvious clinical worsening

Relapse and recurrence typically emerge slowly after discontinuing treatment, and their association can easily be missed by patients and clinicians

Abrupt treatment-discontinuation carries markedly increases early relapse/recurrence risk and may induce very early physiological symptoms of withdrawal

Ironically, treatment-discontinuation is most likely when most needed if early symptoms encourage discontinuation, setting off a nonadherence-recurrence cycle

A central challenge in managing treatment with psychotropic drugs is the limited sharing of understanding of aims and expectations between clinicians and patients

Clinician factors

Excessively brief, superficial clinical assessments, limited follow-up, reliance on simplistic, checklist oriented diagnosis and assessment, and faith in technical solutions to complex human problems

A tendency toward "allopathic compulsion" can encourage aggressive, sometimes reckless and nonrational treatment and polytherapy, even though driven by a wish to help

Growing numbers of drugs/patient require reconsidering diagnosis and clinical strategy

The latest and most aggressively marketed treatment is not necessarily the best option

A light touch is often very helpful: when in doubt, back off!

Closer attention to psychosocial and personal matters is very likely to increase the success of treatment

In short, good clinical psychopharmacology does not exist without good clinical psychiatry

understanding and interpretations of the process, which often are very dissimilar between patient and clinician (Fig. 5.1). From a patient's perspective the concept that a medicine may prove to be helpful is highly theoretical and may or may not be accepted on the recommendation of even a trusted clinician. Typically, patients experience adverse effects of psychotropic drugs almost immediately, and some may persist, as clinical benefits emerge relatively slowly and gradually. The patient may be the last to perceive and acknowledge beneficial effects, if ever, and often much later than the clinician, family members, or friends.

Lapses in treatment, especially as treatment persists for relatively long times following recovery from an acute episode of illness, are to be anticipated routinely. They can be only partial or occasional, or sometimes may involve complete non-acceptance [11]. Often such lapses in adherence to recommended or prescribed treatment are not discussed and remain unknown to the clinician, sometimes with severe illness recurrences as the outcome. A crucial paradox is that lapses in treatment, rather than being perceived by patients as a risk factor for illness-recurrence, are often accompanied by improved subjective states owing to relief of adverse effects, even though these may seem minor and not important clinically. Such cycles of stopping treatment and feeling better can lead to further discontinuation of

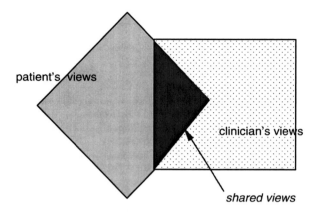

Fig. 5.1 A central challenge in clinical psychopharmacology. A critical challenge for effective therapeutic use of psychotropic drug treatments is to increase the overlap (*middle*) between patient's (*left*) and clinician's (*right*) views of the problem, proposed treatment plan, expected benefits, limitations, adverse effects, and contributions of each to a more successful outcome

treatment followed by potentially serious adverse clinical consequences, especially if clinical monitoring and supervision are not closely and thoughtfully maintained. With further irony, treatment is often discontinued when most needed, in association of emerging symptoms of an impending relapse or recurrence of illness with less perceptive and self-protective judgment. Sometimes such responses can be anticipated from understanding of individual psychopathological tendencies, but too often, not until a major illness-recurrence. As was discussed in Chapters 2–4, rapid or abrupt discontinuation of virtually all classes of psychotropic agents (antipsychotics, mood-stabilizers, antidepressants, anxiolytic-sedatives) is often followed by an excess of relapses or recurrences of the illness being treated, and much more rapidly that following slowly and gradually decreased dosing. Abrupt discontinuations without the clinician's knowledge are very common. Erratic treatment adherence may also contribute to more subtle long-term clinical instability. Suggestions for improving treatment-adherence are provided in Table 5.3.

Various clinician-associated factors also contribute to suboptimal or unsuccessful treatment. These include tendencies that are characteristic of modern clinical psychiatry and were discussed in the preceding section. Notably, even brief encounters with patients can be used to get to know their individual strengths, vulnerabilities, attitudes about the treatment process, and patterns of symptomatic change during periods between major episode recurrences. Instead, brief, routine, and relatively superficial clinical encounters oriented to routine checklist questions are likely to risk unfavorable outcomes. In addition, the tendency to overvalue psychotropic medication as the fundamental basis of treatment can contribute to allopathic compulsion, or a tendency to use more treatment, drug-combinations, and higher doses as if these were rational and powerful interventions of proven value. Complex regimens involving multiple drugs are also much more expensive and can increase risks

Table 5.3 Improving medication-adherence

Simplify regimens (fewer agents and dosing intervals)

Repeated, ongoing patient-education

Acknowledge and negotiate "minor" adverse effects

Monitor high-risk circumstances especially closely (e.g., lithium, clozapine, anticonvulsants, any new treatment, potential pregnancy)

Provide more personal interaction and support

Most disorders respond only partially to medication alone: consider the use of formal psychotherapy of proven value (e.g., CBT, IPT)

Consider support groups of other similar patients who can share experience

Long-acting agents are generally under-utilized

from additive adverse-effect burdens and drug-interactions. Costs of psychotropic drugs vary markedly among drug-types, but most especially between brand name and generic agents (Fig. 5.2).

The very complexity of treatment with multiple drugs, especially in differing dosing cycles, itself, can compromise the accuracy and success of treatment, particularly in patients who are confused or otherwise less able to follow a complex treatment regimen [14]. A further truism is that increasingly complex treatment regimens can be a manifestation of incomplete or erroneous understanding of a clinical problem. Often, trying a different approach or relying more on talking and rehabilitative efforts than drugging proves helpful. It can be appropriate even to share frustrations and disappointments with patients, while seeking to elicit their views of what has been happening and their essential contributions to the shared responsibility of developing a successful treatment plan [11, 16].

Conclusions

A central theme of this book is that modern psychotropic drug treatments are extraordinarily useful and have reshaped modern clinical, experimental, and theoretical psychiatry and psychiatry education and training. Nevertheless, they are limited in effectiveness and carry risks of sometimes clinically significant and occasionally medically serious or even fatal adverse effects. They have become severely overvalued, probably driven largely by their apparent simplicity and ability to appear to increase the "efficiency" of psychiatric treatment, with substantial control of costs. Instead, this view is quite misleading, and can lead to severely compromised, narrow, and misdirected treatment, particularly in fragmented, under-organized, and inconsistent systems of care that are characteristic of contemporary American medicine.

The pharmaceutical industry has been severely criticized in recent years, with accusations of an excessive profit motive, uncritical and unbalanced promotional activities, sometimes hyperbolic claims for expensive, new drugs, and enticement of clinicians into potentially compromising circumstances related to money [2, 13].

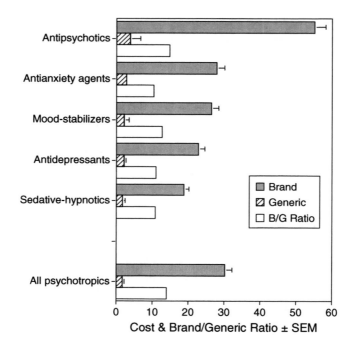

Fig. 5.2 Costs of brand name vs. generic psychotropic drug types and their ratio. Costs are relative, as ratio for each agent considered to the least costly agent of all drugs considered. Note that the average brand/generic cost ratio for all types of psychotropic agents averaged 14-fold, and that the largest cost-difference was for antipsychotics (15-fold) and least for sedative-hypnotics (11-fold difference) considered separately from anxiolytic drugs. Data are based on an analysis by Rosen and Baldessarini [26]

Some of these criticisms may be warranted, but many activities arise from pursuing business aims that may or may not be relevant to the clinical needs of patients. Arguably, clinical research, including experimental therapeutics broadly defined, has been reduced largely to drug trials aimed at licensing and marketing of pharmaceutical products. This is a circumstance in modern academic psychiatry that is much to be lamented and changed. However, it may not be appropriate to criticize the pharmaceutical industry for this trend. Improvements may come about through combined efforts of federal research and regulatory agencies, private foundations, university-based researchers, and other clinical investigators, as well as journal editors. Currently, research arising from fascination with modern brain-imaging, genetic, and other biomedical technologies has a dominant position in research support and publication, whereas the many questions pertaining to optimized clinical applications of the products of the psychopharmaceutical industry remain relatively neglected.

Among many needs for further progress in psychiatric therapeutics, several can be highlighted as tasks for future efforts. These include continued efforts to devise more effective and tolerable treatments for all major mental illnesses, including continued efforts to make use of insights and therapeutic hypotheses arising from both basic and clinical biomedical research, while also remaining open to the value of the unexpected. There is also a need to address aspects of psychiatric disorders that are not well captured by the use of standard, symptom checklists developed largely for convenience and statistical utility in treatment trials, including improved and credible measures of what are usually categorized as the relatively vague concepts of functional status and quality-of-life. Also needed are critical assessments of long-term risks and potential benefits, again including functional as well as symptomatic outcomes. Most psychiatric disorders are recurrent or chronic, making better, less biased and ambiguous, designs of long-term treatment trials aimed at testing for prophylactic effectiveness or sustained, long-term benefits and superior outcomes essential. Much work is needed to define subgroups of patients of each diagnostic type who respond to or tolerate particular treatments unusually well or poorly, using both clinical and biological measures [30]. Finally, it would be of great value to pursue better integration of what has been learned over the past century of psychiatric research at the clinical, psychological, and rehabilitative levels, in integrated combination with artfully and expertly applied pharmacological treatments.

References

1. American Psychiatric Association. Diagnostic and Statistical Manual of Mental Disorders, fourth revision, text revision. Washington, MD: American Psychiatric Association; 2000.
2. Angell M. The Truth About the Drug Companies. New York: Random House; 2004.
3. Baldessarini RJ. American biological psychiatry and psychopharmacology 1944–1994, Chap. 16. In: Menninger RW, Nemiah JC, editors. American Psychiatry After World War II (1944–1994), a Volume Celebrating the 150th Anniversary of the Founding of the American Psychiatric Association. Washington DC: APA Press; 2000. p. 371–412.
4. Centorrino F, Cincotta SL, Talamo A, Fogarty KV, Guzzetta F, Saadeh MG, et al. Hospital use of antipsychotic drugs: polytherapy. Compr Psychiatry. 2008;49:65–9.
5. Centorrino F, Ventriglio A, Vincenti A, Talamo A, Baldessarini RJ. Changes in medication practices for hospitalized psychiatric patients: 2009 versus 2004. Hum Psychopharmacol. 2010;25:179–86.
6. Chiesa A, Chierzi F, De Ronchi D, Serretti A. Quetiapine for bipolar depression: systematic review and meta-analysis. Int Clin Psychopharmacol. 2012;27:76–90.
7. Conn PJ, Roth BL. Opportunities and challenges of psychiatric drug discovery: roles for scientists in academic, industry, and government settings. Neuropsychopharmacology. 2008;33:2048–60.
8. Cutler NR, Sramek JJ, Murphy MF, Riordan H, Biek P, Carta A. Critical Pathways to Success in CNS Drug Development. New York: Wiley; 2010.
9. Eliot TS. The Rock. London: Faber & Faber Publishers; 1934.
10. Frye MA, Ketter TA, Leverich GS, Huggins T, Lantz C, Denicoff KD, et al. Increasing use of polypharmacotherapy for refractory mood disorders: 22 years of study. J Clin Psychiatry. 2000;61:9–15.
11. Gearing RE, Townsend L, MacKenzie M, Charach A. Reconceptualizing medication adherence: six phases of dynamic adherence. Harv Rev Psychiatry. 2011;19:177–89.
12. Harris G. Deal in place for inspecting foreign drugs. New York: New York Times; 13 August 2011. http://www.nytimes.com/2011/08/13/science/13drug.html?scp=2&sq=Gardiner%20Harris&st=cse.
13. Healy D. The Creation of Psychopharmacology. Cambridge, MA: Harvard University; 2002.

14. Ingersoll KS, Cohen J. Impact of medication regimen factors on adherence to chronic treatment: review of literature. J Behav Med. 2008;31:213–24.

15. Khalsa HMK, Salvatore P, Hennen J, Tohen M, Baldessarini RJ. Suicidal events and accidents in 216 first-episode bipolar-I disorder patients: predictive factors. J Affect Disord. 2008;106:179–84.

16. Lazare A, editor. Outpatient Psychiatry: Diagnosis and Treatment. 2nd ed. Philadelphia, PA: Lippincott-Williams & Wilkins; 1988.

17. Leucht S, Corves C, Arbter D, Engel RR, Li C, Davis JM. Second-generation versus first-generation antipsychotic drugs for schizophrenia: meta-analysis. Lancet. 2009;373:31–41.

18. Lieberman JA. Psychiatric drug development: why so slow? 2010. http://www.medscape.com/viewarticle/721671.

19. Marazziti D, Carlini M, Dell'osso L. Treatment strategies of obsessive-compulsive disorder and panic disorder/agoraphobia. Curr Top Med Chem. 2012;12:238–53.

20. Marron DB, editor. Research and development in the pharmaceutical industry. Washington, DC: Congressional Budget Office (SBO) of the US; 2006. 65pp. http://www.google.com/search?client=firefox-a&rls=org.mozilla%3Aen-US%3Aofficial&channel=s&hl=en&gs_sm=s&gs_upl=3381160361019127116116101818101148178916.21810&q=drug%20development%20research&ct=broad-revision&cd=4&ie=UTF-8&sa=X.

21. Meyers M. Happy Accidents: Serendipity in Modern Medical Breakthroughs. New York: Arcade Publishers; 2007.

22. Mojtabai R, Olfson M. National trends in psychotropic medication polypharmacy in office-based psychiatry. Arch Gen Psychiatry. 2010;67:26–36.

23. Ösby U, Brandt L, Correia N, Ekbom A, Sparén P. Excess mortality in bipolar and unipolar disorder in Sweden. Arch Gen Psychiatry. 2001;58:844–50.

24. Payne K, Gurwitz D, editors. The economics of drug discovery and development. Drug Dev Res. 2010;71(special issue):445–506.

25. Pope Jr HG, McElroy SL, Keck Jr PE, Hudson JI. Valproate in the treatment of acute mania. A placebo-controlled study. Arch Gen Psychiatry. 1991;48:62–8.

26. Rosen S, Baldessarini RJ. Acquisition costs of psychopharmacological agents (internal report). Belmont, MA: McLean Hospital Pharmacy Department; 2011.

27. Salvatore P, Baldessarini RJ, Tohen M, Khalsa HMK, Perez Sanchez-Toledo J, Zarate Jr CA, et al. McLean-Harvard International First-Episode Project: two-year stability of DSM-IV diagnoses in 500 first-episode psychotic disorder patients. J Clin Psychiatry. 2009; 70:458–66.

28. Salvatore P, Baldessarini RJ, Tohen M, Khalsa HM, Sanchez-Toledo JP, Zarate Jr CA, et al. McLean-Harvard International First-Episode Project: two-year stability of ICD-10 diagnoses in 500 first-episode psychotic disorder patients. J Clin Psychiatry. 2011;72:183–93.

29. Seo HJ, Chiesa A, Lee SJ, Patkar AA, Han C, Masand PS, et al. Safety and tolerability of lamotrigine: results from 12 placebo-controlled clinical trials and clinical implications. Clin Neuropharmacol. 2011;34:39–47.

30. Simon GE, Perlis RH. Personalized medicine for depression: can we match patients with treatments? Am J Psychiatry. 2010;167:1445–55.

31. Tohen M, Vieta E, Goodwin GM, Sun B, Amsterdam JD, Banov M, et al. Olanzapine versus divalproex vs. placebo in the treatment of mild to moderate mania: a randomized, 12-week, double-blind study. J Clin Psychiatry. 2008;69:1776–89.

32. Undurraga J, Baldessarini RJ. Randomized, placebo-controlled trials of antidepressants for acute major depression: thirty-year meta-analytic review. Neuropsychopharmacology. 2012;37:851–64.

33. Williams M. Productivity shortfalls in drug discovery: contributions from the preclinical sciences? J Pharmacol Exp Ther. 2011;336:3–8.

34. Wittchen HU, Jacobi F, Rehm J, Gustavsson A, Svensson M, Jönsson B, et al. Size and burden of mental disorders and other disorders of the brain in Europe in 2010. Eur Neuropsychopharmacol. 2011;21:655–79.

35. World Health Organization. International Classification of Diseases, tenth revision (ICD-10). Geneva: World Health Organization (WHO); 1994.
36. Yildiz A, Guleryuz S, Ankerst DP, Ongür D, Renshaw PF. Protein kinase C inhibition in the treatment of mania: double-blind, placebo-controlled trial of tamoxifen. Arch Gen Psychiatry. 2008;65:255–63.
37. Yildiz A, Vieta E, Leucht S, Baldessarini RJ. Efficacy of antimanic treatments; meta-analysis of randomized, controlled trials. Neuropsychopharmacology. 2011;36:375–89.

Index

CPSIA information can be obtained at www.ICGtesting.com
Printed in the USA
LVOW10*2120180915

454783LV00007B/94/P

9 781461 437093